Handbook of Geotechnical Testing

Handbook of Geotechnical Testing

Basic Theory, Procedures and Comparison of Standards

Yanrong Li

Department of Earth Sciences and Engineering, Taiyuan University of Technology, Taiyuan, China

CRC Press
Taylor & Francis Group
Boca Raton London New York Leiden

CRC Press is an imprint of the
Taylor & Francis Group, an **informa** business

A BALKEMA BOOK

CRC Press/Balkema is an imprint of the Taylor & Francis Group, an informa business

First issued in paperback 2021

© 2020 Taylor & Francis Group, London, UK

Typeset by Apex CoVantage, LLC

Library of Congress Cataloging-in-Publication Data
Names: Li, Yanrong (Writer on geology), author.
Title: Handbook of geotechnical testing : basic theory, procedures and comparison
   of standards / Yanrong Li, Department of Earth Sciences, Taiyuan University, Taiyuan, China.
Description: First edition. | Boca Raton : CRC Press/Taylor & Francis Group, [2020] |
   Includes bibliographical references and index.
Identifiers: LCCN 2019043589 (print) | LCCN 2019043590 (ebook) |
   ISBN 9780367340643 (hardcover) | ISBN 9780429323744 (ebook)
Subjects: LCSH: Geotechnical engineering—Materials—Testing—Handbooks, manuals, etc.
Classification: LCC TA706.5 .L525 2020 (print) | LCC TA706.5 (ebook) |
   DDC 624.1/510287—dc23
LC record available at https://lccn.loc.gov/2019043589
LC ebook record available at https://lccn.loc.gov/2019043590

Published by: CRC Press/Balkema
                Schipholweg 107C, 2316 XC Leiden, The Netherlands
                e-mail: Pub.NL@taylorandfrancis.com
                www.crcpress.com – www.taylorandfrancis.com

ISBN: 978-0-367-34064-3 (hbk)
ISBN: 978-1-03-208232-5 (pbk)
ISBN: 978-0-429-32374-4 (eBook)

DOI: https://doi.org/10.1201/9780429323744

Contents

Figures

Tables

Biography

Yanrong Li was born in 1978 in Shanxi Province, China. He received a BEng in hydrogeology and engineering geology from Taiyuan University of Technology in 2000, an MSc in geological engineering from Chengdu University of Technology in 2003, and a PhD in engineering geology from the University of Hong Kong in 2009. He is now a full professor of engineering geology at Taiyuan University of Technology, where he is also a founding director of the Collaborative Innovation Center for Geohazard Process and Prevention. He worked for ten years as a practicing engineer and field geologist in Hong Kong, Australia and China before turning to be a professor. He served as an associate editor of the *KSCE Journal of Civil Engineering* from 2012 to 2014 and now serves as an editorial board member of the *Bulletin of Engineering Geology and the Environment*. He is a Chartered Geologist (CGeol) of the Geological  Society of London and a Professional Member (MIMMM) of The Institute of Materials, Minerals and Mining. His research interests include loess geology and surface processes, engineering properties of soils and rocks, and geohazard investigation and mitigation.

Foreword

Standardization is a fundamental process for all human endeavors that are to be repeated and require comparable outcomes. Activities carried out and products manufactured using standardized methods can be trusted to have the same quality. Standardization is particularly important for testing as it allows the results to be attributed only to the relevant properties of the tested material or medium. Consequently, tests conducted by different laboratories on the same material/medium are expected to yield identical results, which in turn provide a basis for development of extensive databases for the properties determined. More importantly, test results are often used in design of structures and decision making, which requires familiarity with and confidence in the methods used to produce them.

It is therefore not surprising to see numerous national standards: however, many test methods in these standards involve significant variations. Combined with a language barrier, such variations can lead to notably dissimilar results. In this context, I am delighted to see that Dr. Yanrong Li's long and intense efforts have led to a comprehensive test manual: *Handbook of Geotechnical Testing: Basic Theory, Procedures and Comparison of Standards*. This book presents the basic theories of soil and rock mechanics and geotechnical laboratory testing methods in an easily comprehensible form, and integrates the testing methods recommended by Chinese and international standards. It also includes a detailed summary of the international industry standard system, history of its development and of the Chinese and other widely used national standards, including those of the United Kingdom, the United States, Singapore, South Africa, Australia and Japan. The flow charts and tables make this book highly accessible. Highlighting differences and similarities among the national standards around the world will be particularly helpful when comparing the test results obtained with different standards. Similarly, the appendices provided in this book are of great value; summary of geotechnical parameters and the pairs of Chinese-English definitions for all relevant terms may be a vital source for the Chinese users.

This book will no doubt help raise the awareness of the differences between the standards and make the comparisons easier and more accurate. In this sense, it has the content and presentation style to play the vital role of a catalyzer in improving Chinese national standards in geotechnical testing. I highly recommend this book as a reference manual for soil and rock mechanics laboratories, libraries, and the researchers and engineers, specifically those using multiple standards in their projects.

Adnan Aydin
Professor
The University of Mississippi
February 2019

Preface

Geotechnical and geological engineering has wide applications in military, mining, petroleum, natural geohazard and other engineering disciplines that are concerned with stability of the ground and construction occurring on the surface or within the ground. It uses principles of soil mechanics and rock mechanics to investigate subsurface conditions in order to evaluate stability of natural slopes and man-made deposits, assess risks posed by site conditions, design earthworks and structure foundations, and monitor site conditions, earthwork and foundation construction. Determination of the relevant physical, chemical and mechanical properties of ground materials is the key to successfully deliver such projects. Therefore, a book containing both theory of geomaterial testing and up-to-date testing methods is much in demand to college students, academic researchers and practicing technicians and engineers for obtaining reliable and accurate test results. This book is intended primarily to serve this need and aims at the clear explanation, in adequate depth, of the fundamental principles of soil and rock tests.

The book is organized into four parts. Part 1, consisting of the first 3 chapters, provides the fundamental theories on topics directly related to geotechnical testing, including soil classification, earth stress, soil strength, index properties for engineering classification of rocks, rock strength and deformability, rock mass classification, strength of rock mass etc. Part 2, embracing Chapters 4, 5, 6, 7 and 8, consists of a deep review of the international standard systems (BSI, ASTM, Chinese GB, etc.) for soil and rock tests in the field of geotechnical and geological engineering. It also provides the basic principles for test selection and test preparation, and in form of flowchart, the procedures for conducting soil and rock tests. Part 3, consisting of Chapters 9 and 10, elaborates on an in-depth comparison and discussion of the international standards of different origins for each test. There are also suggestions for improving standards currently in effect. Part 4 is a series of appendices. Appendices I and II present comprehensive summaries of reference values (average, maximum, minimum and range) of basic physical, mechanical and chemical properties of common soil and rocks, based on statistical analysis of the data retrieved from published literature and unpublished reports. Appendix III presents a full spectrum of glossary relevant to geotechnical testing with accurate definition. Appendix IV includes a full collection of dimensions of physical quantities used in geotechnical testing. A series of tables are provided for unit conversion of these physical quantities.

It is intended that the book will serve as a useful source of reference for professionals in the field of geotechnical and geological engineering. It can work as a one-stop knowledge warehouse to build a basic cognition of material tests the readers are working on. It helps college students bridge the gap between class education to engineering practice, and helps academic researchers guarantee reliable and accurate test results. It is also useful for training new technicians and providing a refresher for veterans. Engineers contemplating the ICE, IOM3 and other certification exams will find this book an essential test preparation aid. It is assumed that the reader has no prior knowledge of the subject but has a good understanding of basic mechanics.

The author wishes to record his thanks to the various publishers, organizations and individuals who have given permission of the use of figures and tables of data, and to acknowledge his dependence on those authors whose works provided sources of material. Extracts from BSI, ASTM, GB are reproduced by permission of the corresponding organization. The author thanks Professor Runqiu Huang from SKLGP, Professor Adnan Aydin from the University of Mississippi (USA), Professor Yongxin Xu from the University of Western Cape (South Africa) and Professor Zhen Guo from Taiyuan University of Technology for reading various versions of this book. The author also thanks a number of graduate students who aided in the data collection and preparation of the charts. The journey for pursuing this book would not have been possible without perpetual support and encouragement of my family. My special thanks are owed to my wife Ms. Xiaohong Deng and my sons Haoqian and Haoze for their understanding, concern, inspiration and loving support.

Yanrong Li
Professor of Earth Sciences and Engineering
Taiyuan University of Technology
July 2018

Part I

Theory of geotechnical testing

The theory of soil mechanics was formulated in Europe in the 18th century. The large-scale construction of industrial plants, urban buildings and railways aimed at meeting the needs of the development of capitalist industrialisation and the outward expansion of the market resulted in many problems related to soil mechanics. In 1773, Coulomb created the famous formula for the shear strength of sand. In 1856, Darcy studied the permeability of sand and proposed Darcy's law. Boussinesq determined the theoretical solution of stresses and deformation for an elastic semi-infinite space under a vertical concentrated load in 1885. In 1922, Fellenius proposed the slope stability analysis method. The establishment and development of these theories laid the foundation for the formation of the discipline of soil mechanics. In 1925, Terzaghi summed up previous research, proposed a one-dimensional consolidation theory and elaborated the principle of effective stress. He published the first monograph of *Soil Mechanics*, which marked the formation of modern soil mechanics.

Thereafter, scholars such as Casagrande, Taylor and Skempton began to conduct extensive research on the shear strength, deformation, permeability, stress–strain relationship and failure mechanism of soil. Basic theories of soil mechanics had been gradually applied to solve engineering problems under different conditions. In 1936, the first international conference on soil mechanics and basic engineering was held. Roscoe and others created and published the famous Cambridge elastic–plastic model in 1963, which marked the period when people's understanding of soil properties and related research went beyond the framework of the rigid–plastic theory model and entered a new stage.

Rock mechanics developed slightly later than soil mechanics did. The embryonic period of rock mechanics began in the late 19th century and continued up to the early 20th century. In this period, preliminary theories were derived to solve the problems involving the mechanical calculation of rock excavation. Griffith proposed the theory of brittle failure in 1921 and began to use centrifuges to study the failure of a mine model under simulated gravity loading in 1931. Starting in the 1950s, the construction of high dams, high slopes and large-span and high-side-wall underground buildings around the world introduced new requirements to rock research and thus promoted systematic development in rock mechanics. In April 1956, the term 'rock mechanics' was used for the first time at the professional meeting held at the Colorado School of Mines in the United States. The book *Rock Mechanics*, written by Talobre and published in Paris in 1957, was the earliest systematic monograph in this field. Many papers were then published in relevant publications, and scholars began to form different schools (such as the French School emphasising plasto-elastic theory and the Austrian School emphasising geological structures). In December 1959, the foundation failure of the Malpasset concrete arch dam in France caused the death of approximately 450 people. In October 1963, approximately 2,500 people died in the rubble of a landslide in Longarone, Italy. These engineering disasters greatly promoted the study of rock structural planes. The differences in the engineering properties of *in situ* rock masses and rock masses were also revealed. The geological structure and occurrence of rock masses were valued, and 'discontinuity' became the focus of rock mechanics research. The first International Conference on Rock Mechanics was held in Lisbon in 1966. Afterward, nonlinear theory, uncertainty theory and system science theory were gradually applied to rock mechanics.

Soil, which is a type of loose material with pores and weakly connected skeletons and is easily deformed with the reduction in pores after being loaded, is the research subject in soil mechanics. The research subjects in rock mechanics are rocks and rock masses. A rock is a dense solid with strong cementation between particles; the body of a rock contains rock blocks and joint fractures. The deformation of rocks and rock masses after being loaded is the deformation and displacement of rock blocks. This part mainly discusses the special aspect of soil, rocks and rock masses. Their physical and mechanical properties are also described.

Chapter 1

Characteristics of soils

Soil is the accumulation of mineral particles formed by the weathering, transportation, sedimentation and reconstruction of rocks. The weathering process may be physical, chemical and biological. The agent of transportation can be gravity, wind, water and glaciers. The reconstruction modes include tectonics, weathering, chemistry, gravity, radiation, temperature, biological processes and cementation.[1]

In civil engineering, soil is a geologic material that can be broken by hand squeezing or grinding.[2] The characteristics of soils mainly include physical properties (such as basic physical indicators, particle size distribution, consistency and compaction) and mechanical properties (such as compressibility and strength). A comprehensive understanding of these properties is crucial for the usage of soils.

1.1 Basic properties and engineering classification

Soil is a natural deposit composed of solid particles (approximately 50% of the total volume), and voids (filled with water and/or air) between particles. The bonding or stacking of soil particles constitutes the soil skeleton. The soil whose pores are completely filled with water is called saturated soil, and the soil whose pores are only filled with air is called dry soil.[3]

The solid components in soils are formed by weathering, transportation and sedimentation of parent rocks. Soil-forming minerals can be classified into primary and secondary minerals according to the composition of the parent rock. The soil particles composed of primary minerals have the same properties as those of parent rocks and are usually formed by the physical weathering of parent rocks. The nature of soil particles is relatively stable, has no plasticity and is weak in absorbing water. Common primary minerals are quartz, feldspar, mica, hornblende and pyroxene. Soil particles composed of secondary minerals, whose material composition differs from that of parent rocks, are usually formed from the chemical weathering of parent rocks. Particle size is relatively small and exhibits unstable properties. Soil particles can absorb water. As a result of changes in water content, the volume of soil particles can easily expand or contract. The secondary minerals are mainly clay minerals, the most common being kaolinite, illite and montmorillonite. A small amount of organic matter and soluble salts also exist in soil. The test methods for properties of soil can be found in the specifications GB 50123-1999, BS 1377-3 and ASTM D 2974-14.

Soil water exerts a significant effect on the physical and mechanical properties of soil and is a direct cause of many accidents in geotechnical engineering and geological hazards.

A small portion of soil water is tightly adsorbed on the outside lattice of solid particles in the form of bound and free water.

Clay particles present electrification characteristics in water-bearing media which cause the formation of electric fields around them. The bonds present in soil water absorb soil particles caused by the electric field. Bonded water cannot transfer hydrostatic pressure or arbitrarily flow in soil. According to the distance to the surface of particles and the force of the electric field, bound water can be divided into strongly bound water and weakly bound water. The soil water that is unaffected by the electric field around the particles is called free water and can be divided into capillary and gravity water.

The air in soil can be divided into two types, namely, free air and enclosed air (bubble), according to whether the soil is in contact with the atmosphere. The free air in soil can be discharged or absorbed when the soil suffers external forces without affecting the elasticity of the soil. The enclosed air is usually unable to come into contact with the atmosphere. The enclosed air contributes to soil elasticity and reduces soil permeability to some extent.

1.1.1 Physical indices

Soils have a three-phase composition comprising solid particles, pore water and air voids. The characteristics of soils are closely related to these three phases. A three-phase diagram is commonly used to represent the phase relationship of soil, as shown in Figure 1.1. In the figure, 'm' represents mass; 'V' represents volume; and subscripts s, w, a and v represent solids, water, air and pores, respectively. The mass of the air in soil usually defaults to 0. The basic physical indicators of soil include bulk density, water content, specific gravity, void ratio, porosity, degree of saturation, dry density, saturated density and buoyant density. The physical meanings and calculation formulas of each index are shown in Table 1.1.

The density, water content and specific gravity of soil can be obtained through testing, and other indicators can be derived from the conversion of the three indicators. Density reflects the looseness of the soil structure and can be used to calculate self-weight stress, pressure of retaining wall and slope stability. In addition, it can be used to estimate soil-bearing capacity and final settlement, and it serves as an important indicator to control the compaction degree of the construction filling of subgrade or pavement.

Current specifications that can be referenced for density tests include GB 50123-1999, BS 1377-2 and ASTM D7263-09. Water content reflects the content of water in soil. Changes in water content can affect the physical, mechanical and engineering properties of soils, such as consistency, structural strength and stability. Void ratio, liquidity index and saturation can be calculated from water content. Details about water content tests can be found in GB

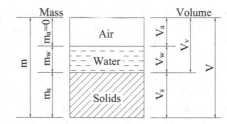

Figure 1.1 Phase diagram of soil

Table 1.1 Common physical property index of soil and basic conversion formula

Name	Symbol	Physical meaning	Expression	Unit	Basic conversion formula	Other conversion formula
Density	ρ	Mass per unit volume of soil	$\rho = \dfrac{m}{V}$	g/cm³	Directly measured by testing	$\rho = \dfrac{G_s(1+\omega)}{1+e}\rho_w$, $\rho = \rho_d(1+\omega)$
Specific gravity	G_s	Ratio of mass of soil particles to mass of pure water at the same volume of 4°C	$G_s = \dfrac{m_s}{V_s \times (\rho_w)_{4°C}}$	-		$G_s = \dfrac{eS_r}{\omega}$
Water content	ω	Ratio of the mass of water contained in the pore spaces of soil to the solid mass of particles in that material; expressed as a percentage	$\omega = \dfrac{m_w}{m_s} \times 100\%$	%		$\omega = \dfrac{\rho}{\rho_d} - 1$
Void ratio	e	Ratio of (1) the volume of voids in soil to (2) the volume of solids in a unit total volume of soil	$e = \dfrac{V_v}{V_s}$	-	$e = \dfrac{G_s\rho_w(1+\omega)}{\rho} - 1$	$e = \dfrac{\rho_s}{\rho_d} - 1 = \dfrac{\rho_s(1+\omega)}{\rho} - 1$, $e = \dfrac{n}{1-n}$, $e = \dfrac{\omega G_s}{S_r}$
Porosity	n	Ratio of (1) the volume of voids in a unit total volume of soil to (2) the unit total volume of that soil; expressed as a percentage	$n = \dfrac{V_v}{V} \times 100\%$	%	$n = 1 - \dfrac{\rho}{G_s\rho_w(1+\omega)}$	$n = 1 - \dfrac{\rho_d}{\rho_s} = 1 - \dfrac{\rho}{\rho_s(1+\omega)}$, $n = \dfrac{e}{1+e}$
Degree of saturation	S_r	Ratio of (1) the volume of water to (2) the volume of voids in a given soil mass; expressed as a percentage	$S_r = \dfrac{V_w}{V_v} \times 100\%$	%	$S_r = \dfrac{\omega G_s\rho}{G_s\rho_w(1+\omega)-\rho}$	$S_r = \dfrac{\omega G_s}{e}$

(Continued)

Table 1.1 (Continued)

Name	Symbol	Physical meaning	Expression	Unit	Basic conversion formula	Other conversion formula
Dry density	ρ_d	Mass of dry soil per unit total volume	$\rho_d = \dfrac{m_s}{V}$	g/cm³	$\rho_d = \dfrac{\rho}{1+\omega}$	$\rho_d = \dfrac{G_s \rho_w}{1+e}$, $\rho_d = \dfrac{\rho}{1+\omega}$, $\rho_d = \dfrac{n S_r}{\omega} \rho_w$
Saturated density	ρ_{sat}	Mass of saturated soil per unit total volume	$\rho_{sat} = \dfrac{m_s + V_v \times \rho_w}{V}$	g/cm³	$\rho_{sat} = \dfrac{\rho(G_s - 1)}{G_s(1+\omega)} + \rho_w$	$\rho_{sat} = \dfrac{G_s + e}{1+e} \rho_w$, $\rho_{sat} = \rho' + \rho_w$, $\rho_{sat} = G_s \rho_w (1-n) + n \rho_w$
Buoyant density	ρ'	Difference between saturated density of soil and density of water	$\rho' = \dfrac{m_s - V_v \times \rho_w}{V}$	g/cm³	$\rho' = \dfrac{\rho(G_s - 1)}{G_s(1+\omega)}$	$\rho' = \dfrac{G_s - 1}{(1+e)} \rho_w$, $\rho' = \rho_{sat} - \rho_w$, $\rho' = (G_s - 1)(1-n) \rho_w$

50123-1999, BS 1377-2, ASTM D2216-10, ASTM D2974-14 and ASTM D4959-16. Specific gravity can be used to calculate void ratios and evaluate soils. Full details of specific gravity tests are given in GB 50123-1999, BS 1377-2 and ASTM D854-14.

Commonly used weight indicators are natural gravity γ, dry gravity density γ_d, saturated gravity density γ_{sat} and buoyant gravity density γ'; they correspond to bulk density ρ, dry density ρ_d, saturated density ρ_{sat} and buoyant density ρ', respectively.

Their relationships can be defined as

$\gamma=\rho g$, $\gamma_d=\rho_d g$, $\gamma_{sat}=\rho_{sat} g$, $\gamma'=\rho' g$ (g is the acceleration of gravity)

The basic definition of density indicates that $\rho_{sat}>\rho>\rho_d>\rho\rho'$, $\gamma_{sat}>\gamma>\gamma_d>\gamma$.

1.1.2 Particle size distribution

In civil engineering, particles with similar properties and sizes within a certain range are usually divided into groups called particle fractions. The particle fraction of soils usually includes clay, silt, sand, gravel, cobble and boulder. The Chinese standard adopts different particle size limits to divide particles into particle fractions (Table 1.2). For example, in Chinese and American standards, the limit between coarse fraction and fine fraction is 0.075 mm; in the British standard, the limit is 0.063 mm.

The proportion of each particle fraction in soil is called particle size distribution, which reflects the composition of the particle fraction in the soil. Particle size composition (particle size distribution) is one of the important properties of soils, and it serves as an indicator for soil classification. The particle size distribution curve of soil is presented as a semilogarithmic plot on the basis of the result of a particle analysis test; the ordinate is the percentage by mass of particles smaller than the size given by the abscissa (Figure 1.2). Specifications that can be referenced for particle analysis tests include GB 50123-1999, BS 1377-2, ASTM D6913-04 (2009) e1 and ASTM D7928-16e1.

The particle size corresponding to any specified value on the 'percentage smaller' scale can be read from the particle size distribution curve. d_{10} is defined as the effective size,

Table 1.2 Fraction classification in Chinese, British and American standards (grain size (mm))

Fraction			GB	ASTM	BS
Fine grain	Clay		$d\leq0.005$		$d\leq0.002$
	Silt	Fine silt	$0.005<d\leq0.075$		$0.002<d\leq0.0063$
		Medium silt			$0.0063<d\leq0.02$
		Coarse silt			$0.02<d\leq0.063$
Coarse grain	Sand	Fine sand	$0.075<d\leq0.25$	$0.075<d\leq0.425$	$0.063<d\leq0.2$
		Medium sand	$0.25<d\leq0.5$	$0.425<d\leq2$	$0.2<d\leq0.63$
		Coarse sand	$0.5<d\leq2$	$2<d\leq4.75$	$0.63<d\leq2$
	Gravel	Fine gravel	$2<d\leq5$	$4.75<d\leq19$	$2<d\leq6.3$
		Medium gravel	$5<d\leq20$	—	$6.3<d\leq20$
		Coarse gravel	$20<d\leq60$	$19<d\leq75$	$20<d\leq63$
Giant grain	Cobble		$60<d\leq200$	$75<d\leq300$	$63<d\leq200$
	Boulder		$d>200$	$d>300$	$d>200$

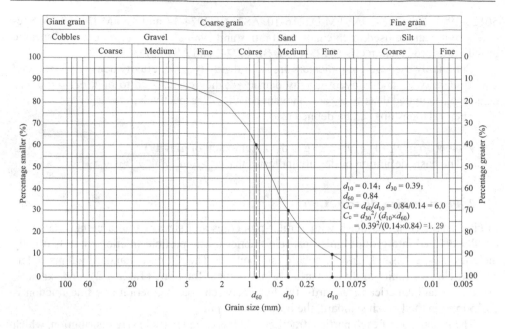

Figure 1.2 Grain size distribution curve[3]

referring to such a particle size that 10% of the particles in a soil are smaller than it. The median size d_{30} and constrained size d_{60}, can be defined in a similar way. The general slope and shape of the distribution curve can be respectively described by the coefficient of uniformity (C_u) and the coefficient of curvature (C_c) which are defined as follows:

$$C_u = \frac{d_{60}}{d_{10}}, \tag{1.1}$$

$$C_c = \frac{(d_{30})^2}{d_{10} \times d_{60}}, \tag{1.2}$$

where C_u reflects the degree of dispersion of soil particles; a high value denotes a large range of particle sizes in the soil. C_c describes the overall condition of the particle distribution curve and reflects the absence of particle size. A large particle size range of soil leads to a uniform particle size, that is, a large range of the particle size distribution curve means a smooth curve, a good particle size distribution of the soil and good engineering properties. The Chinese Standard 'Engineer Classification of Soil' (GB/T50145-2007) stipulates that when gravel or sand satisfies $C_u \geq 5$ and $1 \leq C_c \leq 3$, the particle size distribution is good; when conditions 1.1 and 1.2 cannot be met at the same time, then the particle size distribution is poor.

1.1.3 Relative density of cohesionless soils

Cohesionless soil generally refers to sand and gravel. These types of soil have minimal clay content, zero plasticity and a single-grained structure. The physical state of the two types of soil depends mainly on compaction degree.

When cohesionless soil is in a compacted state, the strength is large, and the compressibility is low; when it is in a loose state, the strength is low, and the compressibility is high. Relative density D_r is usually used to measure the compactness of cohesionless soil. Relative density is defined as

$$D_r = \frac{e_{max} - e_0}{e_{max} - e_{min}},$$ (1.3)

where D_r is the relative density of soil, e_{max} is the void ratio of soil in its loosest possible state, e_{min} is the void ratio of soil in its densest possible state and e_0 is the natural void ratio.

The evaluation criteria for the relative density of cohesionless soil are as follows:

$$0 < D_r \leq 1/3 \, \text{loose},$$

$$1/3 < D_r \leq 2/3 \, \text{moderately compacted},$$

$$2/3 < D_r \leq 1 \, \text{compacted}.$$

The difference between the maximum and minimum void ratios of soil also reflects the advantages and disadvantages of particle size distribution. A large difference indicates poor particle size distribution, whereas a small difference implies a good particle size distribution. The physical state of sand can be determined by its relative density, which provides the basis for geotechnical design, construction and quality control. A high relative density of soil indicates its compressibility and good engineering properties; high compressibility leads to poor engineering properties.

Relative density can also be calculated with the two parameters of minimum dry density ρ_{dmin} and maximum dry density $\rho_{dmin.}$

$$D_r = \frac{(\rho_d - \rho_{dmin})\rho_{dmax}}{(\rho_{dmax} - \rho_{dmin})\rho_d}.$$ (1.4)

Full details on the maximum and minimum dry density tests are given in GB 50123-1999, BS 1377-4, ASTM D4253-16 and ASTM D4254-16.

1.1.4 Consistency of cohesive soils

Consistency is the ability of cohesive soil to resist deformation or damage under external force at a certain water content.[4] The consistency states of cohesive soil when its ability against external forces gradually weakens are solid (the soil volume does not change with the change in water content), semisolid (the soil does not produce large deformation under external force but is easily broken), plastic state (the soil can be pinched into any shape without being broken under external force, and the shape remains unchanged after the external force is removed) and flowing state (the soil cannot maintain its shape with fluidity). The consistency limit is the Atterberg limits of the soil when it changes from one consistency state to another. As the water content in the soil increases, the limit of the range of water content when the soil changes from solid to semisolid is defined as the shrinkage limit (ω_S), that when the soil changes from semisolid to plastic state is defined as the plastic limit (ω_P), and that from plastic state to flowing state is defined as the liquid limit (ω_L). Full details on the determination of Atterberg limits are given in GB 50123-1999, BS 1377-2 and ASTM D4318-10e1.

Table 1.3 Classification of hardness of clayey soil[4]

State	Hard	Hard plastic	Plastic	Soft plastic	Flow plastic
Liquidity index	$I_L \leq 0$	$0 < I_L \leq 0.25$	$0.25 < I_L \leq 0.75$	$0.75 < I_L \leq 1.0$	$I_L > 1.0$

Two important indexes, the liquidity index (I_L) and the plasticity index (I_p), can be obtained from the consistency limit. The plasticity index refers to the range of water content when soil is in its plastic state; it can be defined as

$$I_p = \omega_L - \omega_p. \tag{1.5}$$

The plasticity index is generally expressed by the value with its % omitted. A large plasticity index indicates the strong ability of soil to adsorb much water in its plastic state. It also implies that the soil has high clay content and that the clay minerals in the soil have strong water absorption ability. In engineering practice, fine-grained soil is often classified and evaluated by plasticity index. Cohesive soil is defined as such that the soil possesses no more than 50% of particles greater than 0.075, and the soil has a plasticity index greater than 10.[3–5]

The liquidity index I_L reflects the hardness–softness of clay (Table 1.3), and it can be defined as

$$I_L = \frac{w - w_p}{w_L - w_p} = \frac{w - w_p}{I_p}. \tag{1.6}$$

The activity 'A' of soil can be calculated from the plasticity index (the ratio of the plasticity index I_p of the cohesive soil to the percentage of the colloidal content in the soil 'm', as defined in Equation 1.7). The activity reflects the activation of the minerals in the cohesive soil.

$$A = \frac{I_p}{m}, \tag{1.7}$$

where 'm' is the percentage of colloidal content ($d < 0.002$ mm) in the soil.

Clay can be divided into the following three categories according to the degree of activity:

$A < 0.75$ non-activated clay,

$0.75 < A < 1.25$ normal clay,

$A > 1.25$ activated clay.

1.1.5 Thixotropy of soils

Soil structure refers to the size, shape, characteristics of surface relationship and the arrangement of soil particles. When undisturbed soil is disturbed, its structure is damaged or destroyed, and it suffers a considerable loss of strength. The structure of soil strongly influences its mechanical properties, and this property can be referred to as the strength of the soil structure.[6] The index reflecting the structural properties of soil is sensitivity. Soil sensitivity is defined as the ratio of the undrained strength in the undisturbed state to the undrained strength in the remoulded state at the same water content. Sensitivity reflects the structural strength of clay soil, and it can be expressed as

$$S_t = \frac{q_u}{q_u'},$$ (1.8)

where S_t is the sensitivity of clay soil, q_u is the undrained strength (kPa) of clay soil in the undisturbed state and q_u' is the undrained strength (kPa) of clay soil in the remoulded state.

Clay soil can be divided into the following three categories according to sensitivity value:

$S_t > 4$ soil of high sensitivity,

$2 < S_t \leq 4$ soil of medium sensitivity,

$S_t \leq 2$ soil of low sensitivity.

In foundation construction, the application of soil with high sensitivity should entail strengthening its protection around the foundation trench to avoid the trampling of surrounding pedestrians, which in turn causes the reduction of soil strength.

For clay soils with high sensitivity, their strength will partially recover after remoulding, stopping any disturbance and standing for a short period. Under the condition of constant water content, clay soil softens because of disturbances (reduced strength), and softened clay soil becomes hard with the extension of standing time (increased strength). This property is called the thixotropy of clay soil. The thixotropy of soil can be utilised during construction. For example, when piling, soil is disturbed, resulting in the reduction of its strength; in this case, piles can be easily driven. When piling stops, a portion of the strength of the soil is restored after a period of standing, resulting in the increase of the bearing capacity of piles.

1.1.6 Soil compaction

Fills are applied to many types of construction projects, such as foundations, subgrade, earth banks and dams, especially high earth-rock dams. The volume of stone and earthworks required reaches millions or tens of millions of square meters, and high-quality artificial fills are needed.

When filling, soil can be compacted by beating, vibration or rolling to increase the soil's strength, reduce its compressibility and permeability and ensure the stability of the foundation and geotechnical buildings. Compaction refers to the soil under the action of compaction energy. Soil particles overcome interparticle resistance, and displacement occurs; in this condition, the pores in the soil are reduced, and the density increases.

Practices and experiences have shown that compacting fine-grained soil requires the use of a sniper machine or a rolling machine for a good compaction effect. Controlling the water content of the soil is also necessary. If the water content is high or low, a good compaction result cannot be obtained. When compacting coarse-grained soil, vibrating equipment shall be utilised, and watering shall be comprehensively applied to the process. The two different methods indicate that fine- and coarse-grained soils have different compacting properties.

For fine-grained soils, compaction is mainly reflected by the maximum dry density ρ_{dmax} and optimum water content ω_{opt}.

At a certain compaction effort, the relationship between the dry density and the water content of soil is a parabolic-like curve. The maximum dry density of soil is denoted as ρ_{dmax}, and the corresponding water content is called the optimum water content, denoted as ω_{opt}, as shown in Figure 1.3.

If the water content of soil is less than the optimum value, the dry density of the soil increases after compaction by increasing the water content. In another scenario, when the water content of soil is greater than the optimum value, the dry density of the soil decreases after compaction by increasing the water content, as shown in Figure 1.3.

The maximum dry density and optimum water content of soil are not constant values. A large function of compaction leads to a small optimum water content and a large maximum dry density. For the same type of soil, a high dry density results in a small void ratio; hence, the maximum dry density corresponds to the minimum pore ratio of the soil. At a certain water content, soil compaction is at its maximum density, all the air from the pores is removed, and the soil is saturated. Theoretically, the maximum dry density that can be achieved at a certain water content is the dry density of the saturated soil at that water content. The relationship between the dry density and the water content of saturated soil can be

expressed as $\rho_d = \dfrac{\rho_w G_s}{1 + \omega_{sat} G_s}$, as shown by the saturation line in Figure 1.3, where ω_{sat} is the

saturated water content of the sample. The optimum water content and maximum dry density can be determined by a soil compaction test. Full details on the tests are given in specifications GB 50123-1999, BS 1377-4, ASTM D698-12e2 and ASTM D1557-12e1.

The compaction result for coarse-grained soils is shown in Figure 1.4. This figure depicts that the compaction curve of coarse-grained soil is different from that of cohesive soil. Moreover, the maximum dry density is obtained at a water content of zero or at a high water content.

In the filling of coarse-grained soils, sprinkling water continuously is a necessary procedure to compact the soils at a high water content. The filling standard of coarse-grained soils

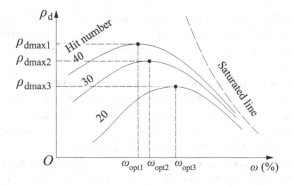

Figure 1.3 Dry density–moisture content relationship of fine-grained soils under different compaction efforts

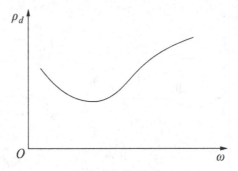

Figure 1.4 Compaction curve of coarse-grained soils

is usually controlled by relative compactness. Compaction tests are not performed in the laboratory, and the British Standard BS 1377-4 only requires a compaction test for coarse-grained soils.

1.1.7 Swelling and shrinkage of cohesive soils

The swelling of cohesive soils (increment in volume) is due to the increase in their water content, and the reciprocal of this property is defined as shrinkage. These properties are collectively called the swelling–shrinkage property. Montmorillonite and illite in cohesive soils have the characteristics of swelling and shrinkage. The volume of montmorillonite is a few times larger than that before suction, and its volume shrinks after losing water. Montmorillonite has a greater swelling and shrinkage ability than illite. Expansive soil refers to clay containing a certain amount of montmorillonite. If the content of montmorillonite in soil is over 5% of its total mass, then that soil shows a good swelling ability.[7]

The indexes concerning swelling include free swelling ratio δ_{ef}, loaded swelling ratio δ_{ep}, nonloaded swelling ratio δ_{ef} and swelling pressure P_e. The specimen to determine free swelling ratio is an artificially prepared specimen which is loose and dry. Free swelling ratio is the ratio of the volume increment to the original volume after the specimen swelling and stabilisation in pure water. The soil specimen to determine loaded swelling ratio is under a confined condition and is subjected to an upper load. Loaded swelling ratio is the ratio of the height increment to the original height after specimen swelling in pure water. Nonloaded swelling ratio is similar to loaded swelling ratio, but the specimen used is not subjected to upper load. Swelling force is the internal stress generated by cohesive soil in contact with water and is numerically equal to the pressure at which a swollen specimen returns to its original volume. These indexes can be measured by a free swelling ratio test, loaded swelling ratio test, nonloaded swelling ratio test and swelling force test. Full details on the test procedures are given in GB 50123-1999. The swelling force test is collected under the British Standard BS 1377-5. ASTM introduces the loaded swelling ratio test, nonloaded swelling ratio test and swelling force test. Full details on test procedures are given in ASTM D4546-14.

The shrinkage indexes of cohesive soil are the linear shrinkage ratio δ_{sf}, volume shrinkage ratio δ_v and shrinkage coefficient λ_n. Linear shrinkage ratio refers to the ratio of the height of the soil during shrinkage to its original height. Volume shrinkage ratio refers to the ratio of the volume change of the soil after shrinkage to its original volume. The volume

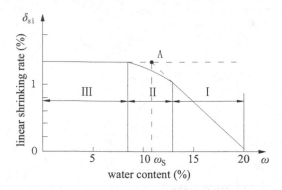

Figure 1.5 Relationship between linear shrinkage rate and water content[9]

change after shrinkage can be determined by subtracting the volume of the drying soil from its original volume. As the water content of the soil decreases, the soil shrinkage process can be roughly divided into three stages (Figure 1.5). In the first stage, the shrinkage ratio of soil is in a linear relationship with the decrease in water content. In the second stage, with the decrease in water content, the shrinkage ratio of soil gradually decreases. In the third stage, the water content continues to decrease, but the soil no longer shrinks, or it shrinks slightly.[8] The shrinkage coefficient is the slope of the first stage (slanted line) of the soil shrinking with the water content.

The soil shrinkage limit ω_S can be determined according to the linear shrinkage ratio and water content of the soil. The line segment of the first stage (linear shrinking stage) and third stage (near horizontal straight-line stage) is extended in Figure 1.5. The water content corresponding to the intersection point E is the shrinkage limit of the soil.

1.1.8 Loess collapsibility

The collapsibility of loess refers to the property of structural failure under the action of self-weight or external load when the loess is immersed in water. Under the load, the loess wetted by water begins to collapse because of the pressure, which is the initial pressure of the collapse.

The collapsible deformation of loess is mainly divided into collapsible deformation and seepage deformation. Collapsible deformation refers to the sudden deformation of loess because of structural failure under load and water immersion. The collapsibility coefficient (δ_s) is an index reflecting the collapsible deformation of loess and is defined as

$$\delta_s = \frac{h_p - h_p'}{h_0},\tag{1.9}$$

where h_0 is the original height of the specimen, h_p is the height of the loaded soil being stable after deformation and h_p' is the height of the loaded soil being stable after deformation because of being wetted by water.

The collapsibility coefficient is also an index for judging whether the loess has collapsibility. When $\delta_s \geq 0.015$, the loess is collapsible; when $\delta_s < 0.015$, the loess is noncollapsible.

Seepage deformation refers to the deformation of the loess caused by salt leaching and the pores in the soil under the long-term action of load and seepage water. The seepage deformation coefficient (δ_{wt}) is an index reflecting the seepage deformation of loess and is defined as

$$\delta_{wt} = \frac{h_z - h_s}{h_0}, \tag{1.10}$$

where h_0 is the original height of the specimen, h_z is the height of the soil after deformation caused by self-weight and h_s is the height of the soil after seepage deformation under a certain range of pressure.

After being wetted by water, the loess that is collapsible under the action of self-weight is denoted as self-weight collapsible loess; the collapsible loess under the action of additional pressure is called non-self-weight collapsible loess. The judging index is the self-weight collapsibility coefficient (δ_{zs}), which is defined as

$$\delta_{zs} = \frac{h_z - h_z'}{h_0} \tag{1.11}$$

where h_0 is the original height of the specimen, h_z is the height of the soil after self-weight deformation and h_z' is the height of the soil wetted by water after self-weight deformation.

The indexes reflecting the collapsibility of loess can be determined by a collapsibility test, and full details on the test are given in GB 50123-1999. The loess collapsibility test is mainly carried out to determine the initial pressure of the collapse, the collapsibility coefficient under the specified pressure and the self-weight collapsibility coefficient for a structure's foundation. For a hydraulic structure, the loess collapsibility test is mainly used to measure the corresponding wetness during the construction and operation phases. The collapsibility index includes the initial collapse pressure, the collapsibility coefficient at a specified pressure, the self-weight collapsibility coefficient and the seepage deformation coefficient.

1.1.9 California Bearing Ratio

California Bearing Ratio (CBR) and rebound modulus are major indexes reflecting the mechanical properties of soils. In current methods for highway design in China, CBR is often used as an indicator to determine the strength of soil foundation, and rebound modulus is often selected as one of the parameters in soil foundation design.

The CBR is a penetration test which was developed by the California Department of Transportation of America to evaluate the mechanical strength of base course materials. The CBR rating was developed to measure the load-bearing capacity of soils used for building roads. CBR refers to the ratio of applied force to a standard force when a cylindrical plunger ($\varnothing 50 \times 100$ mm) penetrates a soil material to 2.5 (or 5) mm; it is expressed as a percentage[10] and defined as

$$CBR_{2.5}(\%) = \frac{p}{7000} \times 100,$$

$$CBR_{5}(\%) = \frac{p}{10500} \times 100, \tag{1.12}$$

where p is the standard force when the cylindrical plunger penetrates the soil material to 2.5 (or 5) mm, 7000 is the standard force when the cylindrical plunger penetrates the soil material to 2.5 mm, and 10500 is the standard force when the cylindrical plunger penetrates the soil material to 5 mm.

Full details on the CBR tests are given in GB 50123-1999, BS 1377-4 and D1883-16.

Rebound modulus refers to the ratio of the stress generated by the roadbed, pavement and road building materials under load and the corresponding rebound strain. This parameter reflects the ability of soils to resist vertical deformation during the rebound deformation phase. The modulus of resilience is related to water content, dry density, soil type and loading frequency. The influence of water content is the largest; a high water content leads to a small rebound modulus.[11] Rebound modulus can be expressed as

$$E_e = \frac{\pi pD}{4l}\left(1-\mu^2\right),$$ (1.13)

where E_e is the rebound modulus (kPa), p is the unit pressure on the bearing plate, D is the diameter of the bearing plate, l is the rebound deformation under unit pressure and μ the Poisson's ratio of the soil with a value of 0.35.

Full details on the rebound modulus test are given in GB 50123-1999.

1.1.10 Soil permeability

Soil permeability refers to the property of soil to allow water to penetrate. An important index reflecting this property is the soil permeability coefficient (k). When the flow velocity of soil is low, the relationship between the discharge velocity and the hydraulic gradient is in accordance with Darcy's empirical law, namely,

$$v = \frac{q}{F} = ki,$$ (1.14)

where k is the permeability of soil, q is the volume of flowing water per unit time, F is the cross-sectional area of the flowing water, i is the hydraulic gradient and v is the discharge velocity.

The permeability coefficient of soil is the rate of flow under a unit hydraulic pressure gradient ($i=1$). In civil engineering, the permeability coefficient is often used to calculate ground settlement. This coefficient is also related to the design of drainage plans and is used to calculate the discharge rate of groundwater when construction is below the groundwater level. The permeability coefficient is related to soil porosity, particle size and distribution, temperature and other factors. The value can be determined by a permeability test. Full details on the test are given in GB 50123-1999, BS 1377-5 and BS 1377-6.

Groundwater seepage produces a volume force on the soil structure, and such force is defined as seepage force (G_D). Seepage force can be expressed as $G_D = i\gamma_w$, and the unit is kN/m^3. Seepage force is equal to the resistance applied by the soil to the water flow, and the directions of the two forces are opposite. If the direction of the water flow is from top to bottom, then the direction of the seepage force generated by the groundwater flow coincides with the direction of soil gravity, and the soil is compacted. If the direction of the water flow is from bottom to top, then the direction of seepage force is opposite to the direction of gravity. If the sum of the seepage pressure and buoyancy is greater than the gravity of the soil, the soil is brought up by the flow of water, causing soil flow. Soil flow occurs on the condition of

$$i \ge i_{cr} = \frac{\gamma'}{\gamma_w},$$ (1.15)

where i_{cr} is the critical hydraulic gradient, γ' is the buoyant gravity density (kN/m³) and γ_w is the weight of 1 kg water (kN/m³).

If the soil particle size distribution is noncontinuous (absence of some particle sizes in the soil), the water flow takes away the fine particles and leaves the coarse particles, resulting in gush and underground soil holes. These soil holes gradually expand under the action of water flow, ultimately causing in-site collapse.

1.1.11 Engineering classification of soils

Engineering classification of soils is the main task of engineering reconnaissance and an important prerequisite for engineering design.

The general methods for the engineering classification of soils are as follows:

(1) According to geological origin, soils can be divided into residual soils, slope materials, diluvial soils, alluvial soils, aeolian soils and glacial soils;
(2) According to geological time, soils can be divided into ancient deposits, general deposits and recent deposits;
(3) According to particle size distribution and plasticity index, soils can be divided into gravel, sand, silt and clay;
(4) According to organic matter, soils can be divided into inorganic soil, organic soil, peat soil and peat; and
(5) According to regional and engineering properties, soils can be divided into general soil and special soil (such as loess, soft soil, red clay, frozen soil and artificial fills). Different countries adopt different classification methods for soils. Here, we mainly discuss China's 'Standard for soil engineering classification' (GB/T50145-2007). The classification of soils in the United Kingdom and the United States mainly adopts BS5930-99 and ASTM D2487-06, respectively.

The first step in soil classification is to discriminate between organic and inorganic soils. Organic matter is usually black, cyan or dark and exhibits odour, plasticity and sponge sensation. This soil component can be judged by visual inspection, hand touch or its odour. When it cannot be judged by experience, then it can be determined by experiment. Inorganic soils can be further divided into three categories according to the relative content of the grain group: coarse-grained to fine-grained soils, coarse-grained soils and fine-grained soils (Table 1.2).

If the content of coarse particles in the soil is greater than or equal to 15% of the total mass, the soil is classified according to the classification criteria of coarse-grained soils. When the content of coarse particles in the soil is less than 15% of the total mass, the coarse particles are deducted and classified according to the corresponding regulations of coarse-grained soils or fine-grained soils.

Soils in which the mass of coarse particles is greater than 50% of the total mass of the soils are called coarse-grained soils. Coarse-grained soils are divided into two types: gravel and sand. If the content of gravel is greater than that of sand, these coarse-grained soils are called gravelly soils; if the content of gravel is not greater than that of sand, these soils are called sand soils.

Soils in which the mass of fine-grained particles is greater than or equal to 50% of the total mass of the soils are referred to as fine-grained soils. Fine-grained soils shall be classified according to the following standards:

(1) Soils in which the mass of coarse-grained particles is less than 25% of the total mass are called fine-grained soils;
(2) Soils in which the mass of coarse-grained particles is greater than 25% and less than 50% of the total mass are called fine-grained soils with coarse-grained particles; and
(3) Soils in which the organic matter content is less than 10% and not less than 5% are called organic soils.

The labelling of fine-grained soils is as follows. If the soils contain some organic matter, the adjective 'organic matter' should be added before the name of the soils, and the soil code should be followed by O for details. Examples are high-liquid-limit organic clay (CHO) and low-liquid-limit organic matter silt (MLO). If the content of coarse-grained particles in fine-grained soils is 25–50%, then the soils shall be classified as fine-grained soils with coarse-grained particles. If the major content of fine-grained soils is sand, then the soils shall be classified as sand-containing fine-grained soils, and the soil code should be followed by 'S'. Examples are sand-containing low-liquid-limit clay (CLS) and sand-containing high-liquid-limit silt (MHS).

1.2 Stress in soil

The stress in soil mass is the key to analysing the mechanical behavior of soils. Such stress can be divided into spatial and plane stresses. Plane stress is the simplification of space stress in special cases. Stress in soil mass can be divided into self-weight and additional stresses according to causes. Stress can also be divided into effective and pore stresses according to the stress-bearing principle or stress transmission mode of the soil skeleton and the pore spaces (water, air) in soils. For saturated soils, the pore pressure is the pore water pressure.

1.2.1 Three-dimensional stress state

An object is deformed under an external force, and the force that resists the deformation of internal parts is called an internal force. The section method can be used for analysis and calculation given that the internal force maintains the balance among the internal parts of an object. As shown in Figure 1.6, when calculating the internal force on a section in an object, the object can be assumed to be divided into two halves by the section. Half of the balance force system is analysed separately, and the internal force on the target section can be determined. When the object is assumed to be continuous, then the internal force on this section is distributed over the entire surface. The stress is the internal force on this surface. The theoretical formula is

$$\lim_{\Delta S \to 0} \Delta Q / \Delta S = p. \tag{1.16}$$

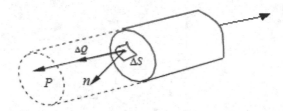

Figure 1.6 Internal force and stress

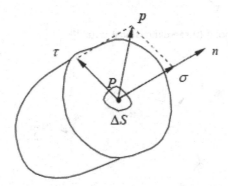

Figure 1.7 Stress is decomposed into normal stress and shear stress[13]

This formula can be understood as the stress p caused by the internal force ΔQ on the unit where the area ΔS tends to be zero. For the convenience of calculation, the average stress p^- is often adopted, and the calculation formula is

$$p^- = \Delta Q / \Delta S. \tag{1.17}$$

Stress p is a vector with the same direction as the internal force ΔQ. The stress changes with the point position because of the changes in the direction of ΔS. Although they are at the same point, the stress changes with the section direction. This property is called the stress state at one point. The position of stress in the object and which point of the stress is differentiated by this point must be determined. Two decomposition methods may be used for the stress vector projecting its component. One is to decompose along the coordinate axis in a given coordinate system. This decomposition method is not often used in practical engineering and is mainly used for theoretical derivation. The other is to decompose the stress perpendicular and parallel to the section direction. The stress decomposed perpendicular to the section is normal stress (σ), and the stress decomposed parallel to the section is shear stress (τ), as shown in Figure 1.7.

A microelement (a regular hexahedron) inside an object is analysed for its stress state. In a rectangular coordinate system, each face of the microelement is subjected to a normal stress and two shear stresses. The six stresses on the opposite sides are equal in magnitude and opposite in direction. Therefore, each microelement can be represented by nine stress components (Figure 1.8). According to the elastic mechanics, the subscript of the normal stress

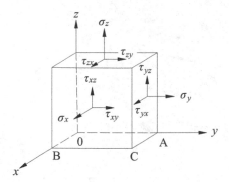

Figure 1.8 Stress state of a point (three-dimensional state)[14]

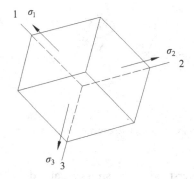

Figure 1.9 Principal stress state of a point[14]

represents the action surface, the first letter in the subscript of the shear stress represents the action surface, and the second angle represents the direction of action. Thus, the stress state of any microelement can be represented by the stress matrix shown in Equation 1.18.

The microelements are in equilibrium; therefore, the three pairs of shear stresses in Equation 1.18 are equal to one another, namely, $\tau_{xy} = \tau_{yx}$, $\tau_{xz} = \tau_{zx}$ and $\tau_{yz} = \tau_{zy}$. In this case, the nine different stress components can be reduced to six. The stress state of the microelement sometimes has a special condition (Figure 1.9), that is, the shear stress on each surface of the microelement is zero, and only the normal stress exists. At this moment, the three normal stresses on the microelement are called the principal stress (maximum principal stress σ_1, intermediate principal stress σ_2 and minimum principal stress σ_3), and the plane that bears the principal stress is called the principal plane.

The three-dimensional stress state can be represented by three circles determined by the principal stresses (σ_1, σ_2, σ_3), as shown in Figure 1.10. The intersection of circle C_1 and the abscissa is A and D; the intersection of circle C_2 and the abscissa is A and B; the intersection point of circle C_3 and the abscissa is B and D. Points A, B and C are at σ_1, σ_2, σ_3 on the abscissa, respectively. Stress circle C_1 reflects the shear stress and normal stress perpendicular to the σ_1 plane. Similarly, stress circles C_2 and C3 respectively represent the stress state on each face perpendicular to σ_2 and σ_3. The stress state of a plane that is not perpendicular

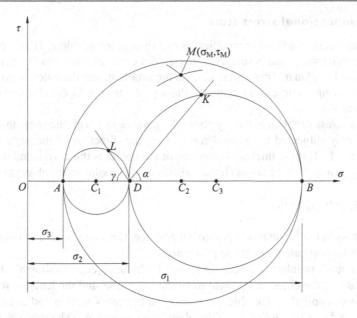

Figure 1.10 Mohr circle in three-dimensional state[15]

to the principal stress axis can be determined by the following method: the angles between its normal axis and the σ_1, σ_2 and σ_3 axes are α, β and γ, respectively. At D point, the angle BDK=α is bisected along the σ axis. For the angle ADL = γ, with C_1 and C_3 as the centers and C_1K and C_3L as the arc radii, the intersection of two arcs M can reflect the stress state of the surface.

Strain can be divided into positive strain ε and shear strain γ. It is dimensionless and is expressed as a percentage. Positive strain is the ratio of the amount of deformation along the normal stress direction of an object subjected to normal stress and the original size. The shear strain is the amount of change between the two intersection angles of the microunit microsection. The ratio of the contraction strain is perpendicular to the direction of force, and the elongation strain parallel to the direction of force when the object is subjected to a tensile force is called Poisson's ratio μ. This ratio is an elastic parameter that reflects the lateral deformation characteristics of an object. The parameters that reflect the elastic deformation of the object include elastic modulus E and shear modulus G. Elastic modulus is the ratio of normal stress to positive strain when the object is subjected to uniaxial tension or compressive stress during the elastic deformation stage. Shear modulus is the ratio of the shear stress (τ) to the shear strain (γ) of the object in the elastic deformation stage.

The relationship between stress and strain, subjected to the action of space stress, can be expressed as

$$[\sigma] = \begin{bmatrix} \sigma_x & \tau_{xy} & \tau_{xz} \\ \tau_{yx} & \sigma_y & \tau_{yz} \\ \tau_{zx} & \tau_{zy} & \sigma_z \end{bmatrix}. \tag{1.18}$$

1.2.2 Two-dimensional stress state

In practice, the stress condition of an object is mostly a space problem. However, when the object has a special shape and is subjected to a special external force condition, the problem can be simplified as planar. This case simplifies the problem, and the calculation results still meet the engineering accuracy requirements. Plane problems can be divided into plane stress and strain problems.

The plane stress problem is mainly about the plate object with the same thickness, and the object is only subjected to external force along the direction of the plate surface. As shown in Figure 1.11, if the thickness direction of the object is the z-axis, and the plate surface is the xoy plane, then the stress characteristics of each point in the object are

$$\sigma_z = 0, \tau_{zx} = 0, \tau_{zy} = 0,$$

that is, the stress in the 'z' direction is zero. In practice, the stress modes of deep-beam and flat-plate dam supports are plane stress problems.

The plane strain problem is mainly for long cylindrical objects. As shown in Figure 1.11, the cross section of the object is assumed to be the xoy plane, and the length direction is the z axis. The force applied by the object is parallel to the cross section and does not change along the length direction. Both ends of the object are constrained. The strain characteristics of the points in a plane strain problem are

$$\varepsilon_z = 0, \gamma_{zx} = 0, \gamma_{zy} = 0,$$

that is, the strain in the z direction is zero. In the plane strain problem, the stress characteristics are as follows: $\tau_{zx} = 0$, and $\tau_{zy} = 0$. The strain in the z direction is not zero. With $\varepsilon_z = 0$ and $\varepsilon_z = \dfrac{1}{E}\left[\sigma_z - \mu\left(\sigma_x + \sigma_y\right)\right], \sigma_z = \mu\left(\sigma_x + \sigma_y\right)$. In engineering practice, the stress state of a gravity-retaining wall can be approximated as a plane strain problem.

When the stress on an object is transformed from a spatial stress problem to a plane stress problem, the number of stresses on any microelement in an object shall be four (Figure 1.12a), which are σ_x, σ_y, τ_{xy} and τ_{yx}, and can be represented by a stress matrix expressed as

$$[\sigma] = \begin{bmatrix} \sigma_x & \tau_{xy} \\ \tau_{yx} & \sigma_y \end{bmatrix}. \tag{1.19}$$

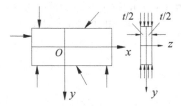

Figure 1.11 Plane stress problem[16]

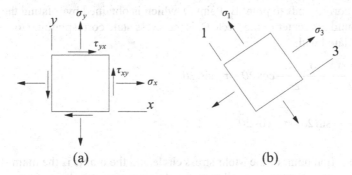

Figure 1.12 Stress state of a point (planar state)[14]

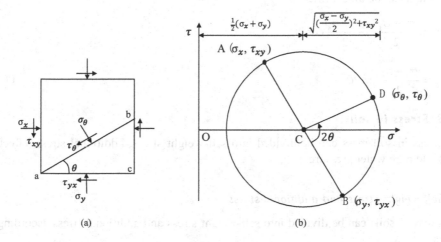

Figure 1.13 Mohr circle in two-dimensional state

When the object is in equilibrium, $\tau_{xy} = \tau_{yx}$ is obtained according to the principle of moment balance. Hence, in the plane state, the stress state of one point can be represented with three different stress components (σ_x, σ_y, τ_{xy}). When the shear stress is zero, the stress state at one point can be represented with two principal stresses (σ_1 and σ_3) (Figure 1.12b).

For plane stress and strain problems, the Mohr stress circle in a two-dimensional state can be used to characterise the stress state at any point. When drawing the Mohr stress circle, the specified compressive stress is positive, and the tensile stress is negative. In the direction of the outer normal, the shearing stress in the counterclockwise rotation is positive; in the clockwise rotation, the shearing stress is negative.

The normal stress is taken as the abscissa, and the shear stress is taken as the ordinate, as shown in Figure 1.13. The stress states of the two faces orthogonal to the microelement are represented with two points (A, B) in the figure. With the intersection of the line and σ axis, C is the center of the circle, the half-length of the line segment is the radius, and the circle is a Mohr stress circle at a certain point reflecting the stress state at this point.

For any inclined surface 'ab' of the microelement (Figure 1.13a), the angle with the 'ac' slope is θ, and the stress state on the 'ac' slope is (σ_y, τ_{yx}). In Figure 1.13b, the stress state of the 'ac' slope corresponds to point B. Point D, which is obtained by rotating the 2θ counter-clockwise around the center of the circle C, is the stress state corresponding to the 'ab' slope; it can be calculated with

$$\sigma_\theta = \frac{\sigma_x + \sigma_y}{2} + \frac{\sigma_x - \sigma_y}{2}\cos 2\theta - \tau_{xy}\sin 2\theta, \tag{1.20}$$

$$\tau_\theta = \frac{\sigma_x - \sigma_y}{2}\sin 2\theta + \tau_{xy}\cos 2\theta. \tag{1.21}$$

The intersection point of the Mohr stress circle and the σ axis is the main stress (σ_1 and σ_3). When the principal stresses σ_1 and σ_3 are used to represent the stress state on any inclined plane (the oblique plane with the angle θ between the maximum principal stress planes), then the formula can be transformed as

$$\sigma_\theta = \frac{\sigma_1 + \sigma_3}{2} + \frac{\sigma_1 - \sigma_3}{2}\cos 2\theta, \tag{1.22}$$

$$\tau_\theta = \frac{\sigma_1 - \sigma_3}{2}\sin 2\theta. \tag{1.23}$$

1.2.3 Stress in soil

The stress in soil mass can be divided into self-weight stress, additional stress, effective stress and pore water pressure.

(1) Self-weight stress and additional stress

The stress of soils can be divided into self-weight stress and additional stress according to causes.

The stress caused by soil weight is called self-weight stress. The stress caused by external load is called additional stress which is the main cause of soil deformation in civil engineering. Before the construction of a building, the stress in the soil is self-weight stress; after the construction, the stress in the soil is the sum of the self-weight stress and additional stress and is called the total stress.

If the ground is horizontal, the soil layer is even, and the natural weight of the soil is γ. In the deep part of z, a microbody is taken. The vertical self-weight stress in the microelement can be expressed as

$$\sigma_{cz} = \gamma_1 z_1 + \gamma_2 z_2 + \cdots = \sum_{i=1}^{n}\gamma_i z_i, \tag{1.24}$$

where σ_{cz} is the self-weight stress of soil mass (kPa), n is the number of soil layers on the microelement, γ_i is the unit weight of layer i, γ is the natural unit weight above the ground water level, γ' is the buoyant gravity density under the ground water level (kN/m³) and z_i is the thickness of layer i (m).

The horizontal stress on the microelement can be expressed as

$$\sigma_{cx} = \sigma_{cy} = K_0\sigma_{cz}, \tag{1.25}$$

where K_0 is the coefficient of earth pressure at rest.

Additional stress is usually calculated with the Buskinsian theoretical formula. The additional stress in soil decreases with increased depth. At the bottom of the foundation, the value of additional stress is equal to the additional stress of the substrate, and the stress distribution starts from the foundation. The additional stress distribution diffuses in the soil. The deeper the soil, the greater the range of stress distribution. The additional stress can be distributed outside the load area.

(2) Effective stress and pore water pressure

The stress in soils can be divided into effective stress (supported by soil particles) and pore stress (supported by water or air in pores) according to the stress of soil particles or the stress transmission mode. Saturated soil is a two-phase soil mass composed of solid particles and pore-filled water, and the two components share the total stress σ. Their relationship can be expressed as

$$\sigma = u + \sigma' \tag{1.26}$$

This expression is called the principle of effective stress, and the deformation (compression) and strength of soil vary depending on the changes in effective stress. When the total stress is constant, a change in the pore water pressure (u) directly causes a change in the effective stress (σ'); consequently, the volume and strength of the soil changes. In the static equilibrium state, the pressure of water around the soil particles from all directions is equal, and thus, no displacement or deformation occurs. The compression of soil particles under hydrostatic pressure is usually neglected because the compressive modulus E in solid soil particles is relatively large.

Pore pressure can be divided into static pore water pressure and excess static pore water pressure. Static pore water pressure refers to the water pressure in the soil below the static water level. The static pore water pressure at any point is equal to the weight of water multiplied by the height of the point from the horizontal plane. The excess pore water pressure refers to the portion of the pore water pressure that exceeds the hydrostatic pressure in the soil. It is caused by the change in the overlying load of the soil and dissipates with the consolidation of the drainage. This excess pressure is generally used to explain the changes in pore water pressure in the consolidation deformation of the soil.

Saturated soils are consolidated and compressed when subjected to pressure, and their internal stress changes with the following process.

(1) When pressure is applied, the total stress is supported by the pore water, thereby forming an excess pore water pressure.
(2) In the consolidation process, soil particles gradually share the overburden pressure. Then, the effective stress increases, and the soil body deforms. Correspondingly, the excess pore water pressure gradually decreases.
(3) At the end of consolidation, the internal excess pore water pressure decreases to zero, and the total soil stress is supported by the static pore water pressure and effective stress.

1.3 Compressibility and consolidation

Soil compression refers to the phenomenon in which the internal pores and volume of the soil are gradually reduced under the action of external force. This phenomenon usually includes three parts: instantaneous compression, consolidation compression and creep.

Soil consolidation refers to the process in which excess water is discharged from the soil, the excess pore water pressure is dissipated, and the soil is compressed and deformed. The compressibility parameters and consolidation coefficients of the soil can be determined by a consolidation test. Full details on the operations of the consolidation test are given in GB 50123-1999, BS 1377-5, BS 1377-6 and ASTM D2435/D2435M-11.

1.3.1 Compressibility

The compression curve is obtained by plotting the compressively stabilised void ratio e against the compressive stress p, as shown in Figure 1.14.

Compression indicators can be obtained from Figure 1.14.

(1) Coefficient of compressibility a

The slope of any secant M_1M_2 on the e-p curve is calculated as follows:

$$a = \frac{e_1 - e_2}{p_2 - p_1} = -\frac{\Delta e}{\Delta p},$$
(1.27)

where:
a = coefficient of compressibility (MPa^{-1}),
Δe = void ratio of soil pressure changing from p_1 to p_2,
Δp = void ratio of compressive stress increasing from p_1 to p_2 (kPa),
e_1 = corresponding soil void ratio when the compressive stress is p_1, and
e_2 = corresponding soil void ratio when the compressive stress is p_2.

As shown in Figure 1.14a, the coefficient compressibility of soil is a value that varies with the level of stress. In civil engineering, the coefficient compressibility a_{1-2} between 100 and 200 kPa is often used to characterise the compressibility of soil. The higher a_{1-2} is, the greater the compressibility of the soil is.

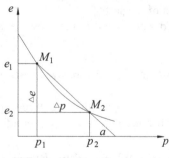

(a) Determination of compression coefficient a from e-p curve

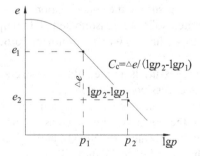

(b) Determine the compression index C_c from the e-lgp curve

Figure 1.14 Soil compression curve[3]

(2) Compression index C$_c$

The slope of the e-lg$p$ curve is calculated as follows:

$$C_c = \frac{e_1 - e_2}{lg p_2 - lg p_1} = -\frac{\Delta e}{lg\left(\dfrac{p_1 + \Delta p}{p_1}\right)}, \tag{1.28}$$

where:
C_c = compression index,
Δe = void ratio of soil pressure changing from p_1 to p_2,
Δp = void ratio of compressive stress increasing from p_1 to p_2 (kPa),
e_1 = corresponding soil void ratio when the compressive stress is p_1, and
e_2 = corresponding soil void ratio when the compressive stress is p_2.

The compression index also reflects soil compressibility. The larger the C_c value is, the higher the compressibility of the soil is. Unlike the coefficient compressibility, the value of the compression index is basically a constant in the higher-pressure range; hence, C$_c$ is more representative than a_{1-2}.

(3) Compression modulus E$_s$

Under lateral restraint, the ratio of vertical stress to vertical strain during soil compression is calculated as

$$E_s = \frac{1 + e_1}{a}, \tag{1.29}$$

where E_s is the oedometer modulus (MPa), e_1 is the initial void ratio and 'a' is the coefficient compressibility (MPa^{-1}).

The compression modulus reflects the ability of the soil to resist deformation under external uniaxial compression. The greater the value is, the lower the soil compressibility is.

(4) Coefficient of volume compression m$_v$

The vertical strain of soil under the unit vertical compressive stress and its value are the reciprocal of compression modulus E_s. The calculation formula is

$$m_v = \frac{1}{E_s} = \frac{a}{1 + e_1}, \tag{1.30}$$

where m$_v$ is the coefficient of volume compression (MPa^{-1}), E_s is the oedometer modulus (MPa), e_1 is the initial void ratio and 'a' is the coefficient of compressibility (MPa^{-1}).

(5) Modulus of soil deformation

Similar to the elastic modulus of an ideal elastomer, the deformation modulus E_0 of soil is the ratio of stress to strain under unconfined conditions. The value reflects the ability of the soil to resist elastoplastic deformation caused by external forces. This ratio can be determined by a loading test or pressure meter test and is often used to represent instantaneous settlement. The relationship between soil deformation modulus and compressive modulus can be expressed as

$$E_0 = E_s \left(1 - \frac{2\mu^2}{1-\mu} \right). \tag{1.31}$$

The Poisson's ratio μ of soil is usually less than or equal to 0.5; therefore, the deformation modulus of soil is always smaller than the compression modulus. The parameters obtained from the soil compression curve can be used to calculate the settlement s of the soil (assuming that the soil is compressed under confined conditions); that is,

$$s = \frac{-\Delta e}{1+e_1} H, \tag{1.32}$$

$$s = \frac{a}{1+e_1} \Delta pH = m_v \Delta pH = \frac{1}{E_s} \Delta pH, \tag{1.33}$$

where Δp is the additional stress loaded above the soil (kPa) and H is the height of the soil before compression.

Equation (1.30) is derived from the change in the void ratio of the soil under the condition in which the soil particles are not deformed. Equation (1.31) is equivalently substituted by Equation (1.30) by each compressibility index. The calculation accuracy of Equation (1.31) is low because compression coefficient a varies under different pressure ranges.

Under the condition of loading, the compressive deformation of soil usually includes two parts: elastic deformation and plastic deformation; after the load is removed, only the plastic deformation is retained (Figure 1.15). In the compression curve, the ring formed by the unloading and recompression process is called a hysteresis loop. In the e-lg$p$ diagram, the area of the hysteresis loop is small. In practice, the hysteresis loop is often treated as a straight line. The slope of the line is called the rebound index C_s and can be calculated from the expression

$$C_s = \frac{e_1 - e_2}{lg p_2 - lg p_1} = -\frac{\Delta e}{lg\left(\dfrac{p_1 + \Delta p}{p_1}\right)}, \tag{1.34}$$

where C_s is the rebound index; p_1, p_2 are the compressive stress at the two ends of the hysteresis loop; and e_1, e_2 are respectively the void ratios corresponding to p_1, p_2.

The recompression curve of the soil indicates that the stress history of the soil exerts a significant effect on its compressibility and that the compressibility is significantly reduced when it is pressed again. The maximum stress experienced by soil is called preconsolidation stress p_c. The ratio of the preconsolidation stress p_c to the effective stress p_0' of the

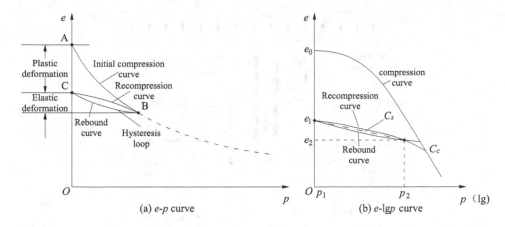

Figure 1.15 Curve of soil rebound versus recompression[4]

soil is called the overconsolidation ratio *OCR*. *OCR* is also an important indicator for soil classification.

> *OCR* > 1 is overconsolidated soil,
>
> *OCR* = 1 is normally consolidated soil,
>
> *OCR* < 1 is underconsolidated soil.

1.3.2 Consolidation

When soil is subjected to stress, the pore water gradually discharges from the saturated soil, and the process by which the volume of the soil gradually decreases is called consolidation of the soil. Terzaghi proposed the theory of one-dimensional consolidation which is used to calculate the deformation of saturated soil layers at any time during the osmotic consolidation process.

Assumptions of Terzaghi's principle:[17]

(1) The soil is fully saturated, homogenous and isotropic.
(2) The soil particles and water are incompressible.
(3) Compression and flow are one-dimensional.
(4) Darcy's law is valid for all hydraulic gradients, and the coefficient of permeability k remains constant.
(5) The change in the void ratio is proportional to the change in the effective stress, that is, $-de/d\sigma' = a$, and the coefficient of volume compression a remains constant.
(6) External load is applied instantaneously.

As shown in Figure 1.16, a homogeneous, isotropic saturated clay layer of thickness H is located above the impervious rock formation and below the sand layer. The soil is consolidated and stabilised under the action of self-weight stress. The continuous uniform load p applied to the ground causes an additional vertical stress that is uniformly distributed along the height inside the soil layer, that is, $\sigma_z = p$. The vertical direction is taken as the positive

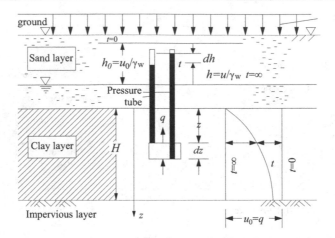

Figure 1.16 Consolidation model of saturated soil[4]

direction, and an element with a depth of z at the clay layer is taken for analysis. Before the ground loading, the water level in the piezometer tube on the top and bottom of the unit body is flushed with the groundwater level. At the moment of loading (at t=0), the water level in the piezometer h_0 ($h_0 = u_0 / \gamma_w$, u_0 is the excess pore water pressure formed at the initial stage of loading, and its value is equal to the additional stress σ_z) increases.

During consolidation, the excess pore water pressure drops to zero, and the water level in the piezometer gradually drops to the groundwater level. At certain times t, the water level in the piezometer at the top of the element is higher than the groundwater level h ($h = u/\gamma_w$, and u is the excess pore water pressure at this time). When seepage occurs, the amount of water in the pores of the element changes, thus causing a change in the pore volume. At this time, the flow rate from the bottom surface of the element is q, and the flow rate from the top surface of the element is $(q + dq)$.

For saturated soils, the net water output of the element is equal to the volume reduction of the unit volume during time dt because the pores are filled with water.

$$dq = m_v d\sigma' dz \tag{1.35}$$

During consolidation, the increase in effective stress is equal to the decrease in pore water pressure because the external load is constant.

$$d\sigma' = d(\sigma - u) = -du \tag{1.36}$$

According to Darcy's law,

$$dq = kidt = k\frac{dh}{dz}dt = \frac{k}{\gamma_w}\frac{du}{dz}dt. \tag{1.37}$$

Equations (1.33), (1.34) and (1.35) indicate that

$$-\frac{\partial u}{\partial t} = \frac{km_v}{\gamma_w}\frac{\partial^2 u}{\partial z^2} = C_v\frac{\partial^2 u}{\partial z^2}. \tag{1.38}$$

Equation (1.36) is the classical differential equation for saturated soils with unidirectional consolidation. In this equation, C_v is the coefficient of consolidation which reflects the rate at which the soil is consolidated. The larger the value is, the faster the consolidation rate is.

The initial and boundary conditions of the operation conditions shown in Figure 1.16 are as follows:

$$t = 0, \ 0 \le z \le H, \qquad u = p,$$
$$0 < t < \infty, \ z = 0, \qquad u = 0,$$
$$0 < t < \infty, \ z = H, \quad q = 0, \ \frac{\partial u}{\partial z} = 0,$$
$$t = \infty, \ 0 \le z \le H, \qquad u = 0.$$

The Fourier series is applied, and the solution that can be obtained is

$$u = \frac{4}{\pi} p \sum_{m=1}^{\infty} \frac{1}{m} \sin\left(\frac{m\pi z}{2H}\right) e^{-m^2 \frac{\pi^2}{4} T_v}, \tag{1.39}$$

where m is a positive odd number (1,3,5...) and T_v is the vertical time factor which is dimensionless and expressed as

$$T_v = \frac{C_v t}{H^2}, \tag{1.40}$$

where H is the maximum drainage distance. The overall thickness of the soil layer is taken under single-sided drainage conditions, and half of the thickness of the soil layer is taken under double-sided drainage conditions.

In engineering applications, the concept of consolidation degree is introduced to simplify the calculation of soil settlement. The degree of consolidation refers to the degree to which the soil is consolidated or the pore water pressure is dissipated after a certain time t under an additional stress. For the soil layer at depth z after the elapse of time t, the degree of consolidation at that point is calculated with

$$U_z = \frac{u_0 - u}{u_0} = 1 - \frac{u}{u_0}, \tag{1.41}$$

where u_0 is the initial excess pore water pressure, the magnitude of which is equal to the additional stress p at that point; and u is the excess pore water pressure at time t.

For further simplification, the average degree of consolidation U of the soil layer is introduced to characterise the average dissipation degree of the excess pore water pressure; it can be expressed as

$$U = 1 - \frac{\int_0^H u \, dz}{\int_0^H u_0 \, dz} = 1 - \frac{\int_0^H u \, dz}{\int_0^H p \, dz} = \frac{\int_0^H \sigma_z' \, dz}{\int_0^H p \, dz}, \tag{1.42}$$

or

$$U = \frac{S_t}{S}, \tag{1.43}$$

where S_t is the amount of foundation settlement after time t and S is the final settlement of the foundation.

When the additional stress is evenly distributed, $\int_0^H u_0 dz = pH$. The pore water stress solution is taken into the average consolidation degree formula, and the expression of the average consolidation degree of the soil layer under the load condition shown in Figure 1.16 can be obtained as

$$U = 1 - \frac{8}{\pi^2} \sum_{m=1}^{\infty} \frac{1}{m^2} e^{-m^2 \frac{\pi^2}{4} T_v} \qquad (m = 1,\ 3,\ 5,\ 7\ldots). \tag{1.44}$$

Equation 1.44 implies that the average degree of consolidation of the soil layer is a single-valued function of time factor T_v which is independent of the magnitude of the applied additional stress. Equation (1.42) is the result of the uniform distribution of additional stress in the soil layer, but the actual load is not a continuous uniform load, as shown in Figure 1.16. The additional stress σ_z generated in the foundation varies along the depth. For example, for a rectangular uniform load, σ_z at the center point is a nonlinear distribution along with the depth. Therefore, the actual initial (excess) pore pressure is nonuniform along the depth, and the calculation formula of the average consolidation degree is also different. For single-sided drainage, the relationship between the average degree of consolidation of the soil under various linear additional stress distributions and the time factor can theoretically be obtained in the same way. Five typical additional linear stress distributions exist, as shown in Figure 1.17. α is a parameter reflecting the shape of the additional stress distribution; it is defined as the ratio of the additional stress σ_z' on the permeable surface to the additional stress σ_z'' on the impervious surface, that is, $\alpha = \sigma_z' / \sigma_z''$. Therefore, different additional stress distributions have different values of α. Moreover, the solutions of the unidirectional consolidation differential equations and the average degrees of consolidation of the soil layers obtained are different. Although the average degree of consolidation of the soil layer is independent of the magnitude of the additional stress, it is related to the value of α, that is, to the distribution of additional stresses in the soil layer.

The relationship between the average consolidation degree of the soil layer and the time factor when the values of α are different is shown in Figure 1.18.

The consolidation time S_t completed by the soil layer at any t time or the time t required for the soil layer to generate a certain settlement amount S_t can be obtained according to Figure 1.19. The final settlement amount S of the soil is known.

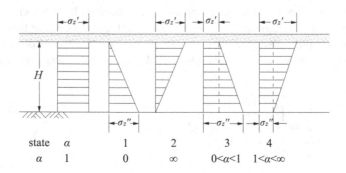

Figure 1.17 Typical stress distribution[4]

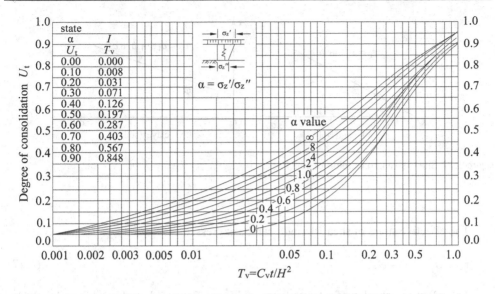

Figure 1.18 Relationship between average degree of consolidation U and time factor T_v.[12, 18]

The content in this section focuses on the case of single-sided drainage. When the soil layer is double-sided drainage, as long as the additional stress in the soil layer is linearly distributed, it is calculated as $\alpha=1$.

1.4 Strength of soils

Soil failure is mainly carried as shear failure. Therefore, the strength theory of soil mainly focuses on the stress state of soil when shear failure occurs. The shear strength of soil refers to the maximum ability of soil to resist shear stress, and it mainly depending on the friction and cohesion between soil particles.

1.4.1 Coulomb's theory

Coulomb proposed Coulomb theory in 1766 on the basis of a series of experiments, also known as the total stress of shear strength equation or the Coulomb equation (Figure 1.19).

$$\text{Sand: } \tau_f = \sigma_f \tan\phi, \tag{1.45}$$

$$\text{Cohesive soils: } \tau_f = \sigma_f \tan\phi + c, \tag{1.46}$$

where τ_f is the shear strength of soil (kPa), σ_f is the normal stress failure on the plane (kPa), ϕ is the angle of shearing resistance and c is the cohesion intercept (kPa).

Equations (1.45) and (1.46) present that the shear strength of cohesive soils is only provided by the friction between particles, $c=0$; the shear strength of cohesive soils is determined by the cohesive force and friction between particles. The shear strength of soil is a linear function of the normal stress σ on the shear plane. When the shear stress $\tau \leq \tau_f$ on the shear plane, the soil does not undergo shear damage.

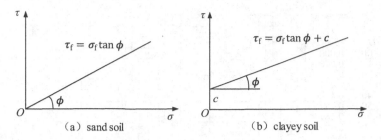

Figure 1.19 Shear strength envelop of soil[19]

The effective stress of shear strength can be expressed as

$$\tau_f = \sigma_f' \tan\phi' + c'. \tag{1.47}$$

The effective cohesion of saturated sand c' is zero. Saturated sand is subjected to vibration load under undrained conditions, and volume reduction causes an increase in internal pore water pressure u. After pore water pressure u increases the total stress σ, according to the effective stress principle $\sigma = u + \sigma'$, the effective stress σ' of the sand becomes zero. At this time, shear strength τ_f is zero, thereby resulting in sand liquefaction.[20]

1.4.2 Mohr-Coulomb criterion

The Mohr circle can be used to represent the stress state at any point within soils, including normal stress and shear stress. The Coulomb equation is also called the shear strength equation. The theory that shear stress reaches the shear strength as the failure criterion is called Mohr-Coulomb failure theory or the Mohr-Coulomb strength criterion. When the Mohr stress circle is below the strength envelope, the soil is in a stable state; when the Mohr stress circle is tangent to the strength envelope, the soil is in a limit equilibrium state (Figure 1.20).

When the soil is in the limit equilibrium state, the relationship between the maximum and minimum principal stresses in the soil is called the limit equilibrium condition of the soil. From the geometric relationship in Figure 1.20,

$$\sigma_{1f} = \sigma_{3f} \tan^2\left(45° + \frac{\phi}{2}\right) + 2c\tan\left(45° + \frac{\phi}{2}\right), \tag{1.48}$$

$$\sigma_{3f} = \sigma_{1f} \tan^2\left(45° - \frac{\phi}{2}\right) - 2c\tan\left(45° - \frac{\phi}{2}\right). \tag{1.49}$$

When the maximum and minimum principal stresses in the soil meet these conditions, the soil is in a state of extreme equilibrium, and the soil is destroyed. The angle θ between the failure surface of the soil and the maximum principal stress surface is

$$\theta = 45° + \frac{\phi}{2}. \tag{1.50}$$

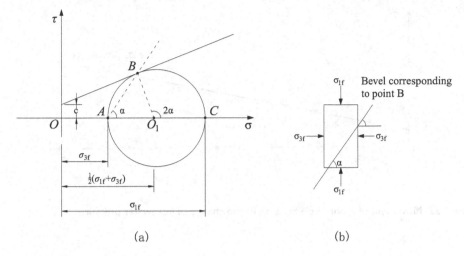

Figure 1.20 Limit stress state of soil

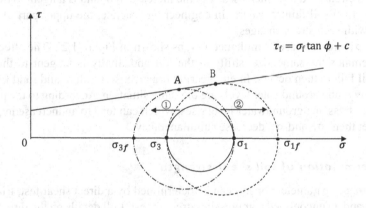

Figure 1.21 Change of stress state of a point in soil

These statements are judged by the total stress, which is the total stress method of the Mohr-Coulomb criterion. When the Coulomb criterion is expressed by effective stress and the Mohr stress circle represents the effective stress circle, it is the effective stress method of the Mohr-Coulomb criterion. The parameters used are the effective positive force σ', effective cohesion c', effective internal friction and angle ϕ'. The formula remains the same.

When the relationship between the Mohr stress circle and the strength envelope is changed from phase separation to tangency, soil failure occurs. The Mohr stress circle can be changed via three ways:

(1) σ_1 remains unchanged, and σ_3 is decreased, as shown by ① in Figure 1.21. The decrease in σ_3 causes the stress circle to gradually increase until it is tangent to the same strength envelope, and then soil failure occurs. In engineering practice, slope excavation, foundation pit excavation or other situations that can cause σ_3 to decrease may cause the soil stress to change according to this situation.

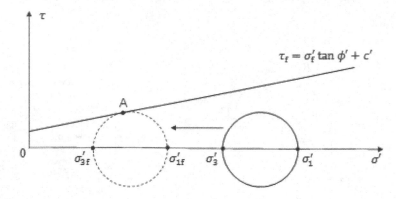

Figure 1.22 Movement of Mohr circle because of the change of porewater pressure

(2) σ_3 remains unchanged, and σ_1 is increased, as shown by ② in Figure 1.21. In this case, the stress circle gradually increases with the increase in σ_1 and is tangent to the intensity line, and then soil failure occurs. In engineering practice, the upper part of the soil is loaded with such stress changes.

(3) σ_1' and σ_3' are decreased simultaneously, as shown in Figure 1.22. The effective stress circle remains the same size, shifts to the left and finally is tangent to the intensity line. Soil failure then occurs. In engineering practice, soil failure and sand liquefaction caused by rising groundwater levels are the same situation. According to the principle of effective stress, the groundwater level rises, which can lead to an increase in pore water pressure; then, σ_1' and σ_3' decrease simultaneously.

1.4.3 Determination of soil shear strength

The shear strength parameters of soil can be determined by a direct shear test, triaxial compression test and an unconfined compressive strength test. Full details on the direct shear test are given in GB 50123-1999, BS 1377-7 and ASTM D3080-11; details on the triaxial compression test are given in GB 50123-1999, BS 1377-6, BS 1377-8, ASTM D2850-15, ASTM D4767-11 and ASTM D7181-11; the references on the unconfined compressive strength test are given in GB 50123-1999, BS 1377-7 and ASTM D2166/D2166M-16.

(1) Direct shear test

Direct shear test method for obtaining shear strength parameters: the shear strength of soil samples under different normal pressures is obtained by a direct shear test. A straight line fitting the plotted points and indicating the state of soil failure is used in the $\sigma - \tau$ coordinate system. The straight line is the Coulomb strength envelope.

The direct shear test suffers from several disadvantages, which include the following:

1) Drainage conditions cannot be controlled, and pore water pressure cannot be measured;
2) During shearing, the effective pressure area of the soil increases with shear displacement, thereby leading to errors in the calculation of compressive stress; and

3) During the failure of the specimen, only an approximation to the state of pure shear is produced in the specimen, and the shear stress on the failure plane is not uniform. Failure occurs progressively from the edges toward the center of the specimen.

Although the direct shear test has many disadvantages, it is easy to prepare because of its simple operation. This test is still widely used in China.[20–21]

(2) Triaxial compression test

The triaxial compressive strength of the soil under different confining pressures can be determined from the triaxial compression specimen. On the σ-τ coordinate axis, a straight shear is tangent to the stress circle, indicating a failure state. This straight line is the Mohr-Coulomb strength envelope.

The triaxial compression test can be classified into an unconsolidated-undrained test (UU), a consolidated undrained test (CU) and a consolidated drainage test (CD) according to test conditions. In practical situations, the consolidation factor is mainly considered the process of dissipation of pore water pressure caused by the total stress change because of construction. This process is related to soil permeability. If the pore water pressure can be rapidly dissipated, then the test does not need to be consolidated; otherwise, the consolidation process is needed.

The unconsolidated-undrained test is applicable to total stress analysis, and the measured strength envelope is a horizontal straight line, that is, the internal friction angle $\phi_u = 0$. In low-permeability soils (such as clay), the measured shear strength can be applied to the strength of the soil when the construction speed is fast, and the pore water pressure does not dissipate significantly; in high-permeability soils (such as sand), it is adapted to the rapid change in total stress, such as an explosion or an earthquake.

The consolidation undrained test can be used for total stress analysis without measurement of pore water pressure and for effective stress analysis with measurement of pore water pressure. Regardless of the total stress method analysis or the effective stress method analysis, the strength envelope measured by the normal consolidation soil is a straight line passing through the origin, that is, $c_{cu} = 0$, but $\phi' > \phi_{cu}$; ϕ' is the effective friction angle under effective internal stress analysis.

The consolidated undrained strength can be applied to the situation in which the soil is balanced but incompletely consolidated in the existing stress system. For some reason, the load is rapidly applied to form an undrained condition. For example, after the foundation or civil construction is completed, the soil itself is consolidated, but the upper load suddenly increases.[22]

In the consolidation drainage test, the total stress is finally converted into effective stress, and the total stress circle is the effective stress circle; thus, it is only suitable for effective stress analysis. The strength envelope measured by the test is usually a diagonal line that does not pass through the origin, and its strength parameters are $\phi_d > \phi_{cu}$, $c_d < c_{cu}$. This parameter can be used to study the bearing capacity or stability of sand foundations or to study the long-term stability problems of clay foundations.

The triaxial test measures the pore water pressure in a specimen. In the consolidation stage, an equal surrounding pressure increment $\Delta\sigma_3$ is applied, and the pore water pressure increment generated at this time is $\Delta u_3 = B \cdot \Delta\sigma_3$, where B is the pore water pressure coefficient under the condition of equal stress. The pore water pressure coefficient B can be used to measure soil saturation. When $B=1$, the soil is saturated; the smaller B is, the smaller the soil saturation is; when $B=0$, the soil is dry soil. In the stage of drainage compression, the confining pressure of the soil is constant, and the top is subjected to the stress increment of $\Delta\sigma_1 - \Delta\sigma_3$. The increment in pore water pressure generated at this stage is $\Delta u_1 = B \cdot A \cdot (\Delta\sigma_1 - \Delta\sigma_3)$. A is the pore water pressure coefficient under partial stress increment and reflects the dilatancy (shrinkage) of the soil. When A is positive, the soil expands by the shear volume, which is referred to as dilatancy. When A is negative, the shear volume of the soil is decreased, which is referred to as shear shrinking.

(3) Unconfined compressive test

The unconfined compressive strength q_u refers to the ultimate strength of the soil against axial stress without lateral restraint. The internal friction angle of the saturated soil under undrained conditions is zero; hence, the strength of the soil is a straight line. The shear strength parameters can be obtained from the unconfined compressive strength of the soil. Shear strength can be expressed as

$$\tau = \frac{q_u}{2}. \tag{1.51}$$

1.4.4 Effective factors on soil shear strength

The physical and chemical properties of soil have a certain influence on its shear strength, and they are mainly reflected in the following five aspects:[5]

(1) Mineral composition: For sand, the higher the quartz mineral content is, the larger the internal friction angle is; the higher the mica mineral content is, the smaller the internal friction angle is. For cohesive soils, the internal mineral composition is different, the charged molecular forces of the soil particles are different, and the cohesive force is also different. If the soil contains various cementing materials, the cohesive force can be increased.
(2) Particle size and distribution: the larger the particle size is, the coarser the soil surface is, and the larger the internal friction angle is. The larger the particle distribution is, the larger the internal friction angle is. The more uniform the particle distribution is, the smaller the internal friction angle is.
(3) Original density of soil: the greater the original density of soil is, the greater the number of contact points between soil particles is, the greater the surface friction between soil particles and the bite force in coarse-grained soil are, and the larger the internal friction angle of soil is. The greater the original density of soil is, the smaller the pores of the soil are, the closer the contact is, and the greater the cohesion is.
(4) Water content of soil: When the water content of the soil increases, the water forms a lubricant on the soil surface, which reduces the internal friction angle. For cohesive soils, an increase in water content thickens the film water and even increases the free water. This condition causes a decrease in shear strength. In fact, landslides generally occur after rain events. When rainwater infiltration increases the water content in hillside soil, the shear strength decreases, which leads to slope instability.
(5) Soil structure: Cohesive soil has structural strength. For example, when the structure of clay soil is disturbed, its cohesive force decreases.

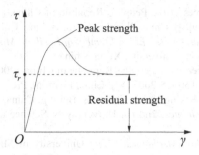

Figure 1.23 Peak and residual strengths of soil[4]

1.4.5 Residual shear strength

During shearing, the strength peaks with the increase in shear displacement. As the process continues, the strength gradually decreases and finally stabilises at a substantially constant value; it is thus called the residual strength of soils τ_r (Figure 1.23). When soils undergo large deformation shear failure, the soil cohesion c is zero generally, and the relationship between residual strength and normal stress can be expressed with

$$\tau_r = \sigma_n \tan\phi_r \tag{1.52}$$

where τ_r is the residual strength of soils (kPa), σ_n is the normal stress on the failure face of soils (kPa) and ϕ_r is the residual internal friction angle.

The factors affecting the residual strength of soils are the shape, size, mineral composition and plasticity index of cohesive soils. The residual strength of soils can be measured by repeated direct shear strength tests. Full details on the test are given in GB 50123-1999 and BS 1377-3.

References

[1] Xie, D.Y., Yao, Y.P. & Dang, F.N. (2008) *Advanced Soil Mechanics*. Higher Education Press, Beijing, China.
[2] Geotechnical Control Office. (1988) *Guide to Rock and Soil Descriptions* (Geoguide 3). Geotechnical Control Office, Hong Kong.
[3] Gong, X.N. & Xie, K.H. (2014) *Soil Mechanics*. China Architecture & Building Press, Beijing, China.
[4] Lu, T.H. (2005) *Soil Mechanics*. Hohai University Press, Nanjing.
[5] Chen, X.Z. & Ye, J. (2013) *Soil Mechanics and Geotechnical Engineering*. Tsinghua University Press, Beijing, China.
[6] Li, G.X. (2005) *Advanced Soil Mechanics*. Tsinghua University Press, Beijing, China.
[7] Qian, J.H. (1996) *Geotechnical Principle and Calculation*. China Water & Power Press, Beijing, China.
[8] Yuan, J.Y. (2003) *Geotechnical Test and Principle*. Tongji University Press, Shanghai.
[9] Technical Committee for Handbook of Engineering Geology. (1992) *Handbook of Engineering Geology*. China Architecture & Building Press, Beijing, China.

[10] Ministry of Water Resources of the People's Republic of China. (1999) *Standard for Soil Test Method* (GB/T 50123-1999[S]). China Planning Press, Beijing, China.

[11] Qin, Q.P. (2005) *The Research of the Effect Element of Resilient Modulus of Pavement Subgrade and Its Correlation.* Chang'an University, Xi'an, China.

[12] Ministry of Water Resources of the People's Republic of China. (2008) *Standard for Engineering Classification of Soil* (GB/T50145-2007 [S]). China Planning Press, Beijing, China.

[13] Xue, Q. (2006) *Elasticity.* Peking University Press, Beijing, China.

[14] Wu, T.H. (1982) *Soil Mechanics.* 2nd ed. University of Science and Technology of Chengdu, Chengdu.

[15] Ren, J.X. (2013) *Rock Mass Mechanics.* China University of Mining and Technology Press, Xuzhou.

[16] Xu, Z.L. (2006) *Elasticity.* 4th ed. Higher Education Press, Beijing, China.

[17] Chen, Z.Y., Zhou, J.X. & Wang, H.J. (1994) *Soil Mechanics.* Tsinghua University Press, Beijing, China.

[18] Huang, C.X. & Wang, Z.Y. (2012) *Soil Mechanics.* Southeast University Press, Nanjing.

[19] Zhao, M.H. (2014) *Soil Mechanics and Foundation Engineering.* Wuhan University of Technology Press, Wuhan.

[20] Knappett, J. (2012) *Craig's Soil Mechanics.* 8th ed. Spon Press, London.

[21] Zhang, K.G. & Liu, S.Y. (2001) *Soil Mechanics.* China Architecture & Building Press, Beijing, China.

[22] Wang, Z.Q. (1986) *Geotechnical Engineering Testing Technology.* China Architecture & Building Press, Beijing, China.

Chapter 2

Characteristics of rocks

In civil engineering, rocks refer to geological materials that cannot be crushed by hand squeezing or be partially scrunched. In geology, rocks pertain to lithified solid materials of igneous, sedimentary, pyroclastic or metamorphic origin.[1] The physical properties of rocks include porosity, density, water absorption and hydraulic properties. Among them, rock porosity exerts important influence on the mechanical properties of rocks. Void spaces in rocks include pores and micro-fissures. Pores affect water flow in rocks. Conversely, the micro-fissures affect the distribution of the deformation and internal stress of rocks. The properties of micro-fissures are as important as the minerals in the rock itself. Micro-fissures can reduce rock strength. Whether the micro-fissures close or not can lead to discrete differences in the rock test results. Griffith's strength theory explains the effects of micro-fissures on tensile strength and the brittle failure of rocks. Commonly used rock failure theories include Mohr strength theory, which mainly deals with the stress analysis of rock shear failure.

2.1 Physical properties

Rocks consist of three phases: solid, liquid and gas. The specific gravity of each phase in a rock has an important influence on the physical properties of the rock. The physical properties of rocks mainly include porosity, density and hydraulic properties, and such features can reflect the strength and weather resistance of rocks.

2.1.1 Porosity

In rocks, the term 'voids' is generally used to refer to pores and fissures. Voids are classified as closed and open according to their compatibility with the outside world. Open voids include large and small open voids. At normal temperature, water can enter large open voids but not small open voids. Only in a vacuum or under 150 atmospheric pressure can water enter small open voids.

Indexes that reflect rock porosity include total void ratio, large open porosity, small open porosity, total open porosity, closed void ratio and void ratio.[2] The commonly mentioned 'rock void' pertains to the total void ratio of the rock.

(1) Total void ratio (n): The ratio (in percentage) of the volume of the void (V_v) to the total volume (V) of the rock specimen. It is calculated as

$$n = \frac{V_v}{V} \times 100\%.$$

(2.1)

(2) Large open porosity (n_b): The ratio (in percentage) of the volume of the large open void (V_{vb}) to the total volume of the test specimen (V). It is calculated as

$$n_{nb} = \frac{V_{vb}}{V} \times 100\%.$$ (2.2)

(3) Small open void ratio (n_a): The ratio (in percentage) of the volume of the small open void in the rock specimen (V_{va}) to the total volume (V) of the test piece. It is calculated as

$$n_l = \frac{V_{va}}{V} \times 100\%.$$ (2.3)

(4) Total open void ratio (n_0): The ratio (in percentage) of the volume of the open void in the rock specimen (V_{v0}) to the total volume (V) of the test piece. It is calculated as

$$n_0 = \frac{V_{v0}}{V} \times 100\%$$ (2.4)

(5) Closed void ratio (n_c): The ratio (in percentage) of the volume of the closed void in the rock specimen (V_{vc}) to the total volume (V) of the test piece. It is calculated as

$$n_c = \frac{V_{vc}}{V} \times 100\%.$$ (2.5)

(6) Void ratio (e): The ratio of the volume of the void (V_v) to the volume of the solid mineral particles (V_s) in the rock specimen. It is calculated as

$$e = \frac{V_v}{V_s} = \frac{V - V_s}{V_s} = \frac{n}{1-n}.$$ (2.6)

Sedimentary rocks are formed by the accumulation of mineral particles, rock fragments or shells. Porosity varies from 0 to 90%, and the typical value of sandstone porosity is 15%. In sedimentary rocks, the void fraction generally decreases as the age of deposition increases. Statistically, chalk is the most porous of all rocks, and in some cases, its porosity can exceed 50%. Some lava (such as pumice) can also exhibit extremely high void rates because of the preservation of volcanic bubbles. The void ratio is usually less than 1% or 2% in fresh igneous rocks, and as the degree of weathering increases, the void ratio can increase to more than 20%.[3]

2.1.2 Density

The densities of rock indexes include grain density, natural bulk density, dry bulk density and saturated bulk density. A rock density index can be measured by a grain density test and a bulk density test. Full details about the grain density and bulk density tests are given in GB/T 50266-2013 and ASTM D4644-16.

(1) Grain density (ρ_s)

The grain density of rocks, also called true density, is the ratio (in g/cm³) of the rock mass m_s to the solid volume V_s. Rock grain density can be calculated as

$$\rho_s = \frac{m_s}{V_s}.$$ (2.7)

(2) Natural bulk density (ρ)

The natural bulk density of a rock refers to the mass of a rock per unit volume of the rock in a natural state, that is, the ratio (in g/cm³) of mass m to volume V. The natural bulk density of a rock can be expressed as

$$\rho = \frac{m}{V}.$$ (2.8)

The natural bulk density of a rock depends on its mineral composition, pore development and water content. This index reflects the merits of rock mechanics to some extent. In general, the greater the natural bulk density, the better the mechanical properties, and vice versa.

(3) Dry bulk density (ρ_d)

The dry bulk density of a rock is the ratio (in g/cm³) of the solid mass m_s in the rock to the volume V of the rock mass. It is calculated as

$$\rho_d = \frac{m_s}{V}.$$ (2.9)

(4) Saturated bulk density of a rock (ρ_{sat})

The saturated bulk density of a rock refers to the ratio (in g/cm³) of the mass of the rock when the pores are filled with water to the volume of the rock. It is calculated as

$$\rho_{sat} = \frac{m_s + V_v \times \rho_w}{V}.$$ (2.10)

The density parameters of rocks are important for engineering practice. In general, rock density is positively correlated with rock strength and elasticity. For oil shale deposits, density can also indicate the value of mineral commodities. In the coal industry, a strong correlation exists between density and ash content and pre-cover depth.

2.1.3 Hydraulic properties

The hydraulic properties of rocks refer to their features underwater or under water immersion conditions. These properties include water absorption, swelling, slake-durability, softness and frost resistance.

(1) Water content and water absorption

The ratio (in percentage) of the mass m_w of the water to the mass m_s of the rock in the natural state is called the water content (water content) ω. It is calculated as

$$\omega = \frac{m_w}{m_s} \times 100\%. \tag{2.11}$$

Full details about the test are given in GB/T 50266-2013 and ASTM D2216-10. Water content is a crucial parameter for weakening rocks. Weak rocks usually contain a large quantity of clay minerals that are easily softened by water. Water content has an important dominating effect on rock strength and deformation.[4]

The capability of rocks to absorb water under certain conditions is called its water absorption capability. Commonly used water absorption indexes include water absorption, saturation rate and coefficient of saturation. The test parameters for the rock water absorption index are given in GB/T 50266-2013 and ASTM D6473-15.

1) Water absorption rate (ω_a)
    The water absorption rate of rock refers to the ratio (in percentage) of the mass m_{w1} of water absorbed by rocks under normal atmospheric pressure and room temperature to the mass of rock solids, and is calculated as

$$\omega_a = \frac{m_{w1}}{m_s} \times 100\%. \tag{2.12}$$

2) Saturated water absorption rate (ω_{sa})
    Saturated water absorption refers to the ratio (in percentage) of the mass m_{w2} of the water absorbed by the rock under high pressure or vacuum to the solid mass of the rock. It is calculated as

$$\omega_{sa} = \frac{m_{w2}}{m_s} \times 100\%. \tag{2.13}$$

3) Coefficient of saturation (η_w)
    The coefficient of saturation is the ratio of the water absorption rate to the saturated water absorption rate and is calculated as

$$\eta_w = \frac{\omega_a}{\omega_{sa}}. \tag{2.14}$$

The water absorption of rocks is closely related to the pores inside the rocks. The water absorption rate of a rock can reflects the development degree of its large open pores. The greater the water absorption rate, the larger the ratio of the large open pores in the rock. The saturated water absorption of a rock indicates the development degree of its open pores; the larger the saturated water absorption, the greater the number of open voids. The coefficient of saturation represents the relative proportional relationship between the open and closed pores in a rock.

The larger the saturation coefficient, the greater the large open pore proportion in the rock, and the lower the number of remaining pores after water absorption at normal temperature and pressure. This kind of rock is susceptible to frost heaving damage because of poor frost resistance.

(2) Swelling

The phenomenon in which the rock volume expands after water absorption and causes structural damage is called rock swelling. Rock swelling is generally caused by clay minerals in the rock (especially montmorillonite). The indexes for evaluating swelling include the axial free swelling ratio (V_H), radial free swelling ratio (V_D), lateral constrained swelling ratio (V_{HP}) and swell pressure (p_e).

Axial free swelling: The rock expands freely in water, and the ratio of axial deformation ΔH to the original height H of the sample is called axial free swelling ratio which is calculated as

$$V_H = \frac{\Delta H}{H} \times 100\%.\tag{2.15}$$

Radial free swelling: The rock expands freely in water, and the ratio of radial deformation ΔD to the sample diameter or side length D is called the radial free swelling ratio which can be calculated as

$$V_D = \frac{\Delta D}{D} \times 100\%.\tag{2.16}$$

Lateral constrained swelling: The rock under lateral restraint expands in water, and the ratio of the axial deformation ΔH_1 to the original height H of the sample is called the lateral constrained swelling ratio which is calculated as

$$V_{HP} = \frac{\Delta H_1}{H} \times 100\%.\tag{2.17}$$

Swell pressure is the maximum pressure applied to maintain the volume of the rock as it expands in water. Full details about the test for rock swelling index are given in GB/T 50266-2013.

(3) Slaking

The enduring slaking index I_{d2} is the main slaking index of rocks. It is measured by the dry–wet cycle test. The said index is the ratio of the dry mass m_r of the sample after two dry–wet cycles to the dry mass m_s of the sample before the test.

$$I_{d2} = \frac{m_r}{m_s} \times 100\%.\tag{2.18}$$

Full details about the test are given in GB/T 50266-2013 and ASTM D4644-16.

(4) Softness

Strength reduction after rock immersion is called the softening of the rock and is measured by the softening coefficient η_c. The softening coefficient refers to the ratio of the compressive strength σ_{cw} of the rock sample in the saturated state to the compressive strength σ_c in the dry state. It is calculated as

$$\eta_c = \frac{\sigma_{cw}}{\sigma_c}. \tag{2.19}$$

Rock softness is related to the hydrophilicity of rocks, their soluble mineral contents and open pores. The coefficient of softness is one of the important indexes for evaluating the mechanical properties of rocks. It is mostly used in hydraulic construction to evaluate the stability of a dam foundation rock mass. That is, when $\eta_c > 0.75$, rock softness is weak, and the engineering geological properties are satisfactory; conversely, when $\eta_c < 0.75$, rock softness is strong, and the engineering geological properties are poor.

(5) Frost resistance

The ability of rocks to resist freeze–thaw damage is called frost resistance, a quality expressed by the coefficient of frost resistance R_d and mass loss rate K_m. The rock frost resistance index can be measured by a rock freeze–thaw test. Full details of the operation are given in GB/T 50266-2013 and ASTM D5312/D5312M-12. The coefficient of frost resistance is the ratio (in percentage) of the dry compressive strength σ_{c2} of the rock specimen after repeated freezing and thawing to the dry compressive strength σ_{c1} before freezing and thawing; it is calculated as

$$R_d = \frac{\sigma_{c2}}{\sigma_{c1}} \times 100\%. \tag{2.20}$$

The mass loss rate refers to the difference between the dry mass before and after freezing and thawing $(m_{s1} - m_{s2})$ and the dry mass m_{s1} before the test, as expressed in percentage and calculated as

$$K_m = \frac{m_{s1} - m_{s2}}{m_{s1}} \times 100\%. \tag{2.21}$$

The main factors that affect rock frost resistance include the thermo-physical properties and strength of rock minerals, the intergranular connections, the development of open pores and water content.

2.1.4 Rock description

The description of rocks is primarily used to provide basic information on their properties. A complete rock description should include strength, color, structure, name, particle size, weathering state and other geological characteristics.[2]

(1) Strength

Rock strength can be described as 'hard' or 'weak', and it has a certain relationship with the point load strength and uniaxial compressive strength index (Table 2.1). The relationship between the uniaxial compressive strength σ_c and the point load strength index $I_s(50)$ is $\sigma_c = 25I_s(50)$.

(2) Color

Rock color is usually presented using three basic parameters: hue, purity and brightness. Hue refers to the basic colors or the mixture of basic colors. Purity pertains to the degree of vividness of color, and brightness denotes the degree of lightness and darkness of color. The description should indicate whether wetting the rock sample reduces its brightness (making the sample darker) but does not change its hue and purity. Therefore, the degree of wetting of the sample should be stated when describing rock color.

(3) Structure and fabric

'Structure' is a broad term that usually refers to the general physical appearance of rocks. It encompasses the size and shape of the particles or crystals (geometrically), the distribution of particle sizes and the degree of crystal development (the relationship between geometric features). This term is mostly applicable to scale features visible to the naked eye in specimens. Rocks that consist of very fine particles must be observed with a microscope.

'Fabric' refers primarily to the arrangement of particles or crystals in a rock. For magmatic rocks and other crystalline rocks, organisation refers to the shape and arrangement direction of the crystalline and amorphous bodies of the rocks. For sedimentary rocks, the fabric focuses on the orientation of individual particles and the location of the particles relative to the cementitious material. In addition, the fabric also includes small structural faces or planes (often referred to as micro-cracks) that pass through the particles or between crystals. The description of microfractures should include their strength, spacing, continuity and direction.

Table 2.1 Classification of rock strengths

Descriptive term	Uniaxial compressive strength (MPa)	Approximate point load strength index values ($I_{s(50)}$) (MPa)
Extremely weak	<0.5	Generally, not applicable
Very weak	0.5–1.25	
Weak	1.25–5	
Moderately weak	5–12.5	0.2–0.5
Moderately strong	12.5–50	0.5–2
Strong	50–100	2–4
Very strong	100–200	4–8
Extremely strong	>200	>8

(4) Weathering and alteration

Weathering can be divided into chemical weathering and physical decomposition, and these two types usually exist simultaneously. Chemical weathering refers to weathering caused by chemical reactions such as hydration, oxidation, ion exchange and dissolution. The chemical weathering of rocks changes their surface color. Therefore, the degree of chemical weathering can be preliminarily judged by the changes in rock color. The degree of chemical weathering of rocks is divided into six levels, from fresh rock to residual soil. Physical decomposition mainly denotes the changes in the surface stress state caused by the thermal expansion and contraction of rocks. Such change leads to rock rupture. The decomposition of rock materials can also be caused or hastened by biological action (such as hydro splitting).

Alteration pertains to changes in properties caused by the circulation of invading hot gases and fluids. Common terms used to describe alteration are kaolinization and mineralisation.

2.2 Deformability of rocks

The constitutive relations of rocks are primarily reflected in the complete stress–strain curve. Rock deformation can be divided into elastic deformation, plastic deformation and viscous deformation according to the stress–strain behavior.[5]

Elasticity: The ability of an object to deform when resisting an external force and to return to its original size and shape once the force is removed is called elasticity. The resulting deformation is called elastic deformation. Objects with elasticity are called elastic bodies, and they can be divided into two types according to their stress–strain behavior: linear elastic body (ideal elastic body) whose stress–strain behavior is linear and non-linear elastic body whose stress–strain behavior is non-linear.

Plasticity: The ability of an object to deform when resisting an external force and its failure to completely return to its original size and shape after unloading is called plasticity. The deformed portion that cannot return presents plastic deformation or permanent deformation. An object that only undergoes plastic deformation under an external force is called an ideal plastic body. The stress–strain curve of the ideal plastic body is a straight horizontal line. When the stress is lower than the yield limit σ_y, the material is not deformed. After the stress reaches σ_y, the deformation increases, and the internal stress remains unchanged.

Viscosity: The ability of an object to deform under stress and for which the deformation cannot be completed instantly while its strain rate increases with the increase of stress is called viscosity. The stress–strain rate curve of an ideal viscous object (Newtonian fluid) is a straight line passing through the origin point.

In addition to mineral composition and structure, the mechanical properties of rocks are also related to environmental factors, such as stress conditions and temperature. Under normal temperature and pressure, rocks are neither ideal elastic bodies nor simple plastic bodies or viscous bodies. Rocks often present elastic–plastic, plastic–elastic, elastic–viscose–plastic or viscoelastic properties.

2.2.1 Stress–strain relationship

Figure 2.1 depicts the complete stress σ–strain ε curve of a rock specimen under a uniaxial compression load. Full details about the test operations are given in GB/T 50266-2013 and ASTM D7012-14.

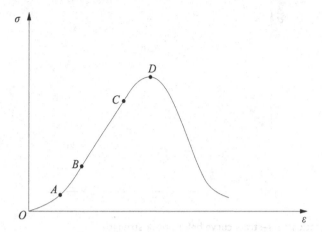

Figure 2.1 Typical curve of stress vs. strain of rock[5]

According to the complete stress–strain curve, the deformation of a rock can be divided into the following four stages:[2,6]

(1) Fissure closing (Section OA): The original open structural surface or micro-cracks in the sample are gradually closed, the rock is compacted, nonlinear deformation occurs, and the stress–strain curve is concave. At this stage, the lateral expansion of the sample is minimal, and its volume decreases as the load increases. The deformation of a fractured rock is more obvious at this stage, whereas the deformation of a hard and less fractured rock is not as obvious.
(2) Elastic compression (Section AC): The stress–strain curve at this stage is at an approximately linear rate. Section AB represents elastic deformation, and BC depicts the stable development of micro-cracks.
(3) Pore structure collapse (Section CD): Point C is the turning point of the rock, or its yield point, from elastic to plastic deformation. The corresponding stress is called yield stress or yield limit. Such stress is two-thirds of the peak strength. After entering this stage, the development of micro-fractures changes qualitatively, and the rupture continues to increase until the sample is completely destroyed. The volume of the sample begins to increase, and the axial strain and volume strain rate increase rapidly. The upper bound stress at this stage is called the peak compressive stress or peak strength of the rock.
(4) Locking (after Point D): The internal structure of the rock block is destroyed, and the sample remains basically intact. At this stage, the cracks develop rapidly, intersect and mutually form a macroscopic fracture surface. Subsequently, the rock block deformation mainly manifests as a block slip along the macro-fracture surface. The bearing capacity of the sample then decreases rapidly with the increase of the deformation, and it generally does not decrease to zero. Thus, the fractured rock block still has a certain bearing capacity.

The stress–strain curves of rocks vary according to their type and nature. Miller divided rocks into six types according to their stress–strain curve characteristics before the peak (Figure 2.2).[6]

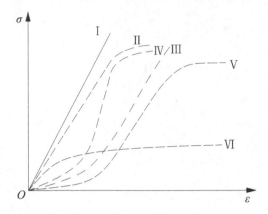

Figure 2.2 Typical rock stress–strain curve before peak strength

Type I: The relationship between pressure and strain is represented as a straight line or until the specimen suddenly breaks. Rocks with such deformation properties (including basalt, quartzite, dolomite and extremely strong limestone) are classified as elastic rocks.

Type II: When the pressure is low, the stress–strain curve is approximately a straight line. When the stress increases to a certain value, the stress–strain curve bends downward, and as the stress increases gradually, the slope of the curve becomes smaller until it is destroyed. Rocks with such deformation properties (including weaker limestone, mudstone and tuff) are called elastic–plastic rocks.

Type III: When the pressure is low, the stress–strain curve bends slightly upward. When the stress increases to a certain value, the stress–strain curve gradually becomes a straight line until damage occurs. Rocks with such deformation properties are called plastic–elastic rocks.

Type IV: When the pressure is low, the stress–strain curve bends upward. When the stress increases to a certain value, the deformation curve becomes a straight line, and the curve bends downward and ultimately shows an S-shaped line. Rocks with such deformation characteristics (including marble and gneiss) are called plastic–elastic–plastic rocks.

Type V: The rock is basically the same as Type IV, and the curve is also an S-shaped line, but the slope of the curve is relatively flat and occurs mostly in rocks with high compressibility, such as schist with stress perpendicular to it.

Type VI: The stress–strain curve begins with a small straight line, follows an inelastic curve portion and continues to creep. Such rocks (including rock salt and soft rock) are called elastic–viscous rocks.

2.2.2 Application of stress–strain curves

In addition to reflecting the constitutive relationships of rocks, the complete stress–strain curve is also applied to the situations listed here:

(1) Prediction of rock burst

As shown in Figure 2.3, the area of the complete stress–strain curve is bounded by the peak strength point C, and the left half of OEC (Area A) represents the strain energy accumulated

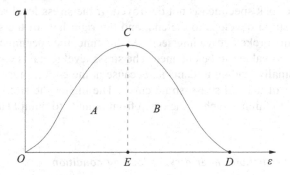

Figure 2.3 Full stress–strain curve used for predicting rock burst[5]

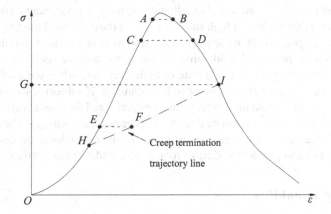

Figure 2.4 Full stress–strain curve for predicting creep damage[2, 6]

in the specimen when the rock reaches peak strength. The right half of DEC (Area B) depicts the energy released by the rock during the destruction process. If B<A, then the rock still has energy remaining after the damage. The rapid release of this part of the energy causes rock burst. If B>A, then the rock has completely released the absorbed strain energy during the failure process, and no rock burst occurs.

(2) Prediction of creep failure

The creep termination trajectory in Figure 2.4 indicates that after the specimen is loaded to a certain stress level, the stress is kept constant, and the specimen creeps. At the appropriate stress level, the creep develops. When the strain reaches a certain value, the creep stops, and the rock specimen reaches a stable state. The creep termination trajectory is the connection of the creep termination point at different stress levels and can be determined by numerous experiments in advance. When the stress level is below point *H*, the stress is kept constant, and the rock specimen does not creep. When the stress level reaches point *E*, the stress is kept constant, and the creep strain develops until point *F* intersects with the creep end trajectory, and the creep stops. Point *G* is the critical point. The stress level remains constant below point *G*, the creep strain develops to intersect the creep termination trajectory, the creep stops, and the rock specimen is not destroyed. If the stress level remains constant at point G, the creep strain develops to the end, and the right half of the stress–strain curve intersects. That is, the broken curve intersects. At this time, the specimen is destroyed. The maximum creep strain value can be obtained. The stress level remains constant above point G and creeps, eventually leading to damage, because at the end, it must intersect the post-destruction segment of the full stress–strain curve. The higher the stress level, the shorter the time from creep to failure, such as the creep from point C to point D and the creep from point A to point B.

(3) Rock damage prediction under a cyclic loading condition

Cyclic loading is often encountered in geotechnical engineering. For example, repeated blasting operations involve applying cyclic loads to surrounding rocks and are dynamic. As most rocks are not linear elastic materials, their loading and unloading paths do not coincide, and each add–unload creates a hysteresis loop, thereby leaving a permanent deformation. Figure 2.5 shows the cyclic loading at high stress levels and the rock breaking shortly. If a cyclic load is applied from point A, then the permanent deformation develops to point B, the peak is intersected with the peak of the full stress–strain curve, and the rock is destroyed. Thus, when the rock engineering itself is in a state of high stress and when the cyclic load occurs again, the rock engineering becomes vulnerable to damage. If subjected to cyclic loading at the stress level at point C, then the loading can be experienced for a relatively long period of time before the rock engineering can be destroyed. Therefore, according to the existing stress level of the rock itself and the magnitude and period of the cyclic load, the complete stress–strain curve is applied to predict the time of rock failure under cyclic loading conditions.

2.2.3 Rock deformability

According to the stress–strain curve of rocks, the deformation parameters such as the deformation modulus and Poisson's ratio of rocks can be determined.

The deformation modulus E is the ratio of the axial compressive stress to the axial strain under uniaxial compression, which is $E=\sigma/\varepsilon$, where σ and ε are the axial stress and strain at any point on the stress–strain curve, respectively.

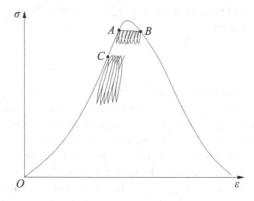

Figure 2.5 Full stress–strain curve predicts failure under cyclic loading conditions[5]

When the stress–strain curve is a straight line, the deformation modulus of the rock is a constant and is numerically equal to the slope of the straight line. As the deformation is mostly elastic deformation, it is also called elastic modulus.

When the stress–strain curve is non-linear, the deformation modulus of the rock is a variable, that is, the modulus on differentiated stress segments. Commonly used examples are the initial modulus E_0, tangent modulus E_t and secant modulus E_s (Figure 2.6).

The initial modulus is the tangent slope at the origin on the curve. The tangent modulus is the slope of the tangent at any point on the curve. The secant modulus refers to the slope of the line connecting the starting point of the curve to another point. The slope of the line connecting the point at $\sigma_c/2$ and the starting point is usually taken.

Poisson's ratio μ is the ratio of the transverse strain ε_x to the longitudinal strain ε_y under uniaxial compression.

$$\mu = \left| \frac{\varepsilon_x}{\varepsilon_y} \right|. \tag{2.22}$$

In practical situations, ε_x and ε_y at $\sigma_c/2$ are commonly used to calculate the Poisson's ratio of rocks.

In addition to the two parameters of elastic modulus and Poisson's ratio, other parameters that reflect the deformation properties of rocks include shear modulus G, Lame constant λ and bulk modulus K_v. These parameters have the following relationship with elastic modulus E and Poisson's ratio μ:

$$G = \frac{E}{2(1+\mu)}, \tag{2.23}$$

$$\lambda = \frac{E\mu}{(1+\mu)(1-2\mu)}, \tag{2.24}$$

$$K_v = \frac{E}{3(1-2\mu)}. \tag{2.25}$$

The elastic parameters described in Equations 2.23, 2.24 and 2.25 are static elastic parameters, that is, the elastic parameters of the material under static load. The elastic

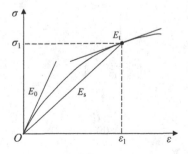

Figure 2.6 Initial, tangent and secant moduli

parameters of the material subjected to dynamic load include rock dynamic elastic modulus E_d, dynamic Poisson's ratio μ_d, dynamic shear modulus G_d, dynamic pull constant λ_d and dynamic bulk modulus K_d. These parameters can be calculated from the longitudinal and transverse wave velocities of rocks. Full details about the rock sound wave velocity test are given in GB/T 50266-2013.

2.3 Rock strengths

Rock strengths commonly adopted in engineering practice include Uniaxial Compressive Strength, point load strength, tensile strength and triaxial compressive strength.

2.3.1 Uniaxial compressive strength

The uniaxial compressive strength (UCS) of rocks σ_c is the maximum compressive stress that rocks can withstand under uniaxial (unconfined) compression conditions. During the test, the specimen is subjected only to a vertical pressure; the lateral pressure is unrestrained, and the deformation of the specimen is not limited. The uniaxial compressive strength of rock is a basic index for determining the mechanical properties of rock mass. It is also an important indicator for the engineering classification of rock mass and for establishing the criteria for rock failure. Such strength can also be adopted to estimate the other strength parameters of rocks.

Full details about the test are given in GB/T 50266-2013 and ASTM D7012-14.

Under the condition of uniaxial compression, the test piece mainly produces the following three types of damage:

(1) Shear failure of X-shaped conjugate slope, $\beta \approx \dfrac{\pi}{4} + \dfrac{\phi}{2}$; (Figure 2.7a);

(2) Single level shear failure, $\beta \approx \dfrac{\pi}{4} + \dfrac{\phi}{2}$; (Figure 2.7b);

(3) Lateral tensile failure. Under axial compressive stress, tensile stress is generated in the transverse direction because of the Poisson effect. Such damage occurs when the transverse tensile stress exceeds the tensile strength of rocks (Figure 2.7c).

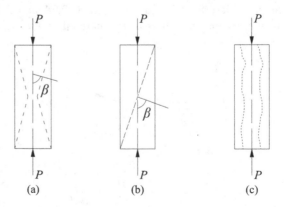

(a) (b) (c)

Figure 2.7 Failure states in uniaxial compression test[5]

The first two types of damage are caused by shear stress on the failure surface that exceeds the shear strength, a phenomenon which is generally called compression shear failure. Accordingly, the third type of damage is sometimes referred to as compression failure.

2.3.2 Point load strength

The test for determining the point load strength of rocks is a simple rock strength test proposed by E. Broch and J. A Franklin in 1972. This index can be referred to in particle distribution and engineering design. Full details about the point load test are given in GB/T 50266-2013 and ASTM D5731-16.

The point load strength index is usually presented as I_s and is expressed as

$$I_s = \frac{P}{D_e^2},\tag{2.26}$$

where P is the load at which the specimen is broken (N), D_e is the diameter of the equivalent rock core (mm), $D_e = \sqrt{\frac{4A}{\pi}}$ and A is the minimum cross-sectional area (mm^2) between the points of loading.

In practice, the core diameter exerts a major impact on the value of the point load strength. The core diameter is changed from 10 mm to 70 mm, and the point load strength can be reduced by 2 to 3 times. Accordingly, the International Society of Rock Mechanics determines the strength index value $I_s(50)$ of the radial load point load test of a 50 mm cylindrical specimen as a standard test value. The point load strength index can also be converted with the Uniaxial Compressive Strength. The conversion equation is

$$\sigma_c = 24I_s(50).\tag{2.27}$$

2.3.3 Tensile strength

The tensile strength of rocks is the maximum tensile stress that rocks can withstand when they are damaged by uniaxial tensile load. It is expressed by σ_t and can be adopted to determine the strength envelope and establish a rock strength criterion. Tensile strength can be identified directly by stretching or indirectly by the Brazilian test. Because of the swiftness and convenience of the Brazilian test, it has become a common method for determining the tensile strength of rocks. Full details about the test are given in GB/T 50266-2013 and ASTM D3967-16. Such tensile strength can be expressed as

$$\sigma_t = \frac{2P}{\pi dt},\tag{2.28}$$

Where σ_t is the tensile strength (Mpa), P is the compressive load (MPa), d is the diameter of the cylinder (mm) and t is the height of the cylinder (mm).

In general, the short cracks present in rocks considerably influence the direct drawing method rather than the Brazilian test method. Thus, the strength value given by the Brazilian

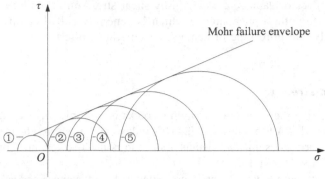

① Mohr circle for specimen failure under uniaxial tensile test
② Mohr circle for specimen failure under uniaxial compression test
③④⑤ Mohr circle for specimen failure during three axial compression test

Figure 2.8 Strength envelope

test is usually higher than that of the direct tensile test, and the difference can sometimes be several times greater.

2.3.4 Shear strength

When the confining pressure is zero or low, a rock usually breaks along a set of inclined fissures. Under high confining pressure, the ductile deformation and strength of the rock increase, and full ductility or plastic flow deformation occurs, along with hardening. The test specimen also exhibits a thick waist barrel-like shape. Full details of the triaxial tests are given in GB/T 50266-2013 and ASTM D7012-14.

The triaxial compression test is mainly used to determine the Mohr strength of rocks. The Mohr strength envelope (Figure 2.8) can be identified through triaxial compression on several specimens in the same coordinate system, drawing the stress state and Mohr circle at the time of the failure of each specimen and supplementing the Mohr circle at the time of the uniaxial compression test and tensile failure test.

2.4 Theory of rock strength

When a rock is destroyed, its stress and strain present a certain relation. The functional relation characterising the stress and strain of rock failure is called the failure criterion or strength criterion. The failure modes of rocks are mainly divided into brittle fracture and ductile fracture. Hard rocks are chiefly characterised by brittle failure in low confining pressure and low temperature environment. Note that hard rocks tend to exhibit ductile failure under high confining pressure and high temperature environment (Table 2.2).

The strength theory of rock mechanics includes maximum positive strain theory, Mohr strength theory, Griffith strength theory and strain energy theory (Table 2.3). Given their wide applicability, Mohr's strength theory and Griffith strength theory are mainly discussed in this work.

Table 2.2 Failure modes of rock under triaxial compression[6]

Strain at failure / %	<1	1~5	2~8	5~10	>10
Destructive form	Brittle failure	Brittle failure	Transition failure	Ductile failure	Ductile failure
Failure of specimens					
Basic type of stress-strain curve					
Failure mechanism	Tension rupture	Rupture with tension	Shear rupture	Shear flow rupture	Plastic flow

Table 2.3 Comparison of the theories of rock strength

Strength theory	Theoretical expression	Failure condition	Scope of application
Maximum normal strain theory	Rupture of the object occurs because the extension strain ε reaches the ultimate strain ε_0.	$\left\| \sigma_3 - \mu(\sigma_1 + \sigma_2) \right\| \geq \sigma_t$ $\sigma_1 \geq \dfrac{1-\mu}{\mu}\sigma_3 + \sigma_c$ σ_c is the rock's uniaxial tensile strength, σ_t is the rock's tensile strength, and μ is the Poisson's ratio.	Only for brittle rocks
Mohr Coulomb strength theory	When a certain functional relationship between shear stress and normal stress on one surface is satisfied, the surface breaks.	$\tau \geq \sigma \tan\varphi + c$ $\sin\varphi \leq \dfrac{\sigma_1 - \sigma_3}{\sigma_1 + \sigma_3 + 2c \cdot \cot\varphi}$ φ is the internal friction angle, and c is the cohesion.	Suitable for rock damage under low confining pressure conditions.
Griffith's strength theory	When the maximum stress on the cracked side wall of the rock is greater than the tensile strength of the rock, the crack expands.	(1) When $\sigma_1 + 3\sigma_3 \geq 0,$ $\dfrac{(\sigma_1 - \sigma_3)^2}{\sigma_1 + \sigma_3} \geq -8\sigma_t.$ (2) When $\sigma_1 + 3\sigma_3 < 0,$ $\left\|\sigma_3\right\| \geq \left\|\sigma_t\right\|.$ σ_t is the rock's tensile strength.	Suitable for the tensile failure of brittle rocks
Shear strain energy theory	Rock failure is due to the plastic flow caused by the shear strain in the unit volume reaching the limit value.	$\dfrac{1}{2}\left[(\sigma_1 - \sigma_2)^2 + (\sigma_2 - \sigma_3)^2 + (\sigma_3 - \sigma_1)^2\right] \geq \sigma_y^2.$ σ_y is yield strength of the material.	Suitable for rocks with ductile failure

2.4.1 Mohr criterion

In 1900, Mohr proposed that the destruction of a material mainly involves shear failure and that the shear stress on the failure surface is a function of the normal stress on that surface. Mohr also made some assumptions about the characteristics of rock failure. He contended that rock strength is independent from the magnitude of the intermediate principal stress σ_2. In addition, the macroscopic fracture surface of a rock is basically parallel to the direction of the intermediate principal stress. Mohr circle can be drawn in a rectangular coordinate system in which the shear stress τ is the vertical axis and normal stress σ is the horizontal axis. In this coordinate axis and with an infinite number of ultimate stress circles, the trajectory line of the breaking stress point is called the Mohr strength line or the Mohr envelope.

As a result of the obvious heterogeneity of rocks, the calculation formula of the Mohr envelope can only be represented with a universal function

$$\tau = f(\sigma). \tag{2.29}$$

The curve of rock strength can be determined by the triaxial compression test. Full details about the test are given in GB/T 50266-2013. If the stress state at any point in the rock falls below the Mohr strength envelope, then the rock will not be destroyed (Case 1, Figure 2.9). If the stress state is tangent to the Mohr strength envelope, then the rock will undergo shear damage (Case 2 in Figure 2.9). If the stress state intersects the Mohr strength envelope, then the rock will be destroyed (Case 3, Figure 2.9).

The Mohr strength envelope is usually presented in three forms: linear envelope, hyperbolic criterion and parabolic criterion.

(1) Linear envelope

The linear strength envelope is also known as the Mohr-Coulomb criterion (Figure 2.10). The Mohr-Coulomb criterion characterises the shear failure of rocks and is suitable for low confining pressures. The angle φ between the line and the σ axis is called the angle of internal friction, and the intercept on the τ axis is the adhesion force c and is expressed as

$$\tau = \sigma \tan \varphi + c. \tag{2.30}$$

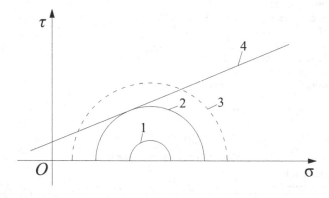

Figure 2.9 Determination of material failure with Mohr envelope 1 – unbroken; 2 – critical; 3 – failed; and 4 – strength envelope.

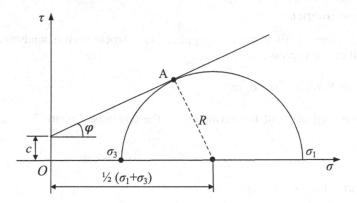

Figure 2.10 Mohr-Coulomb strength

If we assume that the maximum principal stress circle of a micro-body in a rock sample is tangent to the Mohr-Coulomb strength envelope, then such envelope can be ascertained from the geometric relationship shown in Figure 2.10.

$$\sin\varphi = \frac{\sigma_1 - \sigma_3}{\sigma_1 + \sigma_3 + 2c \cdot cot\varphi}. \tag{2.31}$$

Thus, the conditions for determining the failure of rocks using the Mohr-Coulomb strength criterion are

$$\sin\varphi \leq \frac{\sigma_1 - \sigma_3}{\sigma_1 + \sigma_3 + 2c \cdot cot\varphi} \tag{2.32}$$

or

$$\tau \geq \sigma \tan\varphi + c. \tag{2.33}$$

From Formula (2.31), we can obtain

$$\sigma_1 = \sigma_3 \frac{1 + sin\varphi}{1 - cos\varphi} + \frac{2c \cdot cos\varphi}{1 - sin\varphi} \tag{2.34}$$

Under axial compression, $\sigma_3 = 0$, and from Formula (2.31),

$$\sigma_1 = \frac{2c \cdot cos\varphi}{1 - sin\varphi}. \tag{2.35}$$

Thus, the uniaxial compressive strength $\sigma_c = \dfrac{2c \cdot cos\varphi}{1 - sin\varphi}$, setting $\dfrac{1 + sin\varphi}{1 - cos\varphi} = \xi$, leads Formula (2.31) to be expressed with principle stress as

$$\sigma_1 = \sigma_3 \xi + \sigma_c. \tag{2.36}$$

(2) Hyperbolic criterion

The hyperbolic strength criterion curve is applied to hard rocks such as sandstone, limestone and granite. It can be expressed as

$$\tau^2 = (\sigma + \sigma_t)^2 \tan^2\eta + (\sigma + \sigma_t)\sigma_t,$$

(2.37)

where η is the inclination of the asymptote of the envelope (Figure 2.11) and $\tan \eta = \frac{1}{2}\left(\frac{\sigma_c}{\sigma_t} - 3\right)^{1/2}$.

The criterion for rock damage is

$$\tau^2 \geq (\sigma + \sigma_t)^2 \tan^2\eta + (\sigma + \sigma_t)\sigma_t.$$

(2.38)

The calculation formula of $\tan \eta$ indicates that when $\sigma_c/\sigma_t < 3$, $\tan \eta$ will have a void value. Thus, this model does not apply to rocks with $\sigma_c/\sigma_t < 3$.

(3) Parabolic criterion

The quadratic parabolic strength criterion curve is applicable to rocks with weak lithology, such as mudstone and shale. It is expressed as

$$\tau^2 = n(\sigma + \sigma_t).$$

(2.39)

where σ_t is the uniaxial compressive strength of the rock and n is the undetermined coefficient.

According to the geometric relationship shown in Figure 2.12, Formula (2.39) can be converted into

$$(\sigma_1 - \sigma_3)^2 = 2n(\sigma_1 + \sigma_3) + 4n\sigma_t - n^2.$$

(2.40)

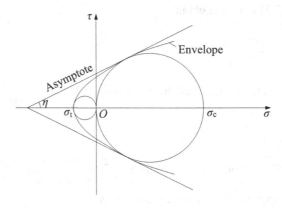

Figure 2.11 Hyperbolic strength[6]

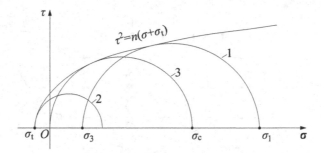

Figure 2.12 Quadratic parabolic strength

1 – Any triaxial compression stress circle

2 – Tensile stress circle

3 – Uniaxial compressive stress circle

Under uniaxial compression, $\sigma_3 = 0$, $\sigma_1 = \sigma_c$, and Formula (2.40),

$$n = \sigma_c + 2\sigma_t \pm 2\sqrt{\sigma_c\left(\sigma_c + \sigma_t\right)} \tag{2.41}$$

By substituting Formula (2.41) into Formula (2.39), the criterion for rock damage is expressed as

$$\tau^2 \geq \left(\sigma_c + 2\sigma_t \pm 2\sqrt{\sigma_c\left(\sigma_c + \sigma_t\right)}\right)\cdot\left(\sigma + \sigma_t\right) \tag{2.42}$$

2.4.2 Griffith theory

Rocks are not an ideal continuous medium, and they contain numerous micro-fissures. Griffith believed that when a brittle material is stressed, the end of the internal micro-fissures generates concentrated stress. When the concentrated stress exceeds the tensile strength of the material, the micro-fissures begins to expand, combine and accommodate. Finally, a macroscopic rupture appears along one or several planes or on the curved surfaces of the rock.

Griffith's theory assumes that the fissures in a rock are flat and elliptical and that they do not affect one another (Figure 2.13).

From Figure 2.13, the tensile stress generated near the internal fissure tip is

$$\sigma_b m = \sigma_y \pm \left(\sigma_y^2 + \tau_{xy}^2\right)^{1/2}, \tag{2.43}$$

where σ_b is the maximum tangential stress around the ellipse, coefficient m is calculated as $m = b/a$ (a is the major semi-axis of the ellipse, and b is the minor semi-axis), σ_y is the normal stress in the direction of vertical fissures and τ_{xy} is the shear stress in the direction parallel to the fissures.

When the tangential stress σ_b around the fissures reaches or exceeds the tensile strength σ_t of the rock, the fissures begin to expand. The σ_t and m of the sidewall of the fissures are difficult to determine. Accordingly, the simplest case can be assumed, that is, the long axis

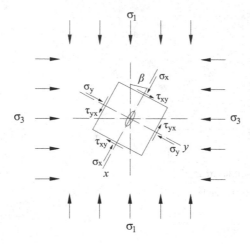

Figure 2.13 Stresses on the periphery of a unit with a crack inclined at an angle β to the maximum principal stress[3]

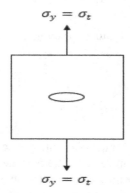

Figure 2.14 Fissure propagation during uniaxial tension

of the fissures is perpendicular to the uniaxial tensile stress to determine the critical value in the tangential stress (Figure 2.14).

(1) The strength criterion can be expressed with σ_y and τ_{xy}:

When the rock is uniaxially stretched to damage, $\tau_{xy} = 0$, $\sigma_y = \sigma_t$, from Formula (2.43). Then,

$$\sigma_b m = \sigma_t + \sigma_t = 2\sigma_t. \qquad (2.44)$$

Thus, the strength criterion can be expressed as

$$2\sigma_t \le \sigma_y \pm \left(\sigma_y^2 + \tau_{xy}^2 \right)^{1/2}. \qquad (2.45)$$

This formula is the Griffith strength criterion, that is, the relation between the normal stress σ_y and the shear stress τ_{xy} when the fissure starts to expand.

(2)   The strength criterion can be expressed with σ_1 and σ_3:
      σ_y and τ_{xy} are replaced with σ_1 and σ_3, respectively. Then,

   1)   When $\sigma_1 + 3\sigma_3 \geq 0$, crack failure angle β is

$$\cos 2\beta = \frac{\sigma_1 - \sigma_3}{2(\sigma_1 + \sigma_3)}. \tag{2.46}$$

Then, the Griffith strength criterion can be expressed as

$$\frac{(\sigma_1 - \sigma_3)^2}{\sigma_1 + \sigma_3} \geq -8\sigma_t. \tag{2.47}$$

   2)   When $\sigma_1 + 3\sigma_3 < 0$, crack failure angle β is

$$\beta = 0.$$

Then, the Griffith strength criterion can be expressed as

$$|\sigma_3| \geq |\sigma_t|. \tag{2.48}$$

According to this criterion, the following formula can be obtained under the uniaxial stress $\sigma_3 = 0$ from (2.47):

$$\sigma_c = -8\sigma_t. \tag{2.49}$$

The Griffith strength criterion is proposed for brittle materials such as glass and steel, and it is therefore only suitable for studying the destruction of brittle rocks. For general rock materials, the Mohr-Coulomb strength criterion is much more applicable than the Griffith counterpart.

References

[1] Geotechnical Control Office. (1988) *Guide to Rock and Soil Descriptions (Geoguide 3)*. Geotechnical Control Office, Hong Kong.
[2] Goodman, R.E. (1989) *Introduction to Rock Mechanics*. 2nd ed. Wiley, Toronto, Canada.
[3] Liu, Y.R. & Tang, H.M. (2009) *Rock Mass Mechanics*. Chemical Industry Press, Beijing, China.
[4] Shen, M.R. & Chen, J.F. (2006) *Rock Mass Mechanics*. Tongji University Press, Shanghai.
[5] Cai, M.F. (2013) *Rock Mechanics and Engineering*. Science Press, Beijing, China.
[6] Ren, J.X. (2013) *Rock Mass Mechanics*. China University of Mining and Technology Press, Xuzhou.

Chapter 3

Characteristics of rock mass

Rock mass usually refers to natural geological rock mass within a certain engineering range which undergoes various geological processes and retains permanent deformation and geological structure traces under the long-term effect of ground stress.[1] In a narrow sense, rock mass is composed of rocks (rock blocks) and discontinuous/structural planes. The structure plays a major role in controlling the mechanics and deformation characteristics of rock mass. Therefore, the properties of the structural plane must be considered in the engineering classification of rock mass.

3.1 Discontinuities in rock mass

The structural plane refers to the planar geological interface formed in the rock mass during the development of geological history. It has a certain extension direction and length and is relatively thin. The tensile strength of the structural plane is close to zero.[2–3] The structural plane includes material differentiation surfaces and discontinuous surfaces such as layers, unconformities, joints, faults and schistosity.

Structural planes are classified into three types according to geological origin: primary, tectonic and secondary structural planes (Table 3.1).

Primary structural planes: Primary structural planes are formed by rock mass in the diagenetic stage. According to the different genesis of rocks, this plane can be classified into sedimentary, magmatic and metamorphic structural planes. The sedimentary rock structural plane is the geological interface formed during diagenesis, and it includes interlayer structural planes such as layers, beddings, inter-deposited sections (unconformity and disconformity surfaces) and primary weak interlayers. The magmatic structural plane is formed by magma intrusion and effusion condensation (e.g. contact surfaces between (a) immersion body and the surrounding rock and (b) dike and the original condensation joint). The metamorphic structural plane is the structural plane formed during the metamorphism of rock mass, and it includes morphological and weak schist interlayers.

Tectonic structural plane: Cleavage, joints, faults and interlayer displacements make up the tectonic structural plane which refers to the plane formed by rock mass under tectogenesis. Cleavage is a planar structure in which the rock can be split into a number of thin slices in a certain direction. The relatively small scale of joints is divided into tension and shear joints depending on the causes of the mechanics. Fault is the discontinuity in which a rock breaks under crustal stress whilst a relative displacement occurs between two fractured blocks. The direction of the displacement is generally parallel to the discontinuous surface. Most faults are formed by shearing, and tectonic rocks are generally present in fault zones.

Table 3.1 Classification of structural planes in rock mass[2]

Genetic type	Geological type	Main features		Characteristic	Engineering geology evaluation
		Occurrence	Distribution		
Primary structural plane	Sedimentary structural plane	Bedding and level; weak intercalation; unconformity and disconformity surfaces; sedimentary discontinuity	Generally consistent with the occurrence of rock formations; interlaminar structural plane	Such marine structural plane formation exhibits stability; the terrestrial type is arranged in a staggered formation for easy pinching.	Structural planes such as layers and weak interlayers are relatively flat; the unconformity and sedimentary sections are mostly composed of clastic mud and are uneven.	Dam sliding and landslides, such as the destruction of the Austin Dam, St. Francis Dam and Malpasai Dam and the massive landslides near the Wayi Reservoir, are caused by such structural planes.
	Magmatic structural plane	The contact surface between the intrusion and the surrounding rocks; the contact between the vein and the rock wall; primary condensation joints	The rock vein is controlled by the tectonic plane whilst the primary joint is controlled by the contact surface of the rock mass.	The contact surface extends farther and is more stable, whereas the original joints tend to be short and dense.	The contact surface with the surrounding rock can have two different characteristics of fusion and fracture. The original joint is generally a cracked surface which is rough and uneven.	Generally, it does not cause large-scale rock mass damage, but sometimes, it can form a slip of the rock mass if it cooperates with the structural fault, such as some local slip of the abutment.
	Metamorphic structural plane	Schistosity	Appearance is consistent with rock formation or structural direction.	The film is short, and the distribution is extremely dense. The weak schist interlayer extends farther and has a fixed level.	The structural plane is smooth and straight, and the schistosity is often closed into a hidden structural plane in the deep part of the rock. The weak schist interlayer has flaky minerals and is scaly.	In shallow metamorphic sedimentary rocks, such as the landslides of ridges and other common slopes, the schist interlayer has an impact on the engineering and stability of underground caves.
Tectonic structural plane		Joint (X-shaped joint, split joint) fault (rushing fault, transverse fault); interlayer shift; feather-like fissures; cleavage	The occurrence is related to the tectonic line, and the interlayer displacement is consistent with the stratum.	The tensile fracture is short, the shear fracture extends farther, and the scale of the compressive fracture is huge, but sometimes, the transverse fault is cut into discontinuities.	The tensile fracture is uneven, often with secondary filling, and is serrated; shear fracture is relatively straight with pin-shaped fissures. Compressive faults have a variety of tectonic rocks which are distributed in strips that often contain fault mud and mylonite.	It has a great influence on the stability of rock mass. Most of the rock mass failure processes mentioned have the coordination effect of structural planes and often cause collapse and roof collapse of slopes and underground works.
Secondary structural plane		Unloading cracks; differentiating fissures; differentiated interlayers; mud interlayers; secondary mud layers	It is controlled by topography and the original structural plane.	The distribution tends to be discontinuous, lenticular, poorly ductile and mainly developed in the surface differentiation zone.	Generally filled with muddy materials; the water-physical property is poor.	Hazards occur on natural and artificial slopes, sometimes affecting dam foundations, dam shoulders and shallow tunnels, but they are generally treated in the foundation during construction.

The interlayer fault zone is a structural plane commonly found in layered rock masses with an occurrence that is generally consistent with the rock formation.

Secondary structural plane: The secondary structural plane refers to the structural plane formed by the external force (e.g. weathering, groundwater and unloading) after the formation of rock mass. This layer includes unloading and weathering fissures, secondary clay layers and muddy interlayers. Unloading fissures occur mostly on the surface of rock masses with free-form surface conditions with common ductility, especially on the banks of deep-cutting gorges. Weathered fissures usually develop along the original interlayers and the original structural plane. These short and dense layers have poor ductility and are limited to a certain depth. The muddy interlayer is formed by the action of water which makes the soft material muddy in the interlayer. The occurrence of the muddy interlayer is similar to that of rock formation, and the degree of muddiness varies depending on groundwater conditions.

The secondary structural plane can be divided according to the extension length of the structural plane, depth of the cut, width of the fracture zone and its own mechanical effects.[2,4,5]

Level I structural plane mainly refers to large or regional faults which generally extend from several kilometres to tens of kilometres, with the fracture bandwidth ranging from several metres to tens or even several hundred metres. The level I structural plane is a large-scale weak structural plane that constitutes an independent medium mechanical unit. Some regional large faults are often associated with modern activities that are related to the stability of the crust in the construction area and thus bring great potential damage to the construction. Construction projects should avoid level I structural planes by implementing effective measures.

Level II structural plane refers to a regional geological interface extending from a few hundred metres to several kilometres with a narrow width (several centimetres to several metres), such as interlayer displacement, unconformity and primary weak interlayer. This structural plane is a weak structural plane that can form a block-fracture boundary which controls several factors that affect project layout, such as the stability of the mountain in the construction area and the deformation and failure mode of the rock mass.

Level III structural plane refers to faults, regional joints, well-extended layers and interlayer shifts with lengths ranging from tens of metres to hundreds of metres and widths ranging from a few centimetres to about one metre. Most level III structural faces belong to hard structural planes, whereas a few belong to weak structural planes. This plane mainly affects or controls the stability of engineering rock masses; it forms block damage of different scales when combined with Levels I and II structural faces.

Level IV structural plane refers to joints, layers, secondary fractures, small faults and developed morphological and physiologic surfaces. The length is generally tens of centimetres to twenty or thirty metres (maximum of ten centimetres for small ones), and the widths range from zero to several centimetres. This plane can form the boundary of rock blocks, destroy the integrity of rock masses and affect the physical and mechanical properties of rock masses and the state of stress distribution. This level includes numerous structural planes with random distribution, thereby affecting the integrity and mechanical properties of rock masses. Aside from being the main content of rock mass classification and structural research, Level IV structural plane is also the key input in the statistical analysis and simulation of structural planes.

Level V structural plane, which is also called the microstructure plane, comprises hidden joints, microlayers, micro-fissures and undeveloped cleavages and textures. The small size, poor continuity and random distribution of this layer reduces the strength of rock mass, mainly affecting the mechanical properties of rocks. If the distribution is dense, rocks can become loose media under weathering.

3.2 Description of rock mass

A rock mass is often described according to strength, color, structure, degree of weathering, name, structural plane and other geological information. Prior to the description, the rock mass should be first divided into units with consistent engineering characteristics. Rock type, weathering degree and structural plane characteristics are usually adopted as the basis for the boundary selection of a rock mass unit. Information about rock mass characteristics should contain the description of geological structures, characteristics of discontinuities and weathering characteristics.[6]

3.2.1 Structure of rock mass

Structure of rock mass refers to the macroscale structure. In geology, construction generally encompasses faults and folds which are often described as large-scale rock structures. Engineers are often concerned about the structural features of small rock masses. Sedimentary rocks are also referred to as 'layered', 'thin-plated' and 'blocky'. For magmatic (and volcanic clastic rocks) and metamorphic rocks, other terms include 'blocky' and 'flow-like', 'leaf-shaped', 'band-shaped' and 'cleavage-shaped'. Construction spacing can be classified as very thick, thick, moderate, thin and very thin.

3.2.2 Characteristics of structural plane

A comprehensive description of the structural plane contains information about occurrence, spacing, continuity, roughness, aperture, filling characteristics and permeability. The survey of structural planes can be divided into two levels: (1) subjective surveys, which only describe the structural planes exerting a significant impact on the project, and (2) objective surveys, which describe all structural planes that intersect the baseline or are located within the defined area.

(1) Occurrence

The appearance of a structural plane can be expressed by the dip and dip angle. A dip is represented by a clockwise angle from the true north direction of the structural plane's tilting direction, whereas a dip angle is expressed by the maximum inclination angle of the structural plane measured in the horizontal direction. Dips and dip angles are commonly measured using compasses and inclinometers. The general expression of the appearance is 'dip direction∠dip angle'.

(2) Spacing

The spacing of structural planes refers to the average spacing in the normal direction of a set of structural planes. The description standards recommended by the International Society of Rock Mechanics are shown in Table 3.2.

Table 3.2 Description of spacing between structural planes[7]

Spacing	Descriptive term
<20 mm	Extremely close-spaced
20–60 mm	Very close-spaced
60–200 mm	Close-spaced
200–600 mm	Medium-spaced
600–2000 mm	Wide-spaced
2000–6000 mm	Very wide-spaced
>6000 mm	Extremely wide-spaced

(3) Continuity

Continuity, which is basically the size of a structural plane, is a crucial description of structural planes. Obtaining the quantity of the three-dimensional shape of a structural plane is difficult. Continuity can only be approximated by measuring the length of the structural plane's traces (i.e. the rock mass exposed to the surface). In the description, specifying whether the structural plane terminates inside the rock or in other structural planes is necessary.

(4) Roughness

The roughness of a structural plane has two parts: corrugation and unevenness. Corrugation, or the fluctuation of structural planes within large-scale ranges (tens of metres long), can be evaluated using wavelength and amplitude. Unevenness refers to the surface irregularities of a structural plane within small-scale ranges (few centimetres to a few metres), as described by Barton's 10-level standard.[8]

(5) Aperture

The degree of aperture is the vertical distance between the rock walls on both sides of an open structural plane. It is usually caused by tensile stress, rinsing and dissolution of filler materials or shearing of a substantially rough structural plane. This characteristic determines the shear strength and hydraulic conductivity of a structural plane. The classification for aperture degrees is shown in Table 3.3.

(6) Characteristics of filling materials

This characteristic describes the filling materials within structural planes. In most cases, parts of the filling materials of a structural plane contains foreign materials which are formed because of the intense weathering in the structural plane. The mechanical strength of the filling material is generally weaker than that of the parent rock. Typical filling materials are soil, weathered or decomposed rock minerals (e.g. quartz, calcite, manganese or kaolin) and breccia formed at the faults or shear zones.

The mineral type, particle size and strength of the filling material should be comprehensively described. The width (maximum, minimum and average), moisture and permeability of the filling material should also be stated.

Table 3.3 Classification of aperture

Descriptive term	Aperture distance between discontinuity walls
Wide	>200 mm
Moderately wide	60–200 mm
Moderately narrow	20–60 mm
Narrow	6–20 mm
Very narrow	2–6 mm
Extremely narrow	0–2 mm
Tight	0 mm

(7) Permeability

The permeability assessment of structural planes is important in engineering practice. The structural plane's permeability is often described as dry, wet (without flowing water) or with existing flow. For the case of existing flow, the amount of flow and flow rate should be recorded. The observation date should also be noted to distinguish dry and rainy seasons.

(8) Fracture state

Core fracture can be described using the following parameters: total core recovery, solid recovery, rock quality index and fracture index. 'Solid core' is the key term in assessing the fracture state. It refers to the core between two natural structural planes. The entire core is not necessarily cylindrical, but at least one complete circular cross section should exist along the axial direction.

Total core recovery is the percentage of the recovered core taken from a certain footage per roundtrip regardless of whether the cross section is complete, partially complete or incomplete.

The rock quality designation (RQD) is the percentage of the total length of the solid core (for lengths greater than 100 mm) per footage per round trip.

The fracture index (FI) refers to the number of structural planes that can be clearly identified per metre of the core. It is measured over the length of the core with a uniform fracture distribution. If the fracture frequency of the core taken from a certain footage per round trip is significantly changed, the fracture index of each part should be separately calculated. When the core is fragmented, FI is described as 'incomplete'.

(9) Rock mass weathering

From the surface to deeper grounds, the weathering intensity of a rock declines. By determining the weathering zone of the rock mass, the rock mass is classified according to the volumetric proportion of the weathering zone. However, the contour of the weathering zone cannot be accurately determined because the three-dimensional features of the rock mass are difficult to obtain. In practice, only the volume percentage can be estimated.

(10) Additional information

When describing rock masses, other information that can help engineers to understand the properties of the rock mass should be recorded. Examples are the geometry of the pores in the carbonate rock, surrounding structural plane, groundwater condition and permeability characteristics.

3.3 Engineering classification of rock mass

In engineering, the purpose of rock mass classification is to comprehensively analyse the geological conditions that affect the stability and the physical and mechanical properties of rock masses. The classification divides rock masses into several categories with different degrees of stability to guide planning, design and construction. Two examples of internationally applied engineering classification systems of rock masses are the rock mass rating (RMR) system proposed by Bieniawski (1974, 1984) and the tunnelling quality index (Q) system proposed by Barton *et al.* (1974).

3.3.1 RMR system

The RMR classification system, also known as the geo-mechanical classification, was proposed by Professor Z.T. Bieniawski (1973) in South Africa. This system is mainly designed for tunnels, mining and foundation construction. The system is developed on the basis of rock strengths (Table 3.4), drill core quality (Table 3.5), joint fracture spacing (Table 3.6), joint characteristics (Table 3.7) and groundwater conditions (Table 3.8). RMR increments corresponding to each parameter are summed to determine the total RMR.

Table 3.4 Rating of rock strength in RMR

Point load strength (MPa)	Uniaxial Compressive Strength (MPa)	Rating
>10	>250	15
4–10	100–250	12
2–4	50–100	7
1–2	25–50	4
Not applicable	10–25	2
	3–10	1
	<3	0

Table 3.5 RQD ratings

RQD (%)	90–100	75–90	50–75	25–50	<25
Rating	20	17	13	8	3

Table 3.6 RMR rankings of joint spacing (most influential set)

Joint spacing (m)	>2.0	0.6–2.0	0.2–0.6	0.06–0.2	<0.06
Rating	20	15	10	8	5

Table 3.7 Ratings of joints in RMR

Description	Rating
Very rough surfaces of limited extent; hard wall rock	30
Slightly rough surfaces; aperture less than 1 mm; hard wall rock	25
Slightly rough surfaces; aperture less than 1 mm; soft wall rock	20
Smooth surfaces, 1–5 mm-thick gouge filling or aperture of 1–5 mm; joints extend more than several metres	10
Open joints filled with more than 5 mm of gouge or open by more than 5 mm; joints extend more than several metres	0

Table 3.8 Rating of groundwater condition in RMR

Inflow per 10 m tunnel length (L·min⁻¹)	Joint water pressure divided by major principal stress	General condition	Rating
None	0	Completely dry	15
<10	<0.1	Damp	10
10–25	0.1–0.2	Wet	7
25–125	0.2–0.5	Dripping	4
>125	>0.5	Flowing	0

Table 3.9 Modification of RMR values for joint orientations

Assessment of influence of orientation on the work	Rating increment for tunnels	Rating increment for foundations
Very favourable	0	0
Favourable	−2	−2
Fair	−5	−7
Unfavourable	−10	−15
Very unfavourable	−12	−25

Table 3.10 RMR classification of rock mass

RMR	100–81	80–61	60–41	40–21	21<
Class	I	II	III	IV	V
Description	Very good rock	Good rock	Fair rock	Poor rock	Very poor rock

Bieniawski (1973) recommended adjusting the sum of the first five ratings to account for favourable or unfavourable orientations because the orientation of joints relative to the work can affect the behavior of the rock.[9] Table 3.9 illustrates the adjustments to account for the adverse effects of joints and fissures on the stability of rock masses. The modified RMR places the rock in one of the five categories defined in Table 3.10.

3.3.2 Q system

The Q system is a tunnel excavation quality classification system proposed by the Norwegian Institute of Geotechnical Engineering (Barton *et al.*, 1974). The classification parameters of the Q system are similar to those of the RMR system. The Q value can be expressed as

$$Q = \left(\frac{RQD}{J_n}\right) \times \left(\frac{J_r}{J_a}\right) \times \left(\frac{J_w}{SRF}\right), \qquad (3.1)$$

where RQD is the rock quality designation, J_n is the number of joint sets (Table 3.11), J_r denotes the roughness of the joints (Table 3.12), J_a is the wall rock condition and/or filling material (Table 3.13), J_w is the water flow characteristic of the rock (Table 3.14) and stress reduction factor (*SRF*) refers to the looseness and stress conditions (Table 3.15).

Table 3.11 Values of J_n[10]

Joint set number	J_n
A. Massive, no or few joints	0.5–1
B. One joint set	2
C. One joint set plus random joints	3
D. Two joint sets	4
E. Two joint sets plus random joints	6
F. Three joint sets	9
G. Three joint sets plus random joints	12
H. Four or more joint sets, random, heavily jointed, 'sugar-cube', etc.	15
J. Crushed rock, earth-like	20

Table 3.12 Values of J_r[10]

Joint roughness number	J_r
(a) Rock–wall contact	
(b) Rock–wall contact before 10 cm sear	
A. Discontinuous joints	4
B. Rough or irregular, undulating	3
C. Smooth, undulating	2
D. Slicken-sided, undulating	1.5
E. Rough or irregular, planar	1.5
F. Smooth, planar	1.0
G. Slicken-sided, planar	0.5
(c) No rock–wall contact when sheared	
H. Zone containing clay minerals thick enough to prevent rock–wall contact	1.0
J. Sandy, gravely or crushed zone thick enough to prevent rock–wall contact	1.0

Table 3.13 Values of J_a[10]

Joint alteration number	ϕ_r (°)	J_a
(a) Rock–wall contact (no mineral fillings, only coatings)		
A. Tightly healed, hard, non-softening and impermeable filling (i.e. quartz or epidote)	-	0.75
B. Unaltered joint walls, surface staining only	25–35	1.0
C. Slightly altered joint walls, non-softening mineral coatings, sandy particles, clay-free disintegrated rock, etc.	25–30	2.0
D. Silty- or sandy-clay coatings, small clay fraction (non-softening)	20–25	3.0
E. Softening or low friction clay mineral coatings (i.e. kaolinite or mica, chlorite, talc, gypsum, graphite and small quantities of swelling clays)	8–16	4.0
(b) Rock–wall contact before 10 cm shear (thin mineral fillings)		
F. Sandy particles, clay-free disintegrated rock, etc.	25–30	4.0
G. Strongly over-consolidated non-softening clay mineral fillings (continuous, 5 mm thickness)	16–24	6.0
H. Medium or low over-consolidation, softening, clay mineral fillings (continuous, but <5 mm thickness)	8–12	8
J. Swelling clay fillings, i.e. montmorillonite (continuous, but >5 mm thickness). Value of J_a depends on the percentage of swelling clay-sized particles, access to water, etc.	6–12	8–12
(c) No rock–wall contact when sheared (thick mineral fillings)		
KLM. Zones or bands of disintegrated or crushed rock and clay (see G, H and J for description of clay conditions)	6–24	6, 8 or 8–12
N. Zones or bands of silty- or sandy-clay, small clay fraction (non-softening)	-	5.0
OPR. Thick, continuous zones or bands of clay (see G, H and J for description of clay conditions)	6–24	10, 13 or 13–20

Table 3.14 Value of J_w[10]

Joint water reduction factor	Water pressure (MPa)	J_w
A. Dry excavations or minor inflow (i.e. <5 L/min locally)	<0.1	1.
B. Medium inflow or pressure, occasional outwash of joint fillings	0.1–0.25	0.66
C. Large inflow or high pressure in competent rock with unfilled joints	0.25–1	0.5
D. Large inflow or high pressure, considerable outwash of joint fillings	0.25–1	0.33
E. Exceptionally high inflow or water pressure at blasting, decaying with time	>1	0.2–0.1
F. Exceptionally high inflow or water pressure continuing without noticeable decay	>1	0.1–0.05

Table 3.15 Value of SRF[10]

Stress reduction factor			SRF
(a) Weakness zones intersecting excavation which may cause loosening of rock mass when tunnel is excavated			
A. Multiple occurrences of weakness zones containing clay or chemically disintegrated rock, very loose surrounding rock (any depth)			10.0
B. Single weakness zones containing clay or chemically disintegrated rock (depth of excavation ≤50 m)			5.0
C. Single weakness zones containing clay or chemically disintegrated rock (depth of excavation >50 m)			2.5
D. Multiple shear zones in competent rock (clay-free), loose surrounding rock (any depth)			7.5
E. Single shear zones in competent rock (clay-free) (depth of excavation <50 m)			5.0
F. Single shear zones in competent rock (clay-free) (depth of excavation >50 m)			2.5
G. Loose, open joints, heavily jointed or 'sugar cube', etc. (any depth)			5.0
(b) Competent rock, rock stress problems			
	σ_c/σ_1	σ_θ/σ_c	SRF
H. Low stress, near surface, open joints	>200	<0.01	2.5
J. Medium stress, favourable stress condition	200–10	0.01–0.3	1.0
K. High stress, very tight structure, usually favourable to stability, may be unfavourable for wall stability	10–5	0.3–0.4	0.5–2.0
L. Moderate slabbing after >1 h in massive rock	5–3	0.5–0.65	5–50
M. Slabbing and rock burst after a few minutes in massive rock	3–2	0.65–1	50–200
N. Heavy rock burst (strain-burst) and immediate dynamic deformations in massive rock	<2	>1	200–400
(c) Squeezing rock: plastic flow of incompetent rock under the influence of high rock pressure		σ_θ/σ_c	SRF
O. Mild squeezing rock pressure		1–5	5–10
P. Heavy squeezing rock pressure		>5	10–20
(d) Swelling rock: chemical swelling activity depending on presence of water			
R. Mild swelling rock pressure			5–10
S. Heavy swelling rock pressure			10–15

The first term of Equation 3.1 is the measure of the joint block sizes, the second factor expresses the shear strength of the block surfaces, and the last factor evaluates the important environmental conditions that influence the behavior of the rock mass. The value of Q ranges from 0.001 to 1000 whilst the quality of the surrounding rock ranges from low to high and can be divided into nine grades (Table 3.16).

Table 3.16 Rock mass classification in Q system[10]

Q	Rock Mass Quality
0.001–0.01	Exceptionally poor
0.01–0.1	Extremely poor
0.1–1	Very poor
1–4	Poor
4–10	Fair
10–40	Good
40–100	Very good
100–400	Extremely good
400–1000	Exceptionally good

The parameters of the Q system can be converted into RMR parameters and vice versa using the connecting relationship proposed by Barton (1993):

$$RMR \approx \lg Q + 50, \tag{3.2}$$

3.4 Strength theory of rock mass

The theory of rock mass strength establishes the criterion for judging rock failure by analysing the stress state of rock mass failure. The empirical models for judging structural plane damage include the joint roughness coefficient–joint compressive strength (JRC-JCS) model, Patton's law and the Gerrard shear formula.[11] This work focuses on the JRC-JCS model, which is the most widely used model in engineering. Models for determining the overall strength of the rock mass include the Hoek-Brown, Bieniawski and Balmer strength criteria, but this study only discusses the Hoek-Brown strength criterion.

3.4.1 JRC-JCS model

The JRC-JCS model is a formula proposed by Barton (1977) for the shear strength of structural planes on the basis of numerous structural plane shear tests.

$$\tau_p = \sigma_n \tan(JRC \lg \frac{JCS}{\sigma_n} + \phi_b), \tag{3.3}$$

where ϕ_b is the basic friction angle, σ_n is the normal force on the structural plane, JRC is the roughness coefficient of the structural plane and JCS is the compressive strength of the wall rock. The JRC-JCS model is used to estimate the peak shear strength of the structural plane under low normal force.

(1) Determination of JCS

The method for determining JCS depends on the weathering degree of the structural plane. For non-weathered or slightly weathered rock masses, the strength of the structural wall rock is approximately equal to that of the rock mass. JCS can adopt the uniaxial compressive strength or point load strength of the rock sample. If the structural wall rock is weathered, then the rebound value R can be measured by the L-type rebound hammer. In addition, the gravity of rock γ is also measured. The JCS is calculated according to the calculation in Figure 3.1 and Equation (3.4).

$$\lg\left(JCS\right) = 0.00088\gamma R + 1.01 \tag{3.4}$$

(2) Determination of ϕ_b

Barton's (1977) research showed that the basic friction angle of the rock mass is mostly between 25° and 35°. The basic friction angle can be determined using the tilt test of the flat surface of the structural plane's wall rock. The method involves obtaining the test rock block from structural plane wall, cutting the block into two halves, removing the rock powder and combining them after air drying. For each type of rock, more than 10 test pieces must be examined. When tilt test data are not available, the residual friction angle of the structural plane ($\phi_b = 30°$) can be adopted.

(3) Determination of JRC

Barton suggested that the value of JRC can be determined using the standard contour curve comparison method and direct shear test back calculation. The standard contour curve comparison method is adopted to artificially compare the contour curve of the surface of the

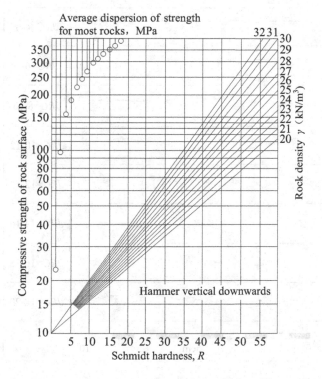

Figure 3.1 Relationship between JCS, Schmidt hardness and rock density[2]

structural plane with the standard contour curve to obtain the value of JRC (Figure 3.2). The direct shear test inverse algorithm calculates the value of JRC by performing the direct shear test to obtain the peak shear strength and the basic friction angle.

$$JRC = \frac{\phi_p - \phi_b}{\lg(JCS / \sigma_n)},$$ (3.5)

where ϕ_p is the peak shear angle ($\phi_p = \arctan[\tau_p/\sigma_n]$).

The common methods used to determine the value of JRC includes fractal theory, statistical parameter method, straight line and trace length ratio method and straight and corrected straight edge method.[12] The following discussions introduce fractal theory.

Joints are measured with different lengths r. The method for measurement is shown in Figure 3.3. If the rock joint has fractal self-similarity, the relationship between the fractal dimension D of the rock joint and the number of measured sizes N can be expressed as

$$N = ar^{1-D}.$$ (3.6)

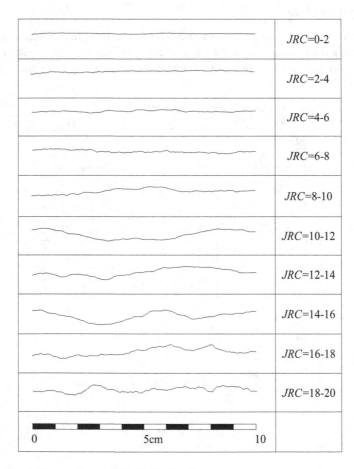

	JRC=0-2
	JRC=2-4
	JRC=4-6
	JRC=6-8
	JRC=8-10
	JRC=10-12
	JRC=12-14
	JRC=14-16
	JRC=16-18
	JRC=18-20

0   5cm   10

Figure 3.2 Standard joint profiles and the JRC values[4]

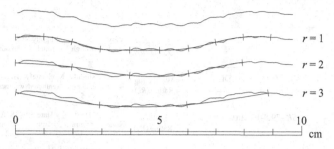

Figure 3.3 Fractal method for measuring joint[4]

By transforming r, D and different values of N can be obtained. Li (2015) calculated D using contour curves and their corresponding JRC values and obtained the following empirical formula:[13]

$$JRC = 520.28(D-1)^{0.7588}.$$ (3.7)

In addition, the JRC-JCS model considers the size effects, which can be corrected as

$$JRC_n \approx JRC_0 \left[\frac{L_n}{L_0}\right]^{-0.02JRC_0},$$ (3.8)

$$JCS_n \approx JCS_0 \left[\frac{L_n}{L_0}\right]^{-0.03JRC_0},$$ (3.9)

where L_0 is the length of the structural plane of the test sample ($L_0 = 100$ mm); L_n is the length of the field structure plane; JRC_0 and JCS_0 are the roughness coefficient of the test structural plane and the strength of the structural wall rock, respectively; and JRC_n and JCS_n are the roughness coefficient of the field structural plane and the strength of the structural wall rock, respectively.

3.4.2 Hoek-Brown criterion

The Hoek-Brown strength criterion is a rock failure criterion proposed by Hoek and Brown (1988) for underground hard rock excavation. The criterion is based on numerous experiments and studies on the properties of jointed rock masses. The mechanical parameters are provided to the rock mass by reducing the mechanical parameters of the intact rock, which can be expressed as

$$\sigma_1' = \sigma_3' + \left(m_b \sigma_{ci} \sigma_3' + s\sigma_{ci}^2\right)^a,$$ (3.10)

where σ_1' is the maximum effective principal stress (MPa) at the time of failure, σ_3' is the minimum effective principal stress (MPa) at the time of failure and σ_{ci} is the uniaxial

Table 3.17 GSI[14]

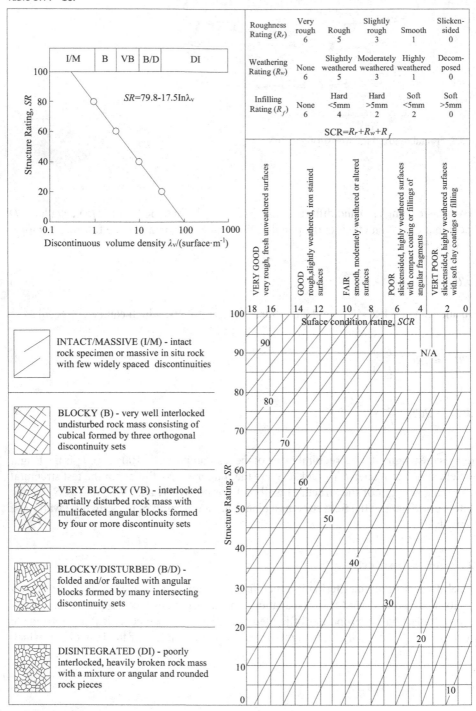

Note: N/A in the figure is the area where the GSI classification method is not app

Table 3.18 Values of m_i of various rocks[14]

Rock type	Class	Group	Texture			
			Coarse	Medium	Fine	Very fine
Sedimentary	Clastic		Conglomerate (21±3)* Breccia (19±5)	Sandstone 17±4	Siltstone 7±2 Greywacke (18±3)	Claystone 4±2 Shale (6±2) Marl (7±2)
	Non-Clastic	Carbonate	Crystalline Limestone (12±3)	Sparitic Limestone (10±2)	Micritic Limestones (9±2)	Dolomite (9±3)
		Evaporite		Gypsum 8±2	Anhydrite 12±2	
		Organic				Chalk 7±2
Metamorphic	Non-foliated		Marble 9±3	Hornfel (19±4) Metasandstone (19±3)	Quartzite 20±3	
	Slightly foliated		Migmatite (29±3)	Amphibolite 26±6	Gneiss 28±5	
	Foliated			Schist 12±3	Phyllite (7±3)	Slate 7±4
Igneous	Plutonic	Light	Granite 32±3 Granodiorite (29±3)	Diorite 25±5		
		Dark	Gabbro 27±3 Norite 20±5	Dolerite (16±5)		
	Hypabyssal		Porphyria (20±5)		Diabase (15±5)	Peridotite (25±5)
	Volcanic	Lava		Rhyolite (25±5) Andesite 25±5	Dacite (25±3) Basalt (25±5)	Obsidian (19±3)
		Pyroclastic	Agglomerate (19±3)	Breccia (19±5)	Tuff (13±5)	

Note: The values in parentheses are estimates.

'*' indicates that the value in the row is for a rock sample perpendicular to the slice plane; the value of m_i will be significantly different if failure occurs along a weakness plane.

compressive strength (MPa) of the rock. m_b, s and a are empirical parameters that reflect the characteristics of the rock mass, and they can be respectively obtained as

$$m_b = \exp\left(\frac{GSI-100}{28-14D}\right)m_i, \tag{3.11}$$

Table 3.19 Values of D[15]

Picture of Rock Masses	Description of Rock Masses	Reference value D
	Well-controlled blasting or TBM excavation produces minimal disturbances.	D=0
	Mechanical or manual excavation of rock mass (no blasting) produces minimal disturbance on the surrounding rock mass. The squeezing problem causes a significant floor arching that is severely disturbed unless a temporary transition (as shown in the picture) is available.	D=0 D=0.5 (No conversion)
	Poor blasting in hard rock tunnels produces severe local damage (the tunnel profile is extended by 2 or 3 m) and rock masses.	D=0.8
	Small-scale blasting causes severe rock damage in geotechnical slope engineering, especially when controlled blasting is adopted (as shown on the left). However, stress relief can cause certain damage.	D=0.7 (Good blasting) D=1.0 (Poor blasting)

Picture of Rock Masses	Description of Rock Masses	Reference value D
	Vast open pit slopes are subject to severe production blasting, and the stresses on the overburden are released, causing disturbances. Some soft rocks can be excavated with little disturbance by crushing and overturning.	D=1.0 (Production blasting) D=0.7 (Machine excavating)

$$s = \exp\left(\frac{GSI - 100}{9 - 3D}\right),$$ (3.12)

$$a = 0.5 + \frac{1}{6}\left[\exp(-GSI/15) - \exp(-20/3)\right],$$ (3.13)

where GSI stands for geological strength index and m_i is the empirical parameter of the rock dimension ($m_i = 1$) which reflects the softness and hardness of the rock. The range of the m_i values is from 0.001 (highly fractured) to 25 (hard and complete rock). Moreover, s reflects the degree of rock fragmentation, with values ranging from 0 to 1 (1 represents a complete rock). D is the disturbance parameter that considers the blasting effect and stress release, and its values range from 0.0 (field undisturbed rock mass) to 1 (disturbed rock mass).

The Hoek-Brown strength criterion can be determined if the rock uniaxial compressive strength (σ_{ci}), GSI, mi and D are known. σ_{ci} can be determined through the rock uniaxial compression test, whereas the GSI can be derived from Table 3.17 on the basis of the rock mass structure grade and the surface condition grade (SCR). The range of m_i can be obtained from Table 3.18. However, exact m_i values must be obtained through laboratory tests, such as uniaxial compression and conventional triaxial compression tests, uniaxial compression and direct tensile tests, uniaxial compression and indirect tensile tests and uniaxial compression and acoustic emission tests. Table 3.19 presents the values for D.

References

[1] Zhang, Y.X. (2004) *Rock Mechanics*. China Architecture & Building Press, Beijing, China.
[2] Liu, Y.R. & Tang, H.M. (2009) *Rock Mass Mechanics*. Chemical Industry Press, Beijing, China.
[3] Shen, M.R. & Chen, J.F. (2006) *Rock Mass Mechanics*. Tongji University Press, Shanghai.
[4] Cai, M.F. (2013) *Rock Mechanics and Engineering*. Science Press, Beijing, China.
[5] Sun, G.Z. (1983) *Basis of Rock Mass Mechanics*. Science Press, Beijing, China.
[6] Geotechnical Control Office. (1988) *Guide to Rock and Soil Descriptions (Geoguide 3)*. Geotechnical Control Office, Hong Kong.

[7] Huang, G.M. (1999) *Description of Jointed Rocks and Engineering Application*. Chengdu University of Technology, Chengdu, China.

[8] Barton, N. & Choubey, V. (1997) The shear strength of rock joints in theory and practice. *Rock Mechanics*, 10(1–2), 1–54.

[9] Bieniawski, Z.T. (1973) Engineering classification of jointed rock masses. *Transactions of the South African Institution of Civil Engineers*, 15(12), 335–344.

[10] Kang, X.B., Xu, M. & Chen, X. (2008) Introduce and application of rock mass quality classification Q-system. *The Chinese Journal of Geological Hazard and Control*, 19(4), 91–95.

[11] Wu, L.H. (2004) *Study on Empirical Strength Criteria*. Chang'an University, Xi'an, China.

[12] Li, H. & Huang, R.Q. (2014) Method of quantitative determination of joint roughness coefficient. *Chinese Journal of Rock Mechanics and Engineering*, 33(s2), 3489–3497.

[13] Li, Y. & Huang, R. (2015) Relationship between joint roughness coefficient and fractal dimension of rock fracture surfaces. *International Journal of Rock Mechanics & Mining Sciences*, 75, 15–22.

[14] Zhu, H.H., Zhang, Q. & Zhang, L.Y. (2013) Review of research progress and applications of Hoek-Brown strength criterion. *Chinese Journal of Rock Mechanics and Engineering*, 32(10), 1945–1963.

[15] Hoek, E. & Carranza-Torres, C. (2002) Hoek-Brown failure criterion - 2002 edition. *Proceedings of the Fifth North American Rock Mechanics Symposium*, 1, 18–22.

[16] Bieniawski, Z.T. (1974) *Geomechanics classification of rock masses and its application in tunnelling–3nd*. Congr. ISRM2, Denver, 27–32.

[17] Bieniawski, Z.T. (1984) *Rock mechanics design in mining and tunnelling*. A.A. Balkema, Rotterdam.

[18] Barton, N.R., Lien, R. & Lunde, J. (1974) Engineering classification of rock masses for the design of tunnel support. *Rock Mechanics and Rock Engineering*, 6(4), 189–236.

[19] Barton, N. & Choubey, V. (1977) The shear strength of rock joints in theory and practice. *Rock Mechanics*, 10(1–2), 1–54.

[20] Li, Y. & Huang, R. (2015) Relationship between joint roughness coefficient and fractal dimension of rock fracture surfaces. *International Journal of Rock Mechanics and Mining Sciences*, 75, 15–22.

Part 2

Test methods for soils and rocks

The physical and mechanical properties of soils and rocks dominate the reliability, service and life span of geotechnical engineering. As one of the important parts of investigation in delivering geotechnical engineering, laboratory tests provide an efficient way to obtain physical and mechanical parameters of soils and rocks.

At present, there are a huge number of domestic and foreign standards and specifications for soils and rock mass, e.g., *Standard for Soils Tests in Laboratory* (GB/T 50123-1999), *Standard for Test Methods of Engineering Rock Mass* (GB/T 50266-2013), *Test Methods of Soils in Highway Engineering* (JTG E40-2007), *Test Methods of Rocks in Highway Engineering* (JTG E41-2005), etc.

Comparison of these industrial standards and specifications highlights several outstanding issues: 1) test format of all current standards and specifications was described too simply and only comprises single section to explain each specific laboratory test for soils and rocks without highlighting key point of each experiment, 2) some content of current standards and specifications is unclear. For example, in *Standard for Soils Test Method* (GB/T 50123-1999) and *Test Method of Soils for Highway Engineering* (JTG E40-2007), there is no clear definition of the judgement method and mass range of "large number of crystal water for determination tests of total mass of soluble salt. This make us difficult to understand whether test results are slightly high or not, 3) test data processing is unsystematic and is not well organized. More than 50% industry standards and specifications merely include formula and some graphs of the data, and sequence of its formula was disorderly arranged. As such, neither quantitative features of test parameters can be shown precisely, nor physical properties of the specimens can be analyzed directly, 4) some current standards and specifications are hard to read and easily lead the misunderstandings. For instance, in *Standard for Soils Test Method* (GB/T 50123-1999) it was pointed out that its test method is applicable for all kinds of soils, including coarse-grained soils, fine-grained soils, organic and frozen soils. Evidently, *Standard for Soils Test Method* (GB/T 50123-1999) as the water content experiment, was not given an accurate description of applicable scope.

On basis of aforementioned situation and problems, we have conducted a deep analysis of 39 laboratory tests for soils and rocks. With systematical consideration of the features of laboratory tests for soils and rocks, the laboratory tests were organized and discussed: (1) experimental preparation including applicable scope and test parameters for soils and rocks as well as requirements of test apparatus and reagents; (2) soils test procedure with flow chart, experimental methods and steps; (3) rock test procedure with flow charts and experimental steps; (4) data analysis and test summary: in this section, computational method and formula for test parameters of soils and rocks were included in data analysis, however, some important problems needed to be carefully considered in the process of test were also provided in the notes. In this chapter, some ambiguous contents are summarized, but method, procedure and notes of laboratory tests for soils and rocks are presented in detail.

Through reading and learning knowledge in this chapter, the practitioners could well understand the basic concept and applicable scope of laboratory tests for soils and rocks; meanwhile, they could also be familiar with apparatus, essential technologies and experimental methods to use them correctly and to improve accuracy and reliability of physical parameters of soils and rocks.

Chapter 4

International standard systems

4.1 Introduction

In the 1940s, the soil test methods began to be standardized. In 1942, the American Association of State Highway and Transportation Officials (AASHO) established standards for test methods, apparatuses and physical properties of soil. In 1948, the United Kingdom issued the first test method (BS1377: 1948) of soil for civil engineering. In 1956, the Ministry of Water Resources of the People's Republic of China issued China's first *Specification of Soil Test*. In 1974, the International Society for Rock Mechanics published the suggested method for rock tests.

The widely applicable standards are those of China, the United Kingdom, the United States, South Africa, Japan, Singapore, Australia, and the International Society for Rock Mechanics. China's standards include national standards (represented by GB/T 50123-1999, GB/T 50266-2013), industry standards (e.g. JTJ 051-93 for transportation industry), and local standards (e.g. the Tianjin standard, DB 29-20-2002). Most of other countries have only one set of national standards, such as the ASTM (American Society for Testing and Materials) in the USA, BS (the British Standards) in the UK, SS (the Singapore Standards), and SANS (the South African National Standards), JIS (Japanese Industrial Standards), AS (Australia Standards), and ISRM Suggested Methods of the International Society for Rock Mechanics.

The purpose of these standards is to regulate soil test to ensure correct assessment of the engineering properties of soils and rocks, and to provide reliable parameters for engineering design and construction. The national standardization systems have been widely investigated. Bian[1] studied the current status of soil standards in China, briefly described the basic regulations of the standard system, and introduced the content of current standards in categories. He pointed out that the naming and classification of standards are somehow confusing in China. Gao[2] put forward that the terminology in some local standards of China are incompatible with those of national and industry standards. Xu[3] reviewed Japan's standardization system and made a brief comparison with China's standardization system. He indicated that both China and Japan's standardization are of government-led. However, the formulating and revising standards is more accessible to the public in Japan than China. An[4] made a summary of South Africa's standardization: clear legislation, timely updating, smooth procedures for revising, and few standards being mandatory. Xia[5] introduced the process of standardization in Singapore and briefed the Singapore standardization system. Ding[6] made a comparison of soil test standards of the United Kingdom, the United States and China, including liquid limit of soils, particle analysis, direct shear test, etc. He pointed out that the test procedures and some specific requirements are quite different.

Understanding the scope of each standard is important for engineers. We have collected the most widely used standards throughout the world, including those of China, the United Kingdom, the United States, Singapore, South Africa, Australia, Hong Kong, and the International Society for Rock Mechanics. In addition, we have collated the classifying strategy and development history of each standard system. We also reviewed the development and revision process of each standard and clarified the correlation among various standards. Finally, the systems of test standards of different countries were compared in terms of classification, formulation, development, and usage of the standards.

4.2 International standards

4.2.1 China's standards

China's soil test standards are divided into national standards, industrial standards, and local standards.[1] The national standard is the first-class standard, and standards at other levels should be compatible with national standard;[1] the industry standard is at the second-level; the local standard is at the third-level for specific geological conditions in a certain region. The latter two are the refinement and concretization of national standards.

As shown in Table 4.1, China's geotechnical test standards are approved and developed by government agencies. With the approval of the Ministry of Construction, the national standards are submitted to the relevant departments for editorial review, and issued by the competent department of national standards, or jointly issued with the Ministry of Construction.[6] The proposal of industrial standards is launched by the national industry authorities and transferred to the nominated academic institutions or practicing companies for drafting. The standard is issued by the competent national authorities, after being reviewed by the editorial board to make sure that the contents meet the requirements of national standards. The local standards are initialized by local competent authority and formulated by relevant professional committees or research institutes. They are issued by regional competent authority. The contents of local standards should be consistent with the relevant industrial and national standards.

The national standard is numbered as 'code + serial number + approval year' (Table 4.1). For example, *Standard for Soil Test Method* has such number as GB/T (code) 50123 (serial number) -1999 (approval year). The letter 'T' in the code section indicates a mandatory standard. Industrial standard is named in the same way as national standard. Naming of local standard is nonuniform, although they all have a code DB indicating they are locally valid.

Revision of the standard is regularly conducted by the government. The standards at other two levels would be updated once the related national standards are renewed. The local standards hold the highest priority, and then industrial and national standards.

4.2.2 ASTM standards

Unlike the classification of Chinese standards, ASTM (American Society for Testing and Materials) standards are published in form of ASTM Standard Annual Book. ASTM standards are divided into 15 categories, covering metals, coatings, plastics, textiles, petroleum, construction, energy, environmental, consumer products, medical services, equipment and electronics, and advanced materials. The ASTM standard is quite complete, rich in content, and has been cited by many countries in the world. The marketing data show that 50% of ASTM standards are sold abroad each year.[7]

Table 4.1 China's standards for geotechnical testing

Title	Code	Class	Valid date	Organization	Editorial board	Scope	Remarks
Standard test methods of geotechnics	GBJ123-88	National	1989.3.1	Ministry of Construction	Ministry of Water Resources and Power	Basic engineering properties test of foundation soil and landfill	In force
	GB/T 50123-1999		1999.10.1		Ministry of Water Resources	materials for industrial and civil construction, transportation, water conservancy and other projects	
Standard test method of rock mass	GB/T 50266-99	National	1999.5.1	Ministry of Construction	Ministry of Power Industry	Rock tests for water conservancy, hydropower, mining, railway, transportation, petroleum, national defense, industrial and civil construction projects	
	GB/T 50266-2013		2013.9.1		Power Enterprise Association	Engineering rock mass test of foundation, surrounding rock, slope and filling materials	In force

(Continued)

Table 4.1 (Continued)

Title	Code	Class	Valid date	Organization	Editorial board	Scope	Remarks
Geotechnical test procedure	SDS01-79	Industrial (Water Conservancy and Power Engineering)	1980	Ministry of Water Resources Power Industry	Nanjing Hydraulic Research Institute	Identification, naming and description of engineering soil	
	SD128-84 SD128-86, SD128-87,		1989	Ministry of Water Resources Department of Energy	Nanjing Hydraulic Research Institute		
	SL 237-1999		1999.4.15	Ministry of Water Resources	Nanjing Hydraulic Research Institute		In force
Test methods for rocks in water resources and hydropower engineering	DLJ204-81, SL2-81	Industrial (Water Conservancy and Power Engineering)	1982	Ministry of Water Resources	Ministry of Power Industry	Rock test of rock foundation, rock slope and underground cavern of first- and second-class buildings in water conservancy and hydropower projects	
	SL 264-2001		2001.4.1		Yangtze River Academy of Sciences	Rock test of water conservancy and hydropower projects	In force

Name	Standard No.	Industry	Date	Issuing body	Drafting organization	Scope	Status
Standard soil test methods for water resources and hydropower engineering	DL/T 5355-2006	Industrial (Power)	2007.5.1	Development and Reform Commission	Chengdu Survey and Design Institute of China Hydropower Consulting Group	Hydropower and water conservancy projects to test the basic engineering properties of foundations, slopes, underground caverns, filling materials, and the control and inspection of construction quality	In force
Test methods for geotechnical engineering	YBJ 42-92 YSJ 225-92	Industrial (Metallurgical and Non-ferrous Metallurgical)	1993.7.1	Former Metallurgical Ministry of Nonferrous Metals	Changsha Institute of Nonferrous Metallurgy Exploration Technology Exploration Institute	Geotechnical test in construction of nonferrous metallurgical industry	
	YS/T 5225-2016		2016.9.1	Ministry of Industry and Information Technology	Changsha Institute of Nonferrous Metallurgy Exploration Technology Exploration Institute		In force

(Continued)

Table 4.1 (Continued)

Title	Code	Class	Valid date	Organization	Editorial board	Scope	Remarks
Geotechnical test methods for railway industry	TB 10102-2004	Industrial (Railway)	2004.4.1	Ministry of Railways	First Survey and Design Institute of Ministry of Railways	Basic properties test of various foundation soils and fillers in railway engineering	
	TB 10102-2010		2010.11.21		China Railway First Survey and Design Institute Group Co., Ltd.	Physical and mechanical properties test of various foundation soils and fillers in railway engineering	In force
Rock test methods for railway industry	TB 10115-98	Industrial (Railway)	1998.7.1	Ministry of Railways	First Survey and Design Institute of Ministry of Railways	Test on Railway engineering foundations, slopes, tunnels and rock used as building materials	
	TB 10115-2014		2015.2.1		China Railway First Survey and Design Institute Group Co., Ltd.	Rock test during railway survey, design and construction	In force

Name	Code	Type	Date	Organization	Institute	Application	Status
Geotechnical test methods for transport industry	JTJ 051-93	Industrial (Transport)	1993.12.1	Ministry of Transport	Science Research Institute of railway	Basic engineering properties test of foundation soil, roadbed soil and other road soils for highway engineering	In force
	JTG E40-2007		2007.10.1		Science Research Institute of railway		In force
Rock test methods for transport industry	JTJ 054-94	Industrial (Transport)	1994.12.1	Ministry of Transport	Second Highway Survey and Design Institute	Rock test of roadbed, pavement, bridge and culvert and tunnel engineering in highway engineering	In force
	JTJ E41-2005		2005.8.1		Second Highway Survey and Design Institute		In force
Geotechnical test regulations	-	Local (Shandong Province)	1993.9.1	Urban and Rural Construction Committee of Shandong Province	Shandong Exploration and Design Association		
Specification of geotechnical engineering	DB 29-20-2002	Local (Tianjin City)	2001.4.1	Urban and Rural Construction Committee of Tianjin	Construction and Design Institute of Tianjin University	Geotechnical investigation, design and construction in construction, municipal and harbor projects in Tianjin	In force

The soil test standards belong to Category D as listed in Table 4.2. The numbering format of ASTM standard follows 'code + type + serial number + approval year', for instance ASTM (code) D (type) 7263 (serial number) - 09 (year). A letter 'M' following the serial number indicates that the standard is in metric unit system, such as D2435M-11. If there is no letter 'M' in the title, the British unit system is employed, for example D7263-09. In some cases, the reapproval year was given in parentheses, e.g. D5312/D5312M-12 (reapproved in 2013). The superscript (ε1, ε2, etc.) after the approval year indicates that the standard has been experienced minor editorial modification without substantial content change, e.g. D4318-10$^{\varepsilon1}$.

Table 4.2 Geotechnical test standards in ASTM

Test	Standard	Code
Particle-size distribution of soils,(d<75 μm)	Standard test method for particle-size distribution (gradation) of fine-grained soils using the sedimentation (hydrometer) analysis	D7928-16[ε1]
Particle-size distribution of soils (d>75 μm)	Standard test methods for particle-size distribution (gradation) of soils using sieve analysis	D6913-04 (Reapproved 2009)[ε1]
Density (unit weight) of soil specimens	Standard test methods for laboratory determination of density (unit weight) of soil specimens	D7263-09
Maximum index density and unit weight of soils	Standard test methods for maximum index density and unit weight of soils using a vibratory table	D4253-16
Minimum index density and unit weight of soils	Standard test methods for minimum index density and unit weight of soils and calculation of relative density	D4254-16
Specific gravity of soil solids	Standard test methods for specific gravity of soil solids by water pycnometer	D854-14
Water content of soil and rock (oven-drying method)	Standard test methods for laboratory determination of water (moisture) content of soil and rock by mass	D2216-10
Water content of soil (direct heating)	Standard test method for determination of water content of soil by direct heating	D4959-16
Limit moisture of soils	Standard test methods for liquid limit, plastic limit, and plasticity index of soils	D4318-10[ε1]
Consolidation properties of soils	Standard test methods for one-dimensional consolidation properties of soils using incremental loading	D2435 / D2435M-11

Test	Standard	Code
Unconsolidated-undrained triaxial compression	Standard test method for unconsolidated-undrained triaxial compression test on cohesive soils	D2850-15
Consolidated-undrained triaxial compression	Standard test method for consolidated-undrained triaxial compression test for cohesive soils	D4767-11
Consolidated-drained triaxial compression	Standard test method for consolidated-drained triaxial compression test for soils	D7181-11
One-dimensional swell or collapse of soils	Standard test methods for one-dimensional swell or collapse of soils	D4546-14
Expansion index of soils	Standard test method for expansion index of soils	D4829-11
Direct shear test of soils	Standard test method for direct shear test of soils under consolidated-drained conditions	D3080 /D3080M-11
Compaction characteristics of soil (standard effort)	Standard test methods for laboratory compaction characteristics of soil using standard effort (12,400 ft-lbf/ft^3 (600 kN-m/m^3))	D698-12$^{\varepsilon2}$
Compaction characteristics of soil (modified effort)	Standard test methods for laboratory compaction characteristics of soil using modified effort (56,000 ft-lbf/ft^3 (2,700 kN-m/m^3))	D1557-12$^{\varepsilon1}$
Maximum dry unit weight and water content range (vibrating hammer)	Standard test methods for determination of maximum dry unit weight and water content range for effective compaction of granular soils using a vibrating hammer	D7382-08
California bearing ratio (CBR) of soils	Standard test method for California bearing ratio (CBR) of laboratory-compacted soils	D1883-16
Unconfined compressive strength of cohesive soil	Standard test method for unconfined compressive strength of cohesive soil	D2166/D2166M-16
Unconfined compressive strength of compacted soil-lime mixtures	Standard test methods for unconfined compressive strength of compacted soil-lime mixtures	D5102-09
Direct shear test of soils under consolidated-drained conditions	Standard test method for direct shear test of soils under consolidated-drained conditions	D3080 /D3080M-11
Organic matter of soils	Standard test methods for moisture, ash, and organic matter of peat and other organic soils	D2974-14

(Continued)

Table 4.2 (Continued)

Test	Standard	Code
pH of soils	Standard test method for pH of soils	D4972-13
Water content of soil and rock by mass	Standard test methods for laboratory determination of water (moisture) content of soil and rock by mass	D2216-10
Splitting tensile strength of rock core	Standard test method for splitting tensile strength of intact rock core specimens	D3967-16
Compressive strength and elastic moduli of intact rock core	Standard test methods for compressive strength and elastic moduli of intact rock core specimens under varying states of stress and temperatures	D7012-14
Point load strength index of rock	Standard test method for determination of the point load strength index of rock and application to rock strength classifications	D5731-16
Direct shear strength tests of rock specimens under constant normal force	Standard test method for performing laboratory direct shear strength tests of rock specimens under constant normal force	D5607-16
Durability of rock under freezing and thawing conditions	Standard test method for evaluation of durability of rock for erosion control under freezing and thawing conditions	D5312/D5312M-12 (reapproved 2013)
Slake durability of rocks	Standard test method for slake durability of shales and other similar weak rocks	D4644-16
Absorption of rock	Standard test method for specific gravity and absorption of rock for erosion control	D6473-15
Rock hardness by rebound hammer method	Standard test method for determination of rock hardness by rebound hammer method	D5873-14

Founded in 1898 by the chemist Dr. Charles B. Dudley in Pennsylvania Railroad, the ASTM International, originally known as the American Society for Testing and Materials, is one of the world's largest non-profit standard development organizations. This organization was renamed ASTM International in 2001.

The process to set up any ASTM standard is shown in Figure 4.1. Anyone (normally a US citizen) may submit a proposal for new standards. The technical committee of ASTM International is responsible for approval of the proposal. Once the proposal is approved, the draft standard would be reviewed by a board with members from academia, industry, and government. All standards issued by ASTM International follow the *ASTM Technical Committee Regulations* and *ASTM Standard Forms and Styles*. ASTM International reviews

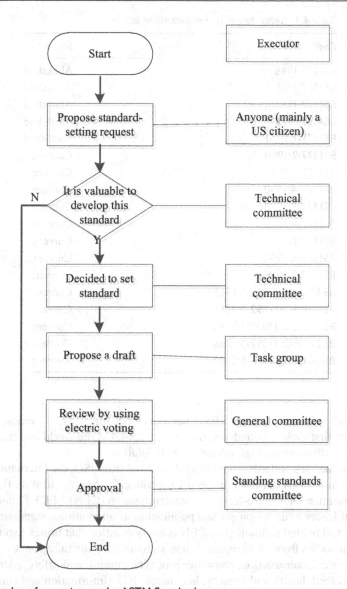

Figure 4.1 Flowchart for setting up the ASTM Standard

standards in force every five years. The technical committee carry out the revision via unscheduled and scheduled (twice a year) meetings.

4.2.3 BSI standards

The UK's soil test standards include not only the BS 1377 series developed in the UK but also the standards from European Codes and ISO. BS 1377 classifies soil tests into nine categories. The indoor test methods are listed in Table 4.3. The coding of standard is formulated as 'BS + serial number + issue year'. The BSI standards is widely applied and cited by other countries in the world.

Table 4.3 Geotechnical test standards in BSI

Code	Status
BS1377:1948	Abolished
BS1377:1961	Abolished
BS1377:1967	Abolished
BS1377:1975	Abolished
BS1377-1:1990	Abolished
BS1377-2:1990	Current
BS1377-3:1990	Current
BS1377-4:1990	Current
BS1377-5:1990	Current
BS1377-6:1990	Current
BS1377-7:1990	Current
BS1377-8:1990	Current
BS1377-1:2016	Current
BS EN ISO 17892-1:2014	Current
BS EN ISO 17892-2:2014	Current
BS EN ISO 17892-3:2015	Current
BS EN ISO 17892-4:2016	Current
BS EN ISO 17892-5:2017	Current

The British Standards Institute (BSI) has developed the BS1377 series. Founded in 1901, BSI is the first national standardization organization in the world and is an unofficial, profitable organization with a high international reputation.

A great number of standards issued by BSI are cited from ISO, the International Electrotechnical Commission (IEC), the European Committee for Standardization (CEN), or the European Committee for Electro-technical Standardization (CENELEC). Founded in 1906, IEC is a global leader in development and publication of international standards for electrical, electronic and related technologies. CEN is an association that brings together national standardization bodies from 34 European countries to develop the European Codes. It serves aerospace, chemical, construction, consumer products, defense and safety, energy, environment, food and feed, health and security, healthcare, ICT (information and communication technology), mechanical materials, stress equipment, services, smart life, and standardization activities in areas such as transportation and packaging. CENELEC was found in 1973 by merging the two former European organizations, CENELCOM and CENEL, and was designated as the official European Organization for Standardization.

The BSI Technical Committee has developed the remaining BS standards. Figure 4.2 shows the development flowchart. Anyone (normally a British citizen) may propose a new standard. The BSI Technical Committee evaluates the draft standard, finalizes and publishes the standard.

BS1377:1948 is the earliest version of BS1377 (Table 4.3). Prior to 1990, the BS1377 was revised in series. After 1990, the standards were revised in category (Table 4.4).

The Technical Committee regularly reviews standards in force through online meetings, conference calls, letters or meetings at least every five years. Publishing errata will correct

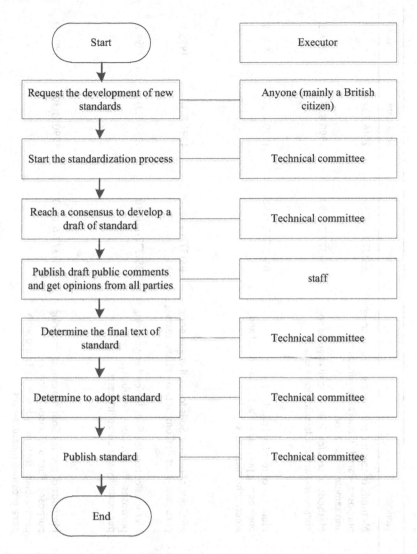

Start	Executor
Request the development of new standards	Anyone (mainly a British citizen)
Start the standardization process	Technical committee
Reach a consensus to develop a draft of standard	Technical committee
Publish draft public comments and get opinions from all parties	staff
Determine the final text of standard	Technical committee
Determine to adopt standard	Technical committee
Publish standard	Technical committee
End	

Figure 4.2 Flowchart for development of BS

unintentional errors. The calibration record is marked on the standard and the standard revision history is shown on the standard front page. If the standard fails review, the standard will be revoked.

4.2.4 ISRM suggested methods

The International Society of Rock Mechanics (ISRM) is a non-profit rock mechanics research society that has published a number of methods for rock tests mainly in form of journals. The methods suggested by ISRM are widely cited worldwide. The main indoor rock test methods suggested by ISRM are listed in Table 4.5.

Table 4.4 Classification of geotechnical tests in BS

Test	Standard	Code	Valid date	Revision
General requirements and sample preparation	Methods of test for soils for civil engineering purposes—Part 1: General requirements and sample preparation	BS1377-1:2016	2016.7.31	
Classification and basic physical properties test of soils (moisture, limit moisture, Shrinkage, density, particle density& particle analysis)	Methods of test for Soils for civil engineering purposes—Part 2: Classification tests	BS1377-2:1990	1990.8.31	1996.5
Chemistry and electrochemistry (organic matter content, burning mass loss, sulfate content, carbonate content, chloride content, soluble content, pH, resistivity and redox potential)	Methods of test for Soils for civil engineering purposes—Part 3: Chemical and electrochemical tests	BS1377-3:1990	1990.8.31	1996.5
Compaction (compaction, relative density, water state value, chalk crushing value and bearing ratio)	Methods of test for Soils for civil engineering purposes—Part 4: Compaction-related tests	BS1377-4:1990	1990.8.31	1995.1
Compressibility, permeability and durability (one-dimensional consolidation, swelling and collapsibility, Permeability of Soils: Falling-head Method disintegration and frost heaving)	Methods of test for Soils for civil engineering purposes—Part 5: Compressibility, permeability and durability tests	BS1377-5:1990	1990.8.31	1994.11
Consolidation and permeability tests in hydraulic cells and with pore pressure measurement	Methods of test for Soils for civil engineering purposes—Part 6: Consolidation and permeability tests in hydraulic cells and with pore pressure measurement	BS1377-6:1990	1990.11.30	1994.11

Test	Title	Standard	Date	
Shear strength tests (total stress) (indoor vane shear, direct shear, ring shear, single-stage and multi-stage loading test UU)	Methods of test for Soils for civil engineering purposes—Part 7: Shear strength tests (total stress)	BS1377-7:1990	1990.6.29	1994.11
Shear strength tests (effective stress) (CU, CD)	Methods of test for Soils for civil engineering purposes—Part 8: Shear strength tests (effective stress)	BS1377-8:1990	1990.10.31	1995.1
Water content	Geotechnical investigation and testing—Laboratory testing of soil Part 1: Determination of water content	BS EN ISO 17892-1:2014	2014.12.31	
Bulk density	Geotechnical investigation and testing—Laboratory testing of soil Part 2: Determination of bulk density	BS EN ISO 17892-2:2014	2014.12.31	
Particle density	Geotechnical investigation and testing—Laboratory testing of soil Part 3: Determination of particle density	BS EN ISO 17892-3:2015	2016.1.31	
Particle size distribution	Geotechnical investigation and testing—Laboratory testing of soil Part 4: Determination of particle size distribution	BS EN ISO 17892-4:2016	2016.12.31	
Oedometer test	Geotechnical investigation and testing—Laboratory testing of soil Part 5: Incremental loading oedometer test	BS EN ISO 17892-5:2017	2017.4.30	

Test methods issued by the ISRM are included in *the Complete ISRM Suggested Methods for Rock Characterization, Testing and Monitoring: 1974-2006*, (commonly known as the Blue Book) and *the ISRM Suggested Methods for Rock Characterization, Testing and Monitoring: 2007-2014*, (commonly known as the Yellow Book). The Blue Book contains suggested methods from 1974 to 2006, and the Yellow Book contains suggested methods from 2007 to 2014. Recent papers related to rock tests were published in *Rock Mechanics and Rock Engineering*.

It can be seen in Table 4.5 that ISRM suggested methods for indoor rock tests are originally published in two journals: *International Journal of Rock Mechanics and Mining Science* (formerly *International Journal of Rock Mechanics and Mining Science and Geomechanics Abstracts*) and *Rock Mechanics and Rock Engineering*.

The method published by ISRM was reviewed and approved by the Commission on Testing Methods of the Rock Mechanics Society. The review procedure for the proposed method is shown in Figure 4.3. Applicants who are interested in developing a method submit method proposal to the committee. If the committee accepts the proposal, the draft by the applicants would be submitted to the chair of the committee for review after being first evaluated by more than three experts in the field. Upon subsequent revisions, if any, the draft is designated as suggested method and published in ISRM journals.

4.2.5 Hong Kong's standards

The soil test standard in Hong Kong is called *Geospec 3: Model Specification for Soil Testing* (2017 ver.), which was issued by Geotechnical Engineering Office (GEO) of Civil Engineering and Development Department (CEDD) of the Hong Kong Special Administrative Region (HKSAR) of China. *Geospec 3* consists of three parts: the first part covers the general technical procedures for planning and supervising laboratory tests, the second and third parts detail the technical procedures for individual tests. *Geospec 3* is mainly applicable in HKSAR.

Geospec 3 was developed by CEDD in accordance with the British Standard *BS1377* series. CEDD was formed by merging the former Civil Engineering Department (CED) and the Territory Development Department (TDD) in July 2004. The GEO carried out technical update in soil test standards and implemented relevant technical guides through publications of CEDD.

4.2.6 Australia's standards

Soil test methods in Australia Standards (AS) have been recognized and widely cited by Australian industry. The standards are divided into different categories according to test type (Table 4.6). The standard is numbered in format of 'AS + serial number + issue year'. Some standard numbers have suffix 'R + year', which means when the standard was reviewed and reconfirmed, such as 4133.4.1-2007 (R2016). The others are followed by "Admt' number + year' where the number stands for times of amendments, and the year represents the time of amendment. For example, AS 1289.3.8.3: 2014 / Amdt 1:2015.

AS was developed by Standards Australian Association, an independent non-governmental and non-profit standard organization. In 1922, the Australian Commonwealth Engineering Standards Association was established. In 1929, it was renamed as the Standards Association of Australia (SAA). In 1988, SAA became the Australian Bureau of Standards. In 1999, it became an Australian Public Company Limited.

Table 4.5 ISRM suggested methods for rock tests

Test	Method	Source	Publication
Density, gravity density, water, content, water absorption, swelling, and disintegration resistance	Suggested methods for determining water content, porosity, density, absorption and related properties and swelling and slake-durability index properties	International Journal of Rock Mechanics and Mining Sciences & Geomechanics Abstracts Vol. 16 No. 2, pp. 141–156	1979.4
Indirect tensile strength of rocks (Brazilian test)	Suggested method for determining mode I fracture toughness using cracked chevron notched Brazilian disc (CCNBD) specimens	International Journal of Rock Mechanics and Mining Sciences & Geomechanics Abstracts Vol. 32, No. I pp. 57–64	1995
Uniaxial compressive strength and deformability of rock	Suggested methods for determining the uniaxial compressive strength and deformability of rock materials	International Journal of Rock Mechanics and Mining Sciences & Geomechanics Abstracts Vol. 16 No. 2, pp. 135–140	1979.4
Sound velocity	Suggested method for determining sound velocity	Rock Mechanics and Mining Science & Geomechanics Abstracts Vol. 15 No. 2, pp. 53–58	1978.4
Triaxial compression	Suggested methods for determination the strength of rock materials in triaxial compression: revised version	International Journal of Rock Mechanics and Mining Sciences & Geomechanics Abstracts Vol. 15 No. 2, pp. 47–51	1978.4
Point load strength	Suggested method for determining point load strength	International Journal of Rock Mechanics and Mining Sciences & Geomechanics Abstracts Vol 22, No. 2, pp. 51–60	1985.4
Shear strength of rock joints	Suggested method for laboratory determination of the shear strength of rock joints: revised version	Rock Mechanics and Rock Engineering Vol. 47 No. I pp. 291	2014.1
Sound velocity	Upgraded ISRM suggested method for determining sound velocity by ultrasonic pulse transmission technique	Rock Mechanics and Rock Engineering Vol. 47, No. I, pp. 255–259	2014.1
Swelling	Suggested methods for laboratory testing of swelling rocks	International Journal of Rock Mechanics and Mining Sciences Vol. 36, No. 3, pp. 291–306	1999.4
Schmidt hammer rebound hardness	Suggested method for determination of the Schmidt hammer rebound hardness: revised version	International Journal of Rock Mechanics and Mining Sciences Vol. 46, No. 3, pp. 627–634	2009.4

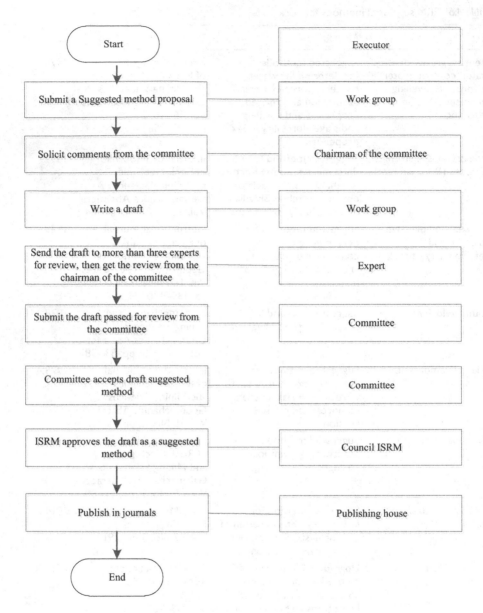

Figure 4.3 Flowchart of setting up an ISRM suggested method

The procedure for issuing and updating a standard in AS is similar to that in ASTM. Anyone (normally an Australian citizen) may make a proposal for developing or revising a standard. If approved, the committee would organize the development or revision. The standards in force are reviewed every five years by the committee in order to reconfirm, modify or even withdraw the existing standards according to the state-of-art technology.

Table 4.6 Geotechnical test methods in AS

Test	Standard	Code
Deformability of rock materials in uniaxial compression (rock strength less than 50 MPa)	Methods of testing rocks for engineering purposes rock strength tests–determination of the deformability of rock materials in uniaxial compression–rock strength less than 50 MPa	AS 4133.4.3.2-2013
Uniaxial compressive strength of rock (strength less than 50 MPa)	Methods of testing rocks for engineering purposes rock strength tests–determination of uniaxial compressive strength–rock strength less than 50 MPa	AS 4133.4.2.2-2013
Uniaxial compressive strength of rock (strength greater than 50 MPa)	Methods of testing rocks for engineering purposes rock strength tests–determination of uniaxial compressive strength of 50 MPa and greater	AS 4133.4.2.1-2007
Point load strength index of rock	Methods of testing rocks for engineering purposes rock strength tests–determination of point load strength index	AS 4133.4.1-2007 (R2016)
Rock porosity and dry density (saturation and caliper techniques)	Methods of testing rocks for engineering purposes rock porosity and density tests–determination of rock porosity and dry density–saturation and caliper techniques	AS 4133.2.1.1-2005 (R2016)
Rock porosity and dry density (saturation and buoyancy techniques)	Methods of testing rocks for engineering purposes rock porosity and density tests–determination of rock porosity and dry density–saturation and buoyancy techniques	AS 4133.2.1.2-2005 (R2016)
Moisture content of rock (oven drying method)	Methods of testing rocks for engineering purposes rock moisture content tests–determination of the moisture content of rock–oven drying method (standard method)	AS 4133.1.1.1-2005 (R2016)
Swelling strain developed in an unconfined rock specimen	Methods of testing rocks for engineering purposes rock swelling and slake durability tests–determination of the swelling strain developed in an unconfined rock specimen	AS 4133.3.1-2005 (R2016)
California bearing ratio of a soil	Methods of testing soils for engineering purposes soil strength and consolidation tests–determination of the California bearing ratio of a soil–standard laboratory method for an undisturbed specimen	AS 1289.6.1.2-1998 (R2013)
Preparation of disturbed soil samples	Methods of testing soils for engineering purposes sampling and preparation of soils–disturbed samples–standard method	AS 1289.1.2.1-1998 (R2013)
Pinhole dispersion classification of a soil	Methods of testing soils for engineering purposes soil classification tests–dispersion–determination of pinhole dispersion classification of a soil	AS 1289.3.8.3:2014/ Amdt 1:2015
Definitions and general requirements of soil	Methods of testing soils for engineering purposes definitions and general requirements	AS 1289.0:2014
Dry density of a soil (water replacement method)	Methods of testing soils for engineering purposes soil compaction and density tests–determination of the field dry density of a soil–water replacement method	AS 1289.5.3.5-1997 (R2013)

(Continued)

Table 4.6 (Continued)

Test	Standard	Code
Compaction control test (Hilf density ratio and Hilf moisture variation)	Methods of testing soils for engineering purposes soil compaction and density tests—compaction control test – Hilf density ratio and Hilf moisture variation (rapid method)	AS 1289.5.7.1-2006
Compaction control test (density index method)	Methods of testing soils for engineering purposes soil compaction and density tests—compaction control test—density index method for a cohesionless material	AS 1289.5.6.1-1998 (R2016)
Dry density/moisture content relation of a soil (standard compactive effort)	Methods of testing soils for engineering purposes soil compaction and density tests—determination of the dry density/moisture content relation of a soil using standard compactive effort	AS 1289.5.1.1:2017
Dry density/moisture content relation of a soil (modified compactive effort)	Methods of testing soils for engineering purposes soil compaction and density tests—determination of the dry density/moisture content relation of a soil using modified compactive effort	AS 1289.5.2.1:2017
Liquid limit of a soil (four-point Casagrande method)	Methods of testing soils for engineering purposes soil classification tests—determination of the liquid limit of a soil – four-point Casagrande method	AS 1289.3.1.1-2009 (R2017)
Liquid limit of a soil (one-point Casagrande method)	Methods of testing soils for engineering purposes soil classification tests—determination of the liquid limit of a soil—one-point Casagrande method (subsidiary method)	AS 1289.3.1.2-2009 (R2017)
Cone liquid limit of a soil	Methods of testing soils for engineering purposes soil classification tests—determination of the cone liquid limit of a soil	AS 1289.3.9.1:2015
Compressive strength of a specimen tested in undrained triaxial compression (with measurement of pore water pressure)	Methods of testing soils for engineering purposes soil strength and consolidation tests—determination of compressive strength of a soil—compressive strength of a saturated specimen tested in undrained triaxial compression with measurement of pore water pressure	AS 1289.6.4.2:2016
Compressive strength of a specimen tested in undrained triaxial compression (without measurement of pore water pressure)	Methods of testing soils for engineering purposes soil strength and consolidation tests—determination of compressive strength of a soil—compressive strength of a specimen tested in undrained triaxial compression without measurement of pore water pressure	AS 1289.6.4.1:2016
California bearing ratio of a soil	Methods of testing soils for engineering purposes soil strength and consolidation tests—determination of the California bearing ratio of a soil—standard laboratory method for a remoulded specimen	AS 1289.6.1.1:2014/ Amdt 1:2017

Test	Standard	Code
Particle density of a soil	Methods of testing soils for engineering purposes soil classification tests–determination of the soil particle density of a soil–standard method	AS 1289.3.5.1-2006
Dry density of a soil (water replacement method)	Methods of testing soils for engineering purposes soil compaction and density tests–determination of the field dry density of a soil–water replacement method	AS 1289.5.3.5-1997 (R2013)
Assignment of maximum dry density and optimum moisture content values	Methods of testing soils for engineering purposes soil compaction and density tests–compaction control test–assignment of maximum dry density and optimum moisture content values	AS 1289.5.4.2-2007 (R2016)
Dry density ratio, moisture variation and moisture ratio of soils	Methods of testing soils for engineering purposes soil compaction and density tests–compaction control test–dry density ratio, moisture variation and moisture ratio	AS 1289.5.4.1-2007 (R2016)
Minimum and maximum dry density of a cohesionless material	Methods of testing soils for engineering purposes soil compaction and density tests–determination of the minimum and maximum dry density of a cohesionless material–standard method	AS 1289.5.5.1-1998 (R2016)
Shrinkage index of a soil (shrink-swell index)	Methods of testing soils for engineering purposes soil reactivity tests–determination of the shrinkage index of a soil–shrink-swell index	AS 1289.7.1.1-2003
Shrinkage index of a soil (loaded shrinkage index)	Methods of testing soils for engineering purposes soil reactivity tests–determination of the shrinkage index of a soil–loaded shrinkage index	AS 1289.7.1.2-2003
Shear strength of a soil	Methods of testing soils for engineering purposes soil strength and consolidation tests–determination of shear strength of a soil–direct shear test using a shear box	AS 1289.6.2.2-1998
Penetration resistance of a soil (9 kg dynamic cone penetrometer)	Methods of testing soils for engineering purposes soil strength and consolidation tests–determination of the penetration resistance of a soil – 9 kg dynamic cone penetrometer test	AS 1289.6.3.2-1997 (R2013)
Penetration resistance of a soil (perth sand penetrometer)	Methods of testing soils for engineering purposes soil strength and consolidation tests–determination of the penetration resistance of a soil–Perth sand penetrometer test	AS 1289.6.3.3-1997 (R2013)
Moisture content of soil (oven drying method)	Methods of testing soils for engineering purposes soil moisture content tests–determination of the moisture content of a soil–oven drying method (standard method)	AS 1289.2.1.1-2005 (R2016)

(Continued)

Table 4.6 (Continued)

Test	Standard	Code
Moisture content of soil (sand bath method)	Methods of testing soils for engineering purposes soil moisture content tests–determination of the moisture content of a soil–sand bath method (subsidiary method)	AS 1289.2.1.2-2005 (R2016)
Moisture content of soil (microwave-oven drying method)	Methods of testing soils for engineering purposes soil moisture content tests–determination of the moisture content of a soil–microwave-oven drying method (subsidiary method)	AS 1289.2.1.4-2005 (R2016)
Moisture content of soil (infrared lights method)	Methods of testing soils for engineering purposes soil moisture content tests–determination of the moisture content of a soil–infrared lights method (subsidiary method)	AS 1289.2.1.5-2005 (R2016)
Moisture content of soil (hotplate drying method)	Methods of testing soils for engineering purposes soil moisture content tests–determination of the moisture content of a soil–hotplate drying method	AS 1289.2.1.6-2005 (R2016)
The total suction of a soil (standard method)	Methods of testing soils for engineering purposes soil moisture content tests–determination of the total suction of a soil–standard method	AS 1289.2.2.1-1998 (R2013)

4.2.7 Japan's standards

The standardization for soil experiments in Japan is similar to that in China and it is proceeded by government agencies. The national standard in Japan is called JIS (Japanese Industrial Standard). JIS covers civil engineering and construction, steel, non-ferrous metals, energy, welding, chemical products, chemical analysis, polymers, textiles, furnaces, daily necessities, mechanical parts, machine tools, precision machinery, general machinery, automotive, aerospace, railway, marine, packaging and transportation, electrical, electronics, household appliances, medical safety appliances, atomic energy and intelligence. Soil testing standards are included in the M (Mine) and A (Architecture) categories (Table 4.7). The standard code is formatted as 'JIS + M/A + series number + year'.[8]

JIS is formulated and revised by Japanese Industrial Standards Committee (JISC). JISC was found on December 6, 1945 as a specifically private institution for standardization work, and serves as the most authoritative agency in Japan for standardization.

Similar to the ASTM, development process of JIS is illustrated in Figure 4.4. Any groups or individual (normally Japanese citizens) may submit the draft standard for JIS to discuss. The approved draft is then transferred to the Ministry of Industry and JISC for reviewing. If JISC considers the draft reasonable, the Minister of Industry will publish it. The departments in charge will revise the draft according to the comments from the public. The draft will be confirmed to be officially involved in JIS and published in the official gazette once it satisfies the quality checking.

Table 4.7 Japanese geotechnical test standards

Test	Code	Standard
Tensile strength of rock	JIS M 0303: 2000	Method of test for tensile strength of rock
Compressive strength of rock	JIS M 0302: 2000	Method of test for compressive strength of rock
Preparation of rock samples	JIS M 0301: 1975	Methods of sampling of rock and preparation of test piece for strength test
Soil density (sand replacement method)	JIS A 1214: 2013	Test method for soil density by the sand replacement method
Load test on soil for road	JIS A 1215: 2013	Method for plate load test on soil for road
Soil compaction (rammer)	JIS A 1210: 2009	Test method for soil compaction using a rammer
Density of soil particles	JIS A 1202: 2009	Test method for density of soil particles
One-dimensional consolidation properties of soils (constant rate of strain loading)	JIS A 1227: 2009	Test method for one-dimensional consolidation properties of soils using constant rate of strain loading
Preparation of disturbed soil samples	JIS A 1201: 2009	Practice for preparing disturbed soil samples for soil testing
Bulk density of soils	JIS A 1225: 2009	Test method for bulk density of soils
Minimum and maximum densities of sands	JIS A 1224: 2009	Test method for minimum and maximum densities of sands
Permeability of saturated soils	JIS A 1218: 2009	Test methods for permeability of saturated soils
One-dimensional consolidation properties of soils (incremental loading)	JIS A 1217: 2009	Test method for one-dimensional consolidation properties of soils using incremental loading
Unconfined compression test of soils	JIS A 1216: 2009	Method for unconfined compression test of soils
California bearing ratio (CBR) of soils	JIS A 1211: 2009	Test methods for the California bearing ratio (CBR) of soils in laboratory
Shrinkage parameters of soils	JIS A 1209: 2009	Test method for shrinkage parameters of soils
Liquid limit and plastic limit of soils	JIS A 1205: 2009	Test method for liquid limit and plastic limit of soils
Particle size distribution of soils	JIS A 1204: 2009	Test method for particle size distribution of soils
Water content of soils	JIS A 1203: 2009	Test method for water content of soils
One-dimensional consolidation properties of soils (constant rate of strain loading)	JIS A 1227: 2009	Test method for one-dimensional consolidation properties of soils using constant rate of strain loading

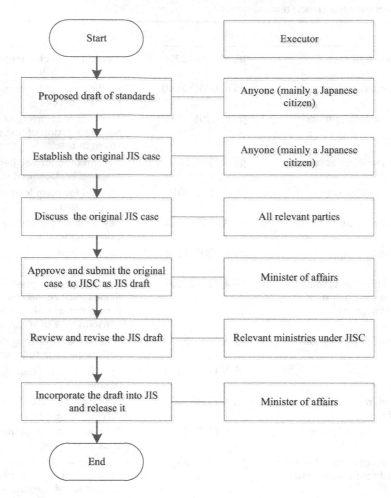

Figure 4.4 Flowchart of standardization in Japan

4.2.8 South Africa's standards

Table 4.8 shows that the locally developed South African National Standards (SANS) for geotechnical tests cover a relatively narrow scope. The ISO or IEC standards are cited whenever the local standards are incompetent.[4]

The SANS was developed by the South African Standard Technical Committee and its development process is shown in Figure 4.5. The South African Bureau of Standards (SABS) initiates the formulation of a standard. The technical committee under SABS is responsible for drafting the standards, soliciting public comments, and modifying the draft according to the comments. The draft will be submitted to the Standards Approval Board for approval and being published.

Table 4.8 Geotechnical test standards in South Africa

Test	Standard	Code
Rock durability	Civil engineering test methods part AG15: determination of rock durability using 10% fact (fines aggregate crushing test) values after soaking in ethylene glycol	SANS 3001-AG15 Ed. 1 (2012)
Wet preparation and particle size analysis	Civil engineering test methods part GR1: wet preparation and particle size analysis	SANS 3001-GR1:2013 (Ed. 1.02)
Liquid limit, plastic limit plasticity index and linear shrinkage	Civil engineering test methods part GR10: determination of the one-point liquid limit, plastic limit, plasticity index and linear shrinkage	SANS 3001-GR10:2013 (Ed. 1.02)
Liquid limit (two-point method)	Civil engineering test methods part GR11: determination of the liquid limit with the two-point method	SANS 3001-GR11:2013 (Ed. 1.02)
Moisture content (oven-drying)	Civil engineering test methods part GR20: determination of the moisture content by oven-drying	SANS 3001-GR20:2010 (Ed. 1.01)
Maximum dry density and optimum moisture content	Civil engineering test methods part GR30: determination of the maximum dry density and optimum moisture content	SANS 3001-GR30:2015 (Ed. 1.02)
Dry density (sand replacement)	Civil engineering test methods part GR35: determination of in-place dry density (sand replacement)	SANS 3001-GR35:2015 (Ed. 1.00)
California bearing ratio	Civil engineering test methods part GR40: determination of the California bearing ratio	SANS 3001-GR40:2013 (Ed. 1.01)

4.2.9 Singapore's standards

Soil test standards used in Singapore consist of the Singapore Standards (SS) issued by the Standards Committee of Singapore and the *Eurocode 7*. SS is classified into recommended standards and mandatory technical standards. The standard number is formatted as 'code + series number'.

The Standards, Productivity and Innovation Commission (SPRING Singapore) has developed the Singapore standards. SPRING Singapore is the statutory committee of the Ministry of Industry and Trade of Singapore, which was formed by combining and

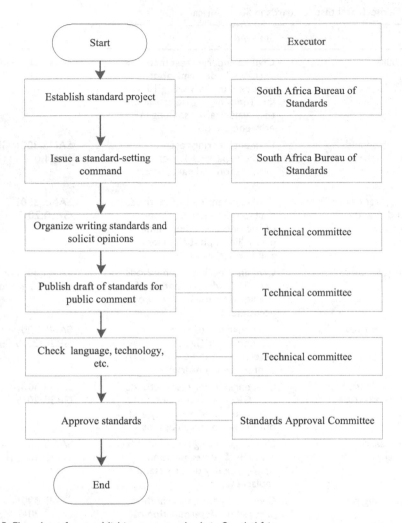

Figure 4.5 Flowchart for establishing test standards in South Africa

reorganizing the former Singapore Productivity and Standards Agency (PSB), the National Productivity Commission (NPB) and the Singapore Standards and Industries Association (SISIR) in 1996. SPRING Singapore was renamed SPRING in April 2002.

The Singapore standard system was established on the basis of international standards. Any proposed Singapore standards must be consulted with all relevant parties prior to publication.

The process of developing Singapore standards is shown in Figure 4.6. Any Singapore citizens may submit a request to Standards Committee for developing new standards. Once the request is accepted, the Standards committee starts to establish the proposed standard, which is then transferred to the Technical Committee for soliciting and reviewing before being published. All Singapore standards must be reviewed every five years to determine if they need revision.[5]

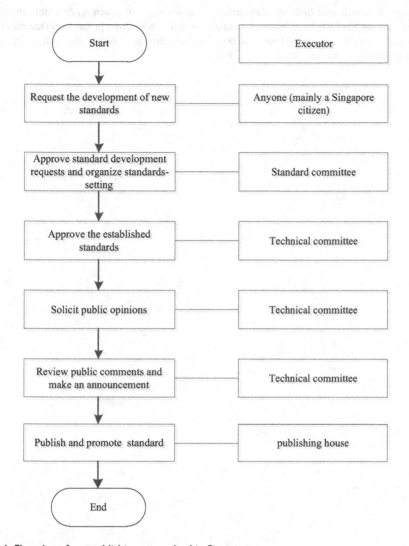

	Executor
Start	
Request the development of new standards	Anyone (mainly a Singapore citizen)
Approve standard development requests and organize standards-setting	Standard committee
Approve the established standards	Technical committee
Solicit public opinions	Technical committee
Review public comments and make an announcement	Technical committee
Publish and promote standard	publishing house
End	

Figure 4.6 Flowchart for establishing a standard in Singapore

References

[1] Biao, Z.Q. (2004) Current status of geotechnical standard specifications in China. *Geotechnical Investigation & Surveying*, (1), 16–20.
[2] Gao, D.Z. (2007) On local standards of geotechnical engineering. *Geotechnical Investigation & Surveying*, (1), 1–6.
[3] Xu, J.Y. (2001) Evaluation of Japan's standardization system. *China Standardization*, (6), 50–50.
[4] An, Y. (2005) South African standardization summary. *China Standardization*, (1), 69–72.
[5] Xia, Y. (2015) *Research on Singapore Standardization System*. China Standardization Forum, Beijing, China.
[6] Ding, Q.B., Xu, Y.B. & Shen, Z.Y. (2010) The major difference among the British standard, American standard and Chinese standard of soil test. *Guangdong Architecture Civil Engineering*, (10), 19–21.

[7] *National Standards and Industry Standards in Engineering Construction.* Available from: www. mohurd.gov.cn[EB/OL], http://www.chinagb.org/Article-58309.html [accessed 9 December 2009].

[8] Technical Committee for Standards. (2002) Introduction to Japan's standardization system. *Printing Quality & Standardization,* (4), 38–39.

Chapter 5

Test preparation

Test preparation is highly important to ensure a successful experiment. This chapter is divided into two sections to detail the applicable scope, parameters to be measured, as well as the apparatus and reagents involved.

5.1 Application scopes and parameters to be measured

According to the Chinese standards, *Standard for Soils Tests* (GB/T 50123-1999) and *Standard for Test Methods of Engineering Rock Mass* (GB/T 50266-2013), Tables 5.1 and 5.2 summarize the application scope and parameters to be measured in each test.

5.2 Apparatus and reagent

The requirements for apparatus and reagent are prerequisites for successful testing. This section summarizes the test apparatuses and reagents for geotechnical testing in soils and rocks. The specifications of devices, and the concentration and amount of reagents are detailed in this section.

5.2.1 Soil tests

(1) Apparatuses for specimen preparation

Apparatus:

   Fine sieve – with hole diameter of 0.5 mm or 2 mm.
   Washing sieve – with hole diameter of 0.075 mm.
   Platform scale and balance – with 5 g, 1 g, 0.5 g, 0.1 g and 0.01 g readability, when having a mass of up to 10 kg, 5 kg, 1 kg, 0.5 kg, and 0.2 kg, respectively.
   Stainless steel cutting ring – with inner diameters of 61.8 mm or 79.8 mm and height of 20 mm; or with inner diameter of 61.8 mm and height of 40 mm.
   Compactor.
   Vacuum cylinder.
   Other devices – soils cutter, wire saw, drying oven, moisturizing cylinder and water spray device, etc.

Table 5.1 Laboratory tests for soils

Test		Application scope	Parameters to be measured
Water content test		Coarse-grained soils, fine-grained soils, organic and frozen soils	1) Mass of wet soil specimens 2) Mass of dry soil specimens
	Cutting ring method	Fine-grained soils without gravel	1) Mass of cutting ring 2) Total mass of cutting ring and soil specimens
Density test	Wax-sealing method	Fragile and irregular-shaped hard soils	1) Mass of the specimens 2) Mass of wax-sealed specimens 3) Mass of wax-sealed specimens in distilled water
Specific gravity of soils solids		Soils with particles smaller than 5 mm	1) Mass of the specimens 2) Total mass of pycnometer, water and specimens 3) Total mass of pycnometer and water
Particle size distribution test	Sieving analysis method	Particle size of soils is in between 0.075 mm and 60 mm	1) Total mass of specimens 2) Mass of soils particles remained on each test sieve
	Densimeter method	Particle size of soils is smaller than 0.075 mm	1) Total mass of the specimens 2) Mass of specimens smaller than a specified particle diameter
Atterberg limits test	Cone penetrometer method	Soils with particles smaller than 0.5 mm; the organic content matter is less than 5 percent	1) Cone penetration 2) Water content
	Rolling method	Soils with particles smaller than 0.5 mm	1) Diameter of soil threads 2) Water content of the soil threads
Relative density test of sand		Soils with particles smaller than 5 mm; and particles of 2–5 mm in size is less than 15% of the total mass of the specimen	1) Minimum index density: volume of soils after depositing in a cylinder and inverting the cylinder 2) Maximum index density: total mass of mold and specimen after vibrating

Compaction test	Light compaction	Soils with particles smaller than 5 mm (cohesive soils)	1) Total mass of specimen and mold after compaction 2) Masses of specimen before and after drying
	Heavy compaction	Soils with particles smaller than 20 mm	
	When adopting three layers of compaction, the largest particle size should be smaller than 40 mm		
California Bearing Ratio (CBR) test		The largest particle size should be smaller than 20 mm. When three layers of compaction is adopted, the largest particle should be smaller than 40 mm	1) Swell during soaking 2) Axial pressure 3) Penetration
Modulus of resilience test		Fine-grained soils with different water content and density	Deformation under each level of pressure
Penetration test	Constant-head method	Coarse-grained soils	1) Difference in head on manometers 2) Volume of water discharged 3) Water temperature
	Variable-head method	Fine-grained soils Saturated clay	Changes of the height of the water head over time
Consolidation test			Heights of specimen during consolidation
Collapse characteristics test of loess		Loess and loess-like soils	Changes of specimen height over time
Triaxial compression test		Fine-grained soils and coarse-grained soils with particles smaller than 20 mm	1) Maximum principle stress 2) Axial deformation

(Continued)

Table 5.1 (Continued)

Test	Application scope	Parameters to be measured
Unconfined compressive strength test	Saturated clay	1) Axial stress 2) Axial deformation
Direct shear test under consolidated-drained conditions	Fine-grained soils	1) Shear stress 2) Shear displacement
Reversal direct shear strength	Clayey soils	1) Shear stress 2) Shear displacement
Tensile strength test	All kinds of soils that can be made in cylinder	1) Diameter of failure surface 2) Axial load 3) Axial deformation
Free swell test of clay	Clay	1) Total mass of the specimen and soils cup 2) Volume of the specimen when expansion ceases
Swelling ratio test	Clay	1) Initial height of the specimen 2) Mass of the specimen 3) Swell of the specimen
Swell pressure test	Undisturbed soils and compacting soils	The balancing load
Shrinkage test	All kinds of soils	Changes of height and mass of the specimen with time
Determination of the pH value	Colorimetric method Electrometric method	– pH value of the suspension
Determination of soluble salt	Measurement of the total mass of soluble salt	All kinds of soils	1) Mass of evaporation pan 2) Total mass of evaporation pan and dried residue

Determination	Method	Condition	Measurement
Measurement of carbonate and bicarbonate			Dosage of sulfuric acid standard solution
Measurement of chloride ion			Dosage of silver nitrate standard solution
Measurement of calcium and magnesium ions			Dosage of EDTA standard solution 1) Dosage of calcium and magnesium to EDTA standard solution 2) Dosage of calcium ion to EDTA standard solution
Measurement of sodium and potassium ions			1) Sodium absorption value 2) potassium absorption value
Measurement of sulfate ion	EDTA complex volume method	Content of sulfate is greater than or equal to 0.025% (equivalent to 50 mg/L) of soils	1) Dosage of calcium and magnesium and barium magnesium mixtures to EDTA standard solution in leaching solution 2) The same volume of barium magnesium mixtures (blank) to EDTA standard solution 3) Dosage of calcium and magnesium to EDTA standard solution in the same volume of leaching solution
	Turbidimetric method	Content of sulfate is smaller than 0.025% (equivalent to 50 mg/L) of soils	Content of sulfate
Determination of organic matter		Organic content is not greater than 15% of soils	Dosage of ferrous sulfate standard solution

Table 5.2 Tests on rocks

Test	Application scope	Parameters to be measured
Water content test		Mass of the specimen before and after drying
Grain density test	All kinds of rocks	1) Mass of rock particles after drying 2) Total mass of pycnometer and liquid and rock particles 3) Total mass of pycnometer and liquid
Point load strength test		1) Dimensions 2) Failure load
Density test — Direct measurement method	All kinds of rocks that can be made into regular shape	1) Diameter of the specimen 2) Height of the specimen 3) Mass of the specimen before and after drying
Density test — Water displacement method	Rocks that are easy to disintegrate, dissolve, have high permeability, dry shrinkage, wet swelling in water and cannot be made into regular specimens	1) Weight of the specimen 2) Weight of wax-sealed specimen 3) Weight of wax-sealed specimen in water
Absorption test	Rocks that do not disintegrate, dissolve or shrink and swell when exposed to water	1) Weight of specimen after drying 2) Weight of specimen after immersion for 48 hours 3) Weight of specimen after compulsory saturation 4) Weight of specimen after compulsory saturation in water
Swelling properties test — The swelling strain developed in an unconfined rock specimen	Rocks that are not easily disintegrated when wet	1) Axial deformation 2) Radical deformation
The swelling strain index for a radially confined rock specimen with axial surcharge		Axial deformation
The swelling pressure index under conditions of zero volume change	All kinds of rocks	Load that keeps the volume constant
Slake durability test	Rocks that are easy to disintegrate in water	Total mass of oven-dried specimen and drum

Test	Applicable rocks	Measured items
Uniaxial compressive strength test		1) Dimensions 2) Failure load
Durability under freezing and thawing conditions		1) Mass of the specimen before and after freezing-thawing cycles 2) Uniaxial compressive strength
The deformability in uniaxial compression test	All kinds of rocks that can be made into cylinder	1) Dimensions 2) Failure Load 3) Axial strain and radial strain
The strength in triaxial compression test		1) Dimensions 2) Failure load
The indirect tensile strength by Brazil test	All kinds of rocks that can be made into regular shape	1) Dimensions of specimen 2) Failure load
The shear strength of rock joints	Discontinuities of all kinds of rocks	1) Shear displacement 2) Shear stress
The sound velocity test by ultrasonic pulse transmission technique	Rocks that can be made into regular shape	1) Height of the specimen 2) Wave velocity
Rebound hardness test	Rocks with no apparent difference in surface and internal quality and no defects	Rebound value

(2) Water content test

Apparatus:

Drying oven (Figure 5.1) – thermostatically controlled oven, capable of maintaining a uniform temperature of 110 ± 5 °C.

Figure 5.1 Drying oven

Balance (Figure 5.2) – with 0.1 g and 0.01 g readability, when having a mass of up to 1000 g and 200 g, respectively.

Figure 5.2 Balance

(3) Density test

1) CUTTING RING METHOD

Apparatus:

Cutting ring (Figure 5.3) – with inner diameter of 61.8 mm or 79.8 mm and a height of 20 mm.

Figure 5.3 Cutting ring

Balance – with 0.1 g and 0.01 g readability, when having a mass of up to 1000 g and 200 g, respectively.

2) WAX-SEALING METHOD

Apparatus:

Wax-sealing apparatus – with a wax melting heater.
Balance – with 0.1 g and 0.01 g readability, when having a mass of up to 1000 g and 200 g, respectively.

(4) Specific gravity of soils solids

Apparatus:

Pycnometer (Figure 5.4) – with volume of 50 mL or 100 mL.

Figure 5.4 Pycnometer

Water bath (Figure 5.5) – capable of maintaining a temperature from room temperature to 100 °C, with an accuracy of ± 1 °C.

Sand bath (Figure 5.6) – capable of maintaining a temperature from 20 °C to 300 °C, with an accuracy of ± 1 °C.

Figure 5.5 Water bath

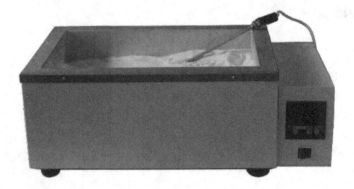

Figure 5.6 Sand bath

Electronic balance (Figure 5.7) – with 0.01 g readability, when having a mass of up to 200 g.

Figure 5.7 Electronic balance

Thermometer – with work temperature ranging from 0 °C to 50 °C and 0.5 °C readability.

(5) Particle size distribution test

1) SIEVING ANALYSIS METHOD

Apparatus:

Test sieves (Figure 5.8) – include coarse sieve (with hole diameters of 60, 40, 20, 10, 5 or 2 mm) and finer sieve (with hole diameters of 2.0, 1.0, 0.5, 0.25 or 0.075 mm).

Figure 5.8 Test sieves

Balance – with 5 g, 1 g, 0.5 g, 0.1 g and 0.01 g readability, when having a mass of up to 5 kg, 1 kg, 0.5 kg, and 0.2 kg, respectively.

Mechanical sieve shaker (Figure 5.9).

Figure 5.9 Mechanical sieve shaker

Other instruments – drying oven, mortar, porcelain plate and brush, etc.

2) DENSIMETER METHOD

Apparatus:

Hydrometer A (Figure 5.10) – with the scale ranging from –5° to 50° and 0.5° readability.
Hydrometer B (20 °C/20 °C) (Figure 5.10) – with the scale ranging from 0.995 to 1.020
and 0.002 readability.

Figure 5.10 Hydrometer

Measuring cylinder (Figure 5.11) – with inner diameter of 60 mm, a volume of 1000 mL,
a height of about 420 mm, a scale ranging from 0 to 1000 mL and accurate of less than
10 mL.

Figure 5.11 Measuring cylinder

Washing sieve – with hole diameter of 0.075 mm.

Washing sieve funnel – the upper diameter of this device should be larger than diameter of washing sieve, whereas its bottom diameter should be smaller than inner diameter of measuring cylinder.

Electronic balance – with 0.1 g and 0.01 g readability, when having a mass of up to 1000 g and 200 g, respectively.

Stirrer with helical vane – with wheel diameter of 50 mm, hole diameter of 3 mm and bar length of 450 mm.

Boiler – with condenser pipe device.

Thermometer – with scale ranging from 0 °C to 50 °C and 0.5 °C readability.

Other device – stopwatch, conical flask (500 mL), mortar, wood pestle and conductivity meter.

Reagents:

4% sodium hexametaphosphate solution (Figure 5.12) – dissolving 4 g sodium hexametaphosphate $(NaPO_3)_6$ into 100 mL water.

Figure 5.12 4%$(NaPO_3)_6$ solution

5% acidic silver nitrate solution (Figure 5.13) – dissolving 5 g silver nitrate $(AgNO_3)$ into 100 mL of 10% nitric acid (HNO_3) solution.

Figure 5.13 5% acidic silver nitrate solution

5% acidic barium chloride solution (Figure 5.14) – dissolving 5 g barium chloride ($BaCl_2$) into 100 mL of 10% hydrochloric acid (HCl) solution.

Figure 5.14 5% acid barium chloride solution

(6) Atterberg limits test

1) CONE PENETROMETER METHOD

Apparatus:

Cone Penetrometer (Figure 5.15) – consists of display screen, release and locking device, sample cup, steel cone with a dial gauge and switch.

Figure 5.15 Cone Penetrometer

Cone (Figure 5.16) – with a mass of 76 g, an angle of 30°.

Figure 5.16 Cone

Specimen cup – with an inner diameter of 40 mm and a height of 30 mm.
Balance – with 0.01 g readability, when having a mass of up to 200 g.

2) ROLLING METHOD

Apparatus:

Ground glass plane – with size of 200 mm×300 mm.
Calipers – with 0.02 mm readability.

(7) Relative density test of sand

1) MINIMUM DRY DENSITY

Apparatus:

Glass graduated cylinder – with a volume of 500 mL or 1000 mL.
Long-stem funnel (Figure 5.17) – with an inner diameter of the stem tube is 1.2 cm and the pipe orifice shall be smoothed.

Figure 5.17 Long-stem funnel

Conical stopper (Figure 5.18) – a cone with a diameter of 1.5 cm welded to the iron pole.

Figure 5.18 Conical stopper

Blade (Figure 5.19) – with the metal sheet welded on the lower end of the copper rod.

Figure 5.19 Blade

2) MAXIMUM DRY DENSITY

Apparatus:

Cylindrical metal mold (Figure 5.20) – with a volume of 1000 mL and 10 cm in diameter; or a volume of 250 mL and 5 cm in diameter. The mold shall be 12.7 cm in height with an extension collar.

Figure 5.20 Cylindrical metal mold

Vibrating fork (Figure 5.21).

Figure 5.21 Vibrating fork

Hammer (Figure 5.22) – with a mass of 1.25 kg, a diameter of 5 cm and the max fall distance of 15 cm.

Figure 5.22 Hammer

(8) Compaction test

In light compaction, the compaction work per unit volume is 592.2 kJ/m³, and that in heavy compaction is 2684.9 kJ/m³.

Apparatus:

Compaction modular (Figure 5.23) and hammer (Figure 5.24) – conforming to the requirements in Table 5.3.

Figure 5.23 Compaction modular

Figure 5.24 Hammer

Table 5.3 Required features of rammer and mold

Method	Diameter of rammer (mm)	Mass of rammer (kg)	Falling distance (mm)	Mold			Extension collar (mm)
				Inside diameter (mm)	Height (mm)	Volume (cm³)	
Standard	51	2.5	305	102	116	947.4	50
Modified	51	4.5	457	152	116	2103.9	50

Balance – with a capacity of 200 g, readable to 0.01 g.

Platform balance – with a capacity of 10 kg, readable to 5 g.

Test sieves – with aperture sizes of 5 mm, 20 mm, 40 mm.

Extruder (Figure 5.25) – Use a hydraulic jack or screw jack. If there is no such device, trowel or spatula can be used to take out the specimen from the mold.

Figure 5.25 Extruder

(9) California bearing ratio test

Apparatus:

The specimen mold (Figure 5.26) – with an inner diameter of 152 mm and height of 166 mm. It shall be assembled with a spacer disc, which is also called bearer with a height of 50 mm and an outside diameter of 151 mm, placed in the bottom of the mold and an extension collar with a height of 50 mm.

Figure 5.26 Mold

Hammer and guide tube – with the bottom diameter of hammer of 51 mm, mass of hammer of 4.5 kg and falling distance of 457 mm).

Test sieve – with hole diameter of 20 mm, 40 mm or 5 mm.

Swell measurement device (Figure 5.27) – mounting a frame to support the dial indicator.

Figure 5.27 Swell measurement device

Perforated swell plate (Figure 5.28) – with an adjustable stem, apertures of the plate shall be smaller than 2 mm.

Figure 5.28 Perforated swell plate

Penetrometer with the following parts: 1) a loading machine (Figure 5.29) and a load-indicating device, the capacity of the load-indicating device shall not be less than 50kN and the minimum penetration rate shall be maintained at 1 mm/min, 2) Penetration piston (Figure 5.30) with a diameter of 50 mm and height of 100 mm and provided with clamp holes for deformation measuring device, 3) Deformation measuring device (Figure 5.31), such as a mechanical dial indicator or an electronic displacement transducer, shall be capable of reading measurements to 0.01 mm.

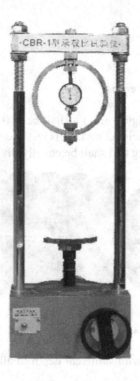

Figure 5.29 Loading machine

Figure 5.30 Penetration piston

Figure 5.31 Deformation measuring device

Annular surcharge weights (Figure 5.32) – shall be 150 mm in diameter, and shall have a center hole of approximately 52 mm. This test needs 4 annular weights and each of them has a weight of 1.25 kg and shall be equally divided into two halves.

Figure 5.32 Annular surcharge weights

Soaking tank (Figure 5.33) – with sufficient depth to allow water around the assembled mold over 25 mm above.

Figure 5.33 Soaking tank

Other devices – platform scale, stripper, etc.

(10) Modulus of resilience test

Apparatus (Lever pressure apparatus method)

Lever pressure apparatus (Figure 5.34) – with maximum pressure of 1500 N. Calibration should be conducted in accordance with requirements of apparatus before testing.

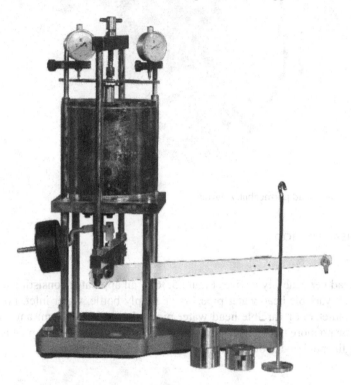

Figure 5.34 Lever pressure device

The specimen mold – conforming to the requirements of apparatus in compaction test. Only a screw hole with an inner diameter of 5 mm and a depth of 5 mm shall be provided at the left of notch plate connected to the column of tamping soleplate for installment of the dial indicator bracket.

Bearing plate – with diameter of 50 mm and a height of 80 mm.

Dial gauge – with measuring range of 2 mm.

Stopwatch – with 0.1 s readability.

(11) Penetration test

1) CONSTANT-HEAD METHOD

Apparatus:

Constant head permeability device (Figure 5.35) – an apparatus consisting of sealing metal cylinder, metal porous disc, gauze, piezometric tube and water supply bottle. Inner diameter of metal cylinder is 10 cm and its height is 40 cm. When using other size cylinder, its inner diameter should be 10 times greater than maximum particle size of the specimen.

Figure 5.35 Constant head permeability device

2) VARIABLE-HEAD METHOD

Apparatus:

Variable head permeability device (Figure 5.36) – an apparatus consisting of penetration container, variable head water pipe, water supply bottle, water inlet. For this device, Inner diameters of variable head water pipes should be uniform, and pipe diameter should be no more than 1 cm. Minimum division value of external pipe is 1.0 mm and the length should be about 2 m.

Figure 5.36 Variable head permeability device

Penetration container – a device consisting of cutting ring, porous stone, water stop gasket, lantern ring, upper cover and lower cover.
Cutting ring – with inner diameter of 61.8 mm and a height of 40 mm.
Porous stone – its coefficient of permeability should be greater than 10^{-3} cm/s.

(12) Consolidation test

Apparatus (Standard consolidation test):

Consolidation container – a device consisting of cutting ring, retaining ring, porous stone, water tank, pressurized upper cover and lantern ring.

Cutting ring – with inner diameters of 61.8 mm or 79.8 mm, and height is 20 mm. It is noted that this device should have certain stiffness and the inner wall should have a high smoothness by coating a thin layer of silicone grease or polytetrafluoroethylene (PTFE).

Porous disc – a device made of alumina or metal material resistant to corrosion and has the greater permeability coefficient than that of the specimen. When using fixed container, diameter of top porous disc should be 0.2–0.5 mm smaller than diameter of cutting ring; when using floating ring container, diameter of upper and bottom porous disc should be equal, and both of them should be smaller than inner diameter of cutting ring.

Loading device (Figure 5.37) – a device capable to apply the specified vertical pressure at all levels in an instant, and there is no impact force. The pressure accuracy should conform to the requirements of the current national standard *"Basic parameters and general technical conditions of geotechnical instruments"* GB/T15406.

Figure 5.37 Loading device

Deformation measurement apparatus – with measuring range of 10 mm and 0.01 mm readability of dial gauge or 0.2% accuracy of displacement detector in the whole measuring range.

Consolidation and pressurization apparatuses – a device should be calibrated regularly with provision of the corresponding calibration curve of equipment distortion. For specific operation see relevant standards.

(13) Collapse characteristics test of loess

Apparatus used in loess collapsibility test should conform to the requirements of consolidation test and have inner diameter of cutting ring of 79.8 mm. Humidity of filter papers and porous stone used in the test should be greater than the natural humidity of the specimen.

(14) Triaxial compression test

Apparatus:

Strain control triaxial apparatus (Figure 5.38) – a device consisting of pressure chamber, axial pressurization equipment, ambient pressure system, back pressure system, measuring system of pore water pressure, axial deformation and measuring system of volume change.

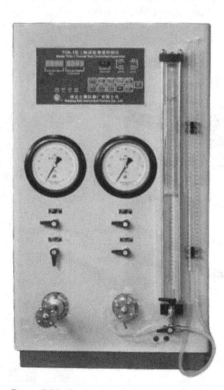 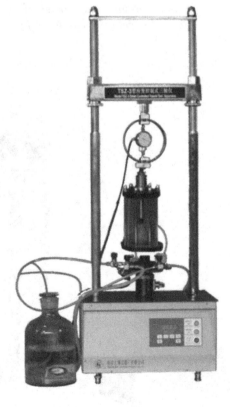

Figure 5.38 Strain-controlled triaxial apparatus

Accessory equipment – a device consisting of compaction apparatus, saturator, soils cutter, undisturbed soils riffle box, soils-cutting plate, film tube and split round film

Balance – with 0.01 g and 0.1 g readability, when having a mass of 200 and 1000 g,
respectively.

Rubber membrane – with diameters of 39.1 mm or 61.8 mm and elastic latex film thickness of 0.1–0.2 mm or with diameter of 101 mm and elastic latex film thickness of
0.2–0.3 mm.

Porous disc – with diameter being equal to that of the specimen and permeability coefficient being greater than that of the specimen. It should be boiled and soaked in water
before use.

(15) Unconfined compressive strength test

Apparatus:

Strain-controlled unconfined compression apparatus (Figure 5.39) – a device consisting
of dynamometer, pressurized framework and lifting appliance.

Figure 5.39 Strain-controlled unconfined compression apparatus

Axial displacement meter – with measuring range of 10 mm, 0.01 mm readability of dial
indicator or 0.2% accuracy of displacement detector in the whole measuring range.

Balance – with 0.1 g readability, when having a mass of up to 500 g.

(16) Direct shear test under consolidated-drained conditions

Apparatus:

Strain-controlled direct shear apparatus (Figure 5.40) – a device consisting of shear box, vertical pressurization equipment, shear transmission, dynamometer and displacement measuring system.

Figure 5.40 Strain-controlled direct shear apparatus

Cutting ring – with inner diameter of 61.8 mm and a height of 20 mm.
Displacement measuring system – with measuring range of 10 mm, 0.01 mm readability of dial indicator or 0.2% accuracy of detector in the whole measuring range.

(17) Reversal direct shear strength test

Apparatus:

Strain-controlled repetitive direct shear apparatus – a device consisting of hear box, vertical pressurization equipment, shear transmission, dynamometer and displacement measuring system, shear variable speed equipment, shear thrust reverser and reversible motor.

Cutting ring – with inner diameter of 61.8 mm and a height of 20 mm.

Displacement measuring system – with measuring range of 10 mm, 0.01 mm readability of dial indicator or 0.2% accuracy of detector in the whole measuring range.

(18) Tensile strength test

Apparatus:

Electronic universal testing machine (Figure 5.41).

Figure 5.41 Universal testing machine

Wire cutting machine (Figure 5.42).

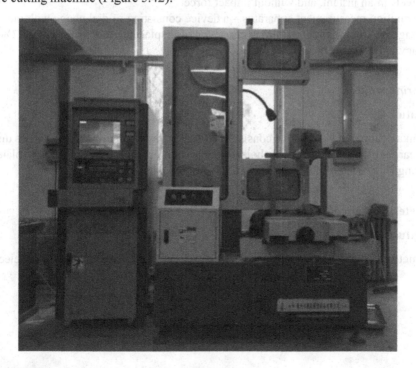

Figure 5.42 Wire cutting machine

Polish sleeve – with diameter of 50 mm and a height of 100 mm.
Specimen clamp.
Soils-cutting device – with variable diameters.
Soils-cutting knife.
Abrasive paper.

(19) Free swell test of clay

Apparatus:

Soils cup – with a volume of 10 mL and inner diameter of 20 mm.
Stemless funnel – with top diameter of 50–60 m, and bottom diameter of 4–5 mm.
Stirrer – a device consisting of straight bar and round plate with hole.
Measuring cylinder – with a volume of 50 mL and accuracy of 1 mL. Its volume and scale need to be calibrated before use.
Balance – with measuring range of 200 g and 0.01 g accuracy.

(20) Swelling ratio test, swell pressure test

Apparatus:

Consolidation container – a device consisting of cutting ring, retaining ring, porous stone, water tank, pressurized upper cover and lantern ring.
Cutting ring – with inner diameter of 61.8 mm or 79.8 mm and a height of 20 mm.
Porous stone – with the larger permeability coefficient than that of the specimen.
Pressurization equipment – a device capable to apply the specified vertical pressure at all levels in an instant, and without impact force.
Deformation measurement apparatus – a device consisting of dial gauge with measuring range of 10 mm and 0.01 mm accuracy or of displacement detector with 0.2% accuracy in the whole measuring range.

(21) Shrinkage test

Apparatus:

Shrinkage apparatus – a device consisting of scale, stay, measuring plate, porous disc and bearing block. The area of the hole accounts for more than 50% of the entire plate area.
Cutting ring – with inner diameter of 61.8 mm and a height of 20 mm.

(22) Determination of the pH value

Apparatus:

pH meter (Figure 5.43) – with glass electrode, calomel electrode and composite electrode.

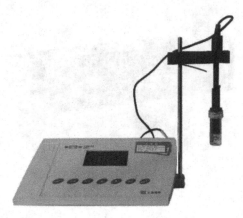

Figure 5.43 pH meter

Sieves – with hole diameter of 2 mm.
Balance – with measuring range of 200 g and 0.01 g of accuracy.
Electric magnetic stirrer (Figure 5.44).

Figure 5.44 Electric magnetic stirrer

Electric oscillator (Figure 5.45).

Figure 5.45 Electric oscillator

Other equipment – drying oven, beaker, jar, glass rod, 1000 mL volumetric flask, filter paper.

Reagents:

Standard buffer solution:

1) pH=4.01 – 10.21 g potassium hydrogen phthalate ($KHC_8H_4O_4$) dried at 105–110 °C was dissolved in distilled water and then solution was transferred into 1000 mL volumetric flask through funnel and finally diluted to 1000 mL.
2) pH=4.01 – 3.53 g sodium phosphate dibasic (Na_2HPO_4) and 3.39 g potassium dihydrogen phosphate (KH_2PO_4) was dried at 105–110 °C, then dissolved in distilled water, then transferred into volumetric flask and finally diluted to 1000 mL with distilled water.
3) pH=9.18 – 3.80 g borax ($Na_2B_4O_7 \cdot 10H_2O$) was dissolved in distilled water without CO_2 and then transferred into 1000 mL volumetric flask through funnel. Finally, the solution was diluted to 1000 mL volumetric flask with distilled water and without CO_2. It should be stored in airtight and moisture-free plastic bottle.

Saturated potassium chloride solution – potassium chloride (KCl) was added into appropriate amounts of distilled water with good stirring until potassium chloride is no longer dissolved.

Note: all reagents are analytical purified chemical reagent.

(23) Determination of soluble salt

I) LEACHING SOLUTION PREPARATION

Apparatus:

Sieve – with hole diameter of 2 mm.
Balance – with measuring range of 200 g and 0.01 g accuracy.
Electric oscillator.
Filtration equipment – with suction flask, porcelain funnel and vacuum pump.
Other equipment – wide mouthed bottle, narrow mouthed bottle, volumetric flask, horn spoon, glass rod and drying oven.

2) MEASUREMENT OF THE TOTAL MASS OF SOLUBLE SALT

Apparatus:

Analytical balance – with measuring range of 200 g and 0.0001 g accuracy.
Water bath and evaporation pan.
Other apparatus – drying oven, dry container, crucible tongs, pipette

Reagents:

15% of hydrogen peroxide solution and 2% of sodium carbonate.

3) MEASUREMENT OF CARBONATE AND BICARBONATE

Apparatus:

Acid burette – with volume of 25 mL and 0.05 mL accuracy.
Analytical balance – with measuring range of 200 g and 0.0001 g accuracy.
Other apparatus – pipette, conical flask, drying oven, volumetric flask.

Reagents:

Methyl orange indicator (0.1%) – 0.1 g methyl orange was dissolved into 100 mL distilled water.
Phenolphthalein indicator (0.5%) – 0.5 g phenolphthalein was dissolved into 50 mL alcohol and diluted with distilled water to 100 mL.
Sulphuric acid standard solution – 3 mL analytical pure concentrated sulfuric acid was dissolved into appropriate amount of water and finally diluted with distilled water to 1000 mL.
Calibration of sulphuric acid standard solution – three samples of 0.1 g anhydrous sodium carbonate with 0.0001 g accuracy were dried at 160–180 °C for 2–4 h before use. Then, 20–30 mL pure water and 2 drops of methyl orange indicator were mixed with three samples in three conical flasks, respectively. Subsequently, they were titrated with the sulfuric acid standard solution until the solution turns from yellow to orange. Finally, the amount of sulfuric acid standard solution was recorded and the exact concentration of the sulfuric acid standard solution was calculated according to the following formula.

$$c(H_2SO_4) = \frac{m(NaCO_3) \times 1000}{V(H_2SO_4) \cdot M(NaCO_3)}$$

Where:
$c(H_2SO_4)$-concentration of sulphuric acid standard solution (mol/L),
$V(H_2SO_4)$-dosage of sulphuric acid standard solution (mL),
$m(Na_2CO_3)$-dosage of sodium carbonate (g), and
$M(Na_2CO_3)$-molar mass of sodium carbonate (g/mol).

The average value with error of less than 0.05 mL should be taken and calculated to be accuracy of 0.0001 mol/L in three parallel titration experiments.

4) MEASUREMENT OF CHLORIDE ION

Apparatus:

Analytical balance – with measuring range of 200 g and 0.0001 g accuracy.
Brown acid burette – with a volume of 25 mL and 0.05 mL accuracy.
Other equipment – pipette, drying oven, conical flask, volumetric flask.

Reagents:

Potassium chromate indicator (5%) – 5 g potassium chromate (K_2CrO_4) was dissolved into appropriate amount of distilled water. Then, silver nitrate standard solution was added the potassium chromate solution dropwise until a brick red precipitation appeared. After standing overnight, it was filtered and the filtrate was diluted to 100 mL with pure water and stored in a dropping bottle.

Silver nitrate $c(AgNO_3)$ standard solution – 3.3974 g analytical pure silver nitrate ($AgNO_3$) was dried at 105–110 °C for 30 minutes before use and it was washed and dissolved with distilled water to 1L volumetric flask. Finally, it was diluted with distilled water to 1000 mL and stored in brown bottle. The concentration of silver nitrate solution is about 0.02 mol/L.

Sodium bicarbonate $c(NaHCO_3)$ solution – 1.7 g sodium bicarbonate was dissolved in distilled water and diluted with distilled water to 1000 mL volumetric flask. The concentration of Sodium bicarbonate solution is about 0.02 mol/L.

Methyl orange indicator

5) MEASUREMENT OF SULFATE ION

EDTA complex volume method

Apparatus:

Analytical balance – with measuring range of 200 g and 0.0001 g accuracy.
Acid burette – with a volume of 25 mL and 0.1 mL accuracy.
Other equipment – pipette, conical flask, volumetric flask, measuring cup, horn spoon, drying oven, mortar, pestle, measuring cylinder.

Reagents:

1:4 hydrochloric acid: mix one concentrated hydrochloric acid and 4 distilled water well by volume.

Barium magnesium mixture – 1.22 g barium chloride ($BaCl_2 \cdot 2H_2O$) and 1.02 g magnesium chloride ($MgCl_2 \cdot 2H_2O$) were dissolved with distilled water into 500 mL volumetric flask and then the solution was diluted with distilled water to 500 mL.

Ammonia buffer solution – 70 g ammonia chloride (NH_4Cl) was put into a beaker and added into appropriate amount of distilled water. After dissolution, it was transferred into 1000 mL measuring cylinder. Then, 570 mL analytical pure concentrated ammonium hydroxide was added into ammonia chloride solution and finally diluted with distilled water to 1000 mL.

Chrome black T indicator: 100 g sodium chloride ($NaCl$) that is dried before use and 0.5 g chrome black T were mixed, grinded and stored in brown bottle.

Zinc primary standard solution – 0.6538 g zinc dust (particle) was dried at 105–110 °C before use and put into beaker, and then 20–30 mL of 1:1 hydrochloric acid solution was gradually and carefully added to the zinc dust. Next, the zinc dust solution was heated until zinc dust was dissolved completely. Finally, it was transferred into a 1000 mL volumetric flask and diluted it with distilled water to 1000 mL. The concentration of zinc primary standard solution is 0.01 mol/L.

EDTA standard solution:

Preparation – 3.72 g Ethylene Diamine Tetraacetic Acid (EDTA) was dissolved in hot distilled water and then transferred into 1000 mL volumetric flask when cold. Finally, it was diluted with distilled water to 1000 mL.

Calibration – three samples of 20 mL zinc primary standard solutions were put in three conical cylinders by pipette, respectively. Then, they were diluted with appropriate amount of water with adding 10 mL ammonia buffer solution, some Chrome black T indicator and 95% alcohol. Next, the solutions were titrated by using EDTA standard solution until they turn to light blue from red. Finally, the amount of standard solution was recorded and the exact concentration of standard solution was calculated according to the following formula:

$$c(\text{EDTA}) = \frac{V(\text{Zn}^{2+})c(\text{Zn}^{2+})}{V(\text{EDTA})}$$

Where:
C(EDTA)-concentration of standard solution (mol/L),
V(EDTA)-dosage of standard solution (mL),
C(Zn^{2+})-concentration of zinc primary standard solution (mol/L), and
V(Zn^{2+})-dosage of zinc primary standard solution (mL).

The average value with error of less than 0.05 mL should be taken and calculated to be accuracy of 0.0001 mol/L in three parallel titration experiments.
95% analytical alcohol.
1:1 hydrochloric acid – the concentrated hydrochloric acid and water was mixed with 1:1 by volume.
5% Barium chloride ($BaCl_2$) solution – 5 g barium chloride ($BaCl_2$) was dissolved into 1000 mL distilled water.

Turbidimetric method

Apparatus:

Photoelectric colorimeter.
Spectrophotometer.
Electric magnetic stirrer.
Measuring spoon – with a volume of 0.2–0.3 cm^3.
Other equipment – pipette, volumetric flask, sieve, drying oven, analytical balance with 0.1 mg readability.

Reagents:

Suspension stabilizer – 30 mL concentrated hydrochloric acid (HCl),100 mL of 95% alcohol, 30 mL distilled water and 25 g sodium chloride (NaCl) were mixed together, and then 50 mL glycerinum was added the solution and mixed again.

Crystallization of barium chloride ($BaCl_2$) – barium chloride was crystallized and sieved to select those of particle size between 0.6 mm–0.85 mm.

Silver sulfate standard solution – 0.1479 g anhydrous sodium sulfate was dried at 105–110 °C before use, dissolved with distilled water and transferred into 1000 mL volumetric flask through a funnel. Finally, the solution was diluted with distilled water to 1000 mL. Concentration of sulfate radical in the solution is 0.1 mg/mL.

6) MEASUREMENT OF CALCIUM AND MAGNESIUM IONS

Apparatus:

Acid burette – with a volume of 25 mL and accuracy 0.1 mL.

Other apparatus – pipette, conical flask, measuring cup, balance, mortar.

Reagents:

2 mol/L sodium hydroxide solution – 8 g sodium hydroxide was dissolved into 100 mL distilled water;

Calcium indicator – 50 g sodium chloride baked before use and 8 g Calcium indicator were put in a mortar and grinded and mixed well. They were stored in a brown bottle and kept in a dryer;

EDTA standard solution – the preparation procedures are the same as the determination (V) of sulfate radical-EDTA complexation volumetric method;

1:4 hydrochloric acid solution – the concentrated hydrochloric acid and water were mixed with 1:4 by volume;

Congo red paper; and

95% ethanol.

7) MEASUREMENT OF SODIUM AND POTASSIUM IONS

Apparatus:

Flame photometer – with accessory apparatus.

Analytical balance – with measuring range of 200 g and 0.0001 g accuracy.

Other apparatus – high temperature furnaces, drying oven, pipette, 1 L volumetric flask, 500 mL volumetric flask, beaker.

Reagents:

Sodium (Na^+) standard solution – 0.2542 g sodium chloride burned at 550 °C before use was dissolved in a little distilled water and transferred into a 1L volumetric flask. Finally, the solution was diluted with distilled water to 1000 mL flask and stored in plastic bottle. The sodium ions (Na^+) concentration of solution is 0.1 mg/mL (100 mg/L).

Potassium standard solution – 0.1907 g potassium chloride (KCl) dried at 105–110 °C before use was dissolved in a little distilled water and transferred into a 1 L volumetric flask. Finally, the solution was diluted with distilled water to 1000 mL flask and stored in plastic bottle. The potassium ions (K^+) concentration of solution is 0.1 mg/mL (100 mg/L).

(24) Determination of organic matter

Apparatus:

Analytical balance – with measuring range of 200 g and 0.0001 g accuracy.
Oil bath pan – with wire cage and vegetable oil.
Heating device – drying oven, electric furnace.
Other apparatus – thermometer with measuring range of 0–50 °C and 0.5 °C accuracy, Test tube, conical flask, burette, small funnel, wash bottle, volumetric flask, and 0.15 mm sieve.

Reagents:

Potassium dichromate standard solution – 44.1231 g potassium dichromate was dried at 105–110 °C and grinded well before use and then was dissolved in 800 mL distilled water. Next, 1000 mL concentrated sulfuric acid was dropped slowly into potassium dichromate solution. After cooling, the solution was transferred into a 2 L volumetric flask and diluted with distilled water to 2 L. Potassium dichromate concentration of solution is 0.075 mol/L.

Ferrous sulfate standard solution – 56 g ferrous sulfate ($FeSO_4 \cdot 7\,H_2O$) (or 80 g ammonium ferrous sulfate) was dissolved in appropriated amount of distilled water and 30 mL of 3 mol/L $c(H_2SO_4)$ solution. Then the solution was diluted with distilled water to 1L and calibrated according to following steps: weigh three parts of 10.00 mL potassium dichromate standard solution. Put them in three conical flask and dilute with distilled water to 60 mL, respectively. Then, 3–5 drops of phenanthroline indicator were added into three parts of potassium dichromate solution, followed by titration with ferrous sulfate standard solution. Record its dosage when the solution suddenly changes to red orange from yellow and green. The average value with error of less than 0.05 mL should be taken in three parallel titration experiments. The concentration of ferrous sulfate standard solution was calculated with 0.0001 accuracy according to the following formula:

$$c\left(FeSO_4\right) = \frac{c\left(K_2Cr_2O_7\right)V\left(K_2Cr_2O_7\right)}{V\left(FeSO_4\right)}$$

Where:
c($FeSO_4$)-concentration of ferrous sulfate (mol/L),
V($FeSO_4$)-dosage of ferrous sulfate titration (mL),
c($K_2Cr_2O_7$)-concentration of potassium dichromate (mol/L), and
V($K_2Cr_2O_7$)-take volume of potassium dichromate (mL).

Phenanthroline indicator – 1.845 g phenanthroline indicator and 0.695 g ferrous sulfate were dissolved into 100 mL distilled water and stored in brown bottle.

5.2.2 Rock tests

(1) Water content test

Apparatus:

Electric heating drying oven – thermostatically controlled oven, capable of maintaining a uniform temperature of 110 ±5 °C.

Balance – with measuring range of 200 g and 0.01 g accuracy.

Drying container.

(2) Grain density test

Apparatus:

Pulverizer, porcelain mortar or agate mortar, magnet block, sieve with bore diameter of 0.25 mm.

Balance – with measuring range of 200 g and 0.01 g accuracy.

Electric heating drying oven – thermostatically controlled oven, capable of maintaining a uniform temperature of 110 ±5 °C.

Thermostatic water tank – with accuracy of less than ±1 °C.

Sand-bath – a thermostatically controlled device with measuring range from 20 °C to 200 °C.

Short-necked pycnometer with a volume of 100 mL.

Thermometer – with measuring range of 0–50 °C and 0.5 °C accuracy.

Other apparatus – drying container, boiling device, vacuum pumping device.

(3) Density test

1) DIRECT MEASUREMENT METHOD

Apparatus:

Electric heating drying oven – thermostatically controlled oven, capable of maintaining a uniform temperature of 110 ±5 °C.

Balance – with measuring range of 200 g and 0.01 g accuracy.

Caliper – with accuracy of 0.01 mm.

Drying container.

Measuring platform.

Other apparatus – drill, rock saw, stone mill, grinding machine.

2) WATER DISPLACEMENT METHOD

Apparatus:

Drill, rock saw, stone mill, grinding machine.

Balance – with measuring range of 200 g and accuracy of 0.001 g.

Electro-optical analytical balance – with measuring range of 200 g and accuracy of 0.01 mg.

Paraffin melting apparatus, electric heating drying oven, drying container, measuring platform.

(4) Absorption test

Apparatus:

Drill, rock saw, stone mill, grinding machine.

Electric heating drying oven – thermostatically controlled oven, capable of maintaining a uniform temperature of 110 ±5 °C.

Balance – with measuring range of 200 g and accuracy of 0.01 g.

Water tank, drying container, measuring platform, paraffin melting apparatus, weighing device in water, caliper.

(5) Swelling properties test

Apparatus:

Drill, rock saw, stone mill.
Measuring platform.
Free swelling rate tester.
Lateral restraint swelling rate tester.
Expansion pressure tester.
Thermometer.

(6) Slake durability test

Apparatus:

Drying oven and drying container.

Balance or platform scale – with measuring range of 15 kg and accuracy of 0.5 g.

Slake durability device (Figure 5.46) – a device consisting of drive equipment, cylindrical sieve drum and water tank. The length and diameter of cylindrical sieve drum is 100 mm and 140 mm, respectively. Diameter of sieve pore is 2 mm.

Figure 5.46 Slake durability apparatus

Thermometer.

(7) Uniaxial compressive strength test

Apparatus:

Drill machine (Figure 5.47), cutting machine (Figure 5.48), stone mill (Figure 5.49), lathe (Figure 5.50).

Figure 5.47 Drilling machine

Figure 5.48 Cutting machine

Figure 5.49 Stone mill

Figure 5.50 Lathe

Measuring platform.
Material testing machine.
Automatic controlled electro-hydraulic servo compression testing machine.

(8) Durability under freezing and thawing conditions

Apparatus:

Balance – with measuring range of 200 g and accuracy of 0.01 g.
Refrigerator or low-temperature freezer – with the lowest work temperature of –24 °C.
Galvanized iron sheet box and iron wire stand.
Drill, rock saw, stone mill, lathe.
Measuring platform.
Material testing machine.

(9) The deformability in uniaxial compression test

Apparatus:

Drill, rock saw, stone mill, lathe.
Measuring platform.
Material testing machine.
Static resistance strain measuring instrument.
Wheatstone bridge, megger, and multimeter.
Resistance strain foil gauge, dial gauge or dial indicator.
Dial gauge stand, magnet stand.

(10) The strength in triaxial compression test

Apparatus:

Drill, rock saw, stone mill, lathe.
Measuring platform.
Triaxial apparatus
Automatic controlled electro-hydraulic servo compression testing machine.

(11) The indirect tensile strength by Brazil test

Apparatus:

Drill, rock saw, stone mill, lathe.
Measuring platform.
Material testing machine.
Microcomputer controlled electric press.

(12) The shear strength of rock joints

Apparatus:

The specimen preparation apparatus.
The specimen saturation and maintenance equipment.
Flat-pushing manipulation direct shear tester
Displacement indicator.

(13) Point load strength test

Apparatus:

Point-load tester (Figure 5.51).

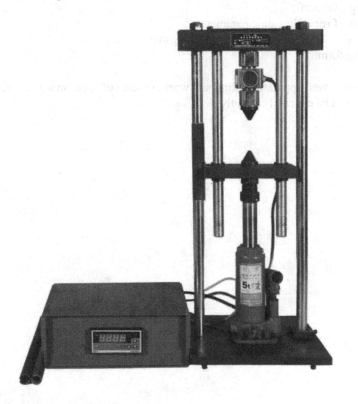

Figure 5.51 Point load apparatus

Vernier caliper (Figure 5.52) – with accuracy of 0.01 mm.

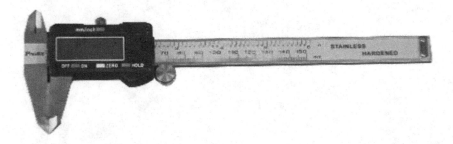

Figure 5.52 Vernier caliper

(14) The sound velocity test by ultrasonic pulse transmission technique

Apparatus:

Drill, stone sawing machine, stone mill, lathe.
Measuring platform.
Instrument of rock ultrasonic parameter.
Transducer of longitudinal wave and shear wave.
Testing platform.

Note: joint properties, apparatus wiring and working state of apparatus and transducer after power-on should be checked carefully.

Chapter 6

Procedures of tests on soils

Laboratory soil test is to determine physical, chemical and mechanical properties of soil. This provides an important guidance for geotechnical engineering design and construction. This chapter includes the preparation of soil specimen and test procedures mainly according to *Standard for Soil Test Method* (GB/T 50123-1999) and relative industry standards, such as highway and water conservation. We prepared flowcharts of test procedures for each test, in order to provide the readers with an overall view of the test.

6.1 Specimen preparation

6.1.1 Preparation

It is required to make a detailed description of undisturbed soil sample prior to cutting it into the required size. For disturbed soil, several steps are normally required: describing, drying, grinding and sieving. Besides this, saturating the specimen may be necessary for certain tests.

Mechanical property test possesses specific requirements for preparation of undisturbed specimen. For instance, density difference of undisturbed specimens in the same group should be less than 0.03 g/cm³. The density of the prepared specimen should be accurate to ±0.01 g/cm³ with respect to the predetermined density. The water content of the prepared specimen should be accurate to ±1% of the predetermined water content.

(1) Requirements and steps for preparation of undisturbed soil specimen

1) Place soil sample drum according to indicated direction, strip wax seal and friction tape, open the soil sample drum and take out the specimen. Inspect the specimen structure to see if the specimen has been disturbed and if soil quality confirms to requirements of the specimen mechanical test.

2) Apply a thin layer of petroleum jelly in the inside of cutting ring. Press the cutter vertically on the sample. Whittle the specimen and soil cutter along the outside of cutter. Level the two ends of the specimen with wire saw or soil cutter according to hardness of the specimen and wipe up the outside of cutting ring. Weigh the total mass of cutting-ring and soil.

3) Take representative specimen and measure its water content. Specimen preparation of specific gravity, particle size analysis and limitation water ratio can be referenced to item 2.

4) Describe the physical properties of specimen prior to cutting. For soft soil having low plasticity and high sensitivity, it should be prepared without disturbing it.

(2) Requirements and steps for disturbed soil specimen preparation

1) Take out the soil sample from sample drum or package. Describe the physical properties of specimen prior to cutting. Then, cut the soil sample into fragments and mix them well. Take representative soil sample and measure its water content.

2) For homogeneous organic soil, representative soil sample with natural water content should be used for particle size analysis and limit water content test. For heterogeneous soil, take the proper amount of soil according to the test items, and dry the specimen until it can be pulverized. The specimen for cohesionless soil should be dried at 105–110 °C. For soil with more than 5% organic matter or containing gypsum and sulphate, it should be dried at 65–70 °C.

3) Put air-dried or oven-dried soil sample on a rubber plate and grind it with wood pestle. For soil sample without sand and gravel, it can be ground with soil breaking device (soil breaking device cannot crush soil particles).

4) For the dispersed coarse-grained soil and fine-grained soil, it shall be sieved according to Table 6.1. Gravel soil containing fine-grained soil should be soaked in water and stirred thoroughly to separate the coarse and fine particles. Then, sieve the specimen according to the requirements of the specific test.

Table 6.1 Test sampling numbers (quantity) and soil sieving standards

Soil types Sampling numbers	Cohesive soil		Cohesionless soil		Sieving-passed standards (mm)
	Undisturbed soil (cylinder) $\varphi10$ cm × 20 cm	Disturbed soil (g)	Undisturbed soil (cylinder) $\varphi10$ cm × 20 cm	Disturbed soil (g)	
Water content		800		500	
Specific gravity		800		500	
Particle size analysis		800		500	
Atterberg limits		500			0.5
Density	1		1		
Consolidation	1	2000			2.0
Loess collapse	1				
Triaxial compression	2	5000		5000	2.0
Swell, Shrinkage	2	2000		8000	2.0
Direct shear	1	2000			2.0
Compaction California Bearing Ratio		Light compaction> 15000 Heavy compaction> 30000			5.0
Unconfined compressive strength	1				
Reversal direct shear strength	1	2000			2.0
Relative density				2000	
Penetration	1	1000		2000	2.0
Chemical analysis		300			2.0
Centrifuge water equivalent		300			0.5

(3) Steps for preparing disturbed soil specimen

Procedures:

Setp-1: 1–2 extra samples should be prepared.

Step 2: The ground air-dried soil sample is passed through a sieve with hole diameter of 2 mm or 5 mm. Thoroughly mix soil sample and measure its water content. Then place the specimen in moist air vat or plastic bag.

Step 3: Calculate the required amount of water in the specimen sample according to the following formula:

$$m_w = \frac{m_0}{1+0.01\omega_0} \times 0.01(\omega_1 - \omega_0) \tag{6.1}$$

Where:

m_w – the amount of water added for the specimen preparation (g),

m_0 – the mass of wet soil (or air-dried soil) (g),

ω_0 – the water content of wet soil (or air-dried soil) (%), and

ω_1 – the water content of the specimen (%).

Step 4: Weigh the air-dried soil sample after sieving and lay it in the enamel pan, spray the water evenly on the soil sample, mix well and keep the sample in the soil container for 24 h (the wetting time of cohesionless soil may be reduced).

Step 5: Measure water content of the specimen (at least two subspecimens).

Step 6: Calculate mass of wet soil according to the following formula:

$$m_0 = (1+0.01\omega_0)\rho_d V \tag{6.2}$$

Where:

ω_0 – the water content of wet soil (or air-dried soil) (%),

ρ_d – dry density of the specimen (g/cm^3), and

V – volume of the specimen (volume of cutting ring) (cm^3).

Step 7: Impactor method and sample-press method can be used for disturbed specimen.

1) Impactor method: put wet soil sample into an impactor apparatus with cutting-ring and compact sample to the desired density.

2) Sample-press method: put wet soil sample into a sample-press apparatus with cutting-ring and compact the sample to the desired density through a piston at a constant pressure.

8. Take out the cutting-ring with the specimen, weight and record the total mass of cutting-ring and the specimen. The samples should be stored in a humidor.

6.1.2 Saturation

Immersion saturation method is used for coarse-grained soil. Capillary saturation method is used for fine-grained soil with permeability coefficient of more than 10^{-4} cm/s; Air exhaust saturation method is used for fine-grained soil with permeability coefficient of less than or equal to 10^{-4} cm/s.

(1) Capillary saturation method

1) Put filter papers on the top of sample and porous disc at the bottom of sample, and then move the sample into frame saturator and tighten the nut.
2) Put saturator with sample in water tank and inject clean water until the surface of specimen will be submerged. Close tank lid to ensure that the specimen is completely saturated (no less than 48 h).
3) Take out cutting-ring and dry its outer wall. Weigh total mass of the specimen and cutting-ring and calculate degree of saturation of the specimen. If the measured degree of saturation is less than 95%, the sample needs to be saturated again.
4) Calculate saturation degree of the specimen according to the following formula:

$$S_r = \frac{(\rho_{sr} - \rho_d)G_s}{\rho_d \cdot e} \tag{6.3}$$

or

$$S_r = \frac{\omega_{sr} G_s}{e} \tag{6.4}$$

Where:
S_r – degree of saturation of the specimen (%),
ω_{sr} – water content of the specimen after being saturated (%),
ρ_{sr} – density of the specimen after being saturated (g/cm³),
G_s – specific gravity of soil grain, and
e – void ratio of the specimen.

(2) Air exhaust saturation method

Step 1: The folded saturator/frame saturator and vacuum degree of saturation device are needed. In the center of clamping plate which is under the folded saturator, porous disc, filter papers, cutting-ring with the specimen, filter papers, porous disc are sequentially placed to the height of bridle rod. Then, cover upper clamping plate of saturator well and tighten the nut in the upper end of bridle rod. Finally, tighten every cutting-ring between upper clamping plate and bottom clamping plate.

Step 2: The specimen is put between vacuum cylinder and lid. Then, tighten lid of the vacuum cylinder, connect vacuum cylinder and air exhauster, and switch on air exhauster. When vacuum manometer reading is close to atmospheric pressure (time for exhausting should not be less than one hour), slightly open the hose clamp to inject clear water into vacuum cylinder with a constant vacuum manometer reading.

Step 3: Stop air exhausting when saturator is submerged by water. Then, open the hose clamp and let air enter into vacuum cylinder. Keep the immersion time of fine-grained soil in water for approximately 10 h to ensure that the sample is well saturated.

Step 4: Open vacuum cylinder and take out cutting-ring with the specimen from saturator. Weigh the total mass of cutting-ring and the specimen and calculate degree of saturation according to Formula 6.3 and Formula 6.4. If the measured degree of saturation is less than 95%, the specimen needs to be saturated repeatedly.

6.2 Tests for physical properties

6.2.1 Water content

Water content of soil is the ratio of the mass of water to the mass of the solid particles in a given mass of soil. This test method is applicable for coarse-grained, fine-grained, organic and frozen soil. Parameters to be measured in this test include the mass of wet soil specimen and the mass of dry soil specimen. The flowchart (Figure 6.1) and the detailed procedures for water content test of soil are present as follows.

Procedures:

Step 1: Weigh and record the mass of the container. Collect representative specimen of about 15 g to 30 g. Put the soil specimen into the container and cover it with the

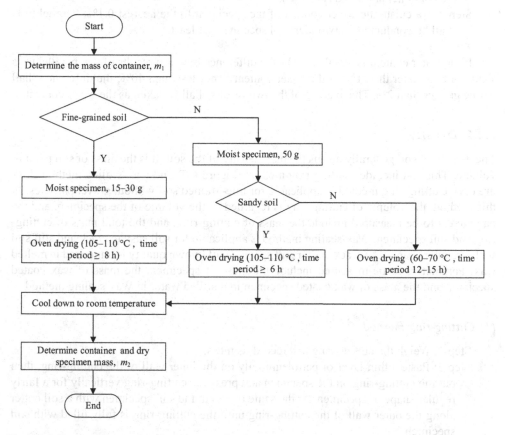

Figure 6.1 Determination of water content of soil

Notes:

[1] According to *Test Method of Soil for Highway Engineering* 2007, the representative soil sample generally should be fine-grained soil.

[2] According to *Test Method of Soil for Highway Engineering* 2007, the specimen should be dried at 60–70 °C for 12–15 h for soil containing gypsum or more than 5% organic compounds.

lid. Then weigh and record the total mass of the specimen and the container. If the
tested soil is organic, sandy or monolithic frozen, the mass of the specimen shall be
around 50 g.

Step 2: Remove the lid. Place the container with lid and specimen in the oven and dry at
105 to 110 °C in order to make sure that the specimen is completely dried.

Soil containing more than 5% organic materials in the dry specimen need to be dried at
65 to 70 °C; Soil containing gypsum need to be dried below 80 °C.

For clay and silt, the drying process shall last at least 8 h; for cohesionless soils, the
drying process shall last at least 6 h in order to make sure the specimen is completely
dried.

Step 3: Take out the container from the oven and put it in a desiccator, cool it down to
room temperature. Then take out the container, weigh and record the total mass of
the container and dried specimen.

Step 4: Calculate the water content of the specimen to the nearest 0.1%. Parallel tests
shall be conducted on two identical specimens at least.

If the water content is less than 40%, the difference between results of two parallel tests
shall not be greater than 1%; if the water content is not less than 40%, the difference shall
not be greater than 2%. The average of the two results shall be taken as the water content.

6.2.2 Density

The density of soil generally means the bulk density of the soil. It is the mass of soil per unit
volume. This test includes cutting-ring method (Figure 6.2) and wax-sealing method (Figure 6.3). Cutting-ring method is applicable for fine-grained soil without gravel particles. In
this method, the volume of cutting-ring is regarded as the volume of the specimen, and the
parameters to be measured include the mass of cutting-ring, and the total mass of cutting-
ring and soil specimen. Wax-sealing method is applicable for soil easy to crack and stiff soil
with irregular shape. The crack and/or stiff specimen of known quality is immersed in melted
wax. Parameters to be measured include the mass of specimen, the mass of wax coated
specimen and the mass of wax coated specimen in distilled water in Wax-sealing method.

(1) Cutting-ring method

Step 1: Weigh the cutting-ring and record its mass.

Step 2: Paste a thin layer of petroleum jelly on the inner wall of the cutting-ring, then
put the cutting-ring on the specimen and press the cutting-ring vertically for a fairly
regular shape of specimen. At the same time, cut the soil specimen with a soil cutter
along the outer wall of the cutting-ring until the cutting-ring is fully filled with soil
specimen.

Step 3: Trim the specimen at both ends of the cutting-ring with the soil cutter and the
wire saw. Wipe up the outer wall of the cutting-ring and weigh the total mass of the
cutting-ring and the specimen. Select a piece of specimen from the remaining specimen and determinate its water content.

Calculate the wet density and the dry density of the specimen and the calculation shall be
accurate to 0.01 g.

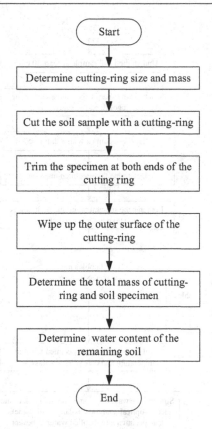

Figure 6.2 Determination of soil density (cutting-ring method)

Parallel test on at least two specimens is a necessity. If the difference is no greater than 0.03 g/cm³, calculate the average of the two calculations as the dry density of the specimen. Otherwise, repeat the test.

For details of another method, please refer to the next episode.

(2) Wax-sealing method

Step 1: Cut at least 30 cm³ representative undisturbed soil specimen. Wipe up the rego-lith and strike off the sharp edges on the surface of specimen then tie it with a thin line. Weigh the specimen and accurate to 0.01 g. Cut a piece of representative test specimen from remaining soil sample and determinate its water content.

Step 2: Hold the line and immerse the specimen into the melted wax that just passed the melting point. After the immersion, take out the specimen immediately. Check the wax mask around the specimen. If there are any bubbles in the wax, puncture the bubbles with a silver needle then fill with liquid wax. Cool the waxed specimen to room temperature. Then weigh and record its mass.

Step 3: Suspend the waxed specimen on the hook of the electro-optical analytical bal-ance and immerse the specimen into distilled water in a container. Make sure that the

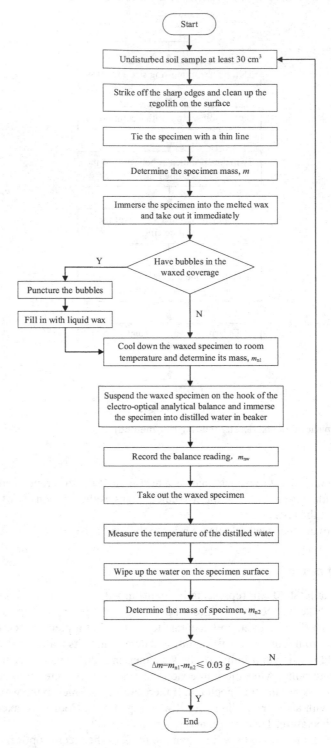

Figure 6.3 Determination of soil density (wax-sealing method)

specimen cannot touch the inner wall of the container. Weigh the mass of specimen in the water then record the measurement.

Step 4: Take out the specimen from water and then measure the temperature of the distilled water. Absorb the water on the surface of the wax and then weigh the mass of wax coated specimen. If the mass of wax coated specimen differs within 0.03 g it means that the wax-sealing is in good condition. Otherwise, repeat the test.

The density of the distilled water under the measured temperature is obtained from the international temperature scale of water density. Calculate the wet density and the dry density of the specimen.

Parallel test should be carried out on another identical specimen. If the two results differ by no greater than 0.03 g/cm³, calculate the average of the two calculations. The average is the dry density of the specimen. Otherwise, repeat the test.

6.2.3 Specific gravity of soil solids

In soil mechanics, specific gravity refers to the ratio of the mass of soil solids to the mass of distilled water of the same volume at 4 °C. It is one of the basic physical indices of soil. For soil with particles smaller than 5 mm in diameter, pycnometer method shall be used. For soil with particles bigger than 5 mm in diameter, floating method or siphon method shall be used. Parameters to be measured include the mass of specimen, total mass of pycnometer, water and specimen and the total mass of pycnometer and water. The pycnometer should be calibrated before the specific gravity test of soil grain. The pycnometer calibration can be conducted by plotting the relationship curve of the temperature against the total mass of pycnometer and water. The specific steps for determining the specific gravity of soil grain by pycnometer method are shown in Figure 6.4. The sample with known mass is initially placed in a pycnometer to measure the total mass. Water is then injected into the pycnometer and heated to be boiled. The mass of the pycnometer and water at the test temperature is obtained from the curve between temperature and mass of pycnometer and water. The specific gravity of water at different temperature can be taken from *CRC Handbook of Chemistry and Physics*. Finally, the specific gravity of soil solids can be calculated according to the formula.

(1) Calibration of pycnometer

Procedures:

Step 1: Wash pycnometer and dry it up, then cool it down at room temperature. Weigh, and record its mass.

Step 2: Fill the pycnometer with air-free distilled water. Plug a stopper in the pycnometer and let the excessive water overflow through the hole of the stopper. Put the pycnometer into a water bath until the pycnometer reaches thermal equilibrium, then record the water temperature. Take out the pycnometer and wipe its outer wall, measure and record the total mass of the pycnometer and water. A parallel test must be conducted for each temperature. The difference of the two measurements shall not be greater than 0.002 g, otherwise repeat the test. Calculate the average of the two measurements for each temperature.

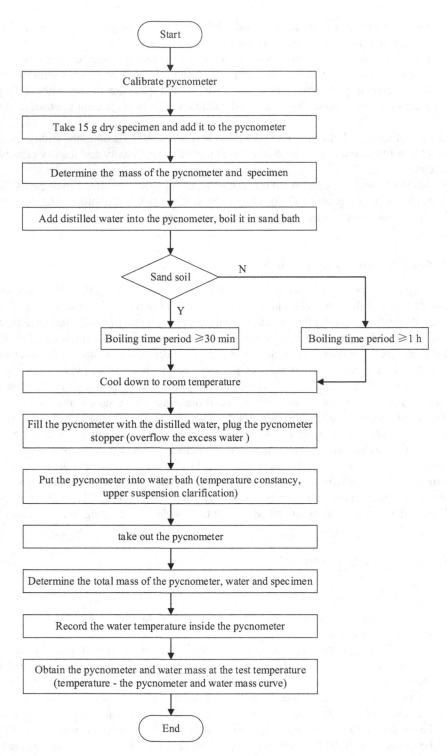

Figure 6.4 Determination of specific gravity of soil particle (pycnometer method)

Step 3: Increase the water temperature in the water bath gradually with intervals of 5 °C. Weigh and record the total mass of pycnometer and water at different temperatures. Plot the relationship curve of the temperature against the total mass of pycnometer and water.

(2) Pycnometer method

Procedures:

Step 1: Weigh the pycnometer and record its mass. Take 15 g dry specimen and add it to the 100 mL pycnometer. Weigh and record the total mass of the pycnometer and the specimen.

Step 2: Add half bottle of distilled water into the pycnometer, shake it up and plug a stopper into the pycnometer. Exhaust the air in the soil by boiling it in a sand bath. The time needed for boiling different samples will be different. Boiling sand sample shall need a duration of at least 30 min. For clay and silt samples the boiling shall need a duration of at least 1 h. Adjust the bath temperature after boiling in case of suspension overflowing from the pycnometer.

Step 3: Take out the pycnometer and cool it down to room temperature. Fill the pycnometer with distilled water. Plug a stopper in the pycnometer while the excess water overflows from the hole of stopper. Put the pycnometer into a water bath until the pycnometer reaches the thermal equilibrium. When the suspension in the upper part of the pycnometer is clarified, take out the pycnometer, wipe its outer wall. Weigh and record the total mass of the pycnometer water and specimen. Record the water temperature inside the bottle.

Step 4: The mass of the pycnometer and water at the test temperature is obtained from the curve between temperature and mass of pycnometer and water. The specific gravity of water at different temperature can be taken from *CRC Handbook of Chemistry and Physics*. Next, calculate the specific gravity of solids according to the formula. A parallel test must be conducted. The difference of the two results shall less than 0.02, otherwise repeat the test. Calculate the average of two calculations as specific gravity of solids.

It should be noted that: When the air in the soil is exhausted, it is usually boiled with distilled water. But for soil containing soluble salts, hydrophilic colloids, or organic matters, pump it in vacuum condition using neutral liquid such as kerosene.

6.2.4 Particle size distribution

Particle size distribution test is to determine the percentage of mass of different size of particles in the soil. This test can be used to evaluate the particle composition of soil and determine the types and engineering properties of soil. Hydrometer method, also known as sedimentation method, is applicable for soil with particles smaller than 75 μm in diameter. Sieving method is applicable for soil with particles larger than 75 μm, in which parameters to be measured include the total mass of specimen and the mass of soil particles remained on each test sieve during sieving. For soil that has more than 10% of particles larger than 75 μm, sieving method shall be first applied. Sieve the soil through the 75 μm wash sieve and then hydrometer method or pipette method shall be conducted. As shown in Figure 6.5, for sieving analysis method, soil particles can be separated by the standard sieves with the different hole diameters. The particle size distribution could be determined by weighing the soil particles with different sizes. In the densitometer method (Figure 6.6), the sample of a certain

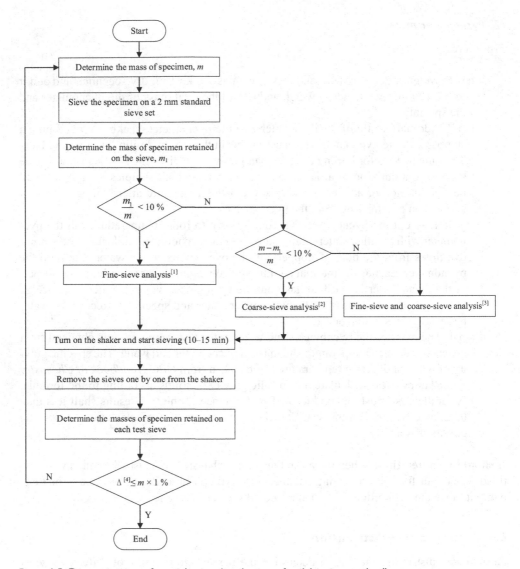

Figure 6.5 Determination of particle size distribution of soil (sieving method)

Notes:

[1] Fine sieve analysis: standard sieves are stacked one by one according to pore diameters, and then add the specimen under the sieve in step 2 to the top sieve.

[2] Coarse sieve analysis: similar to the fine sieve analysis.

[3] Fine sieve and coarse sieve analysis: similar to the fine sieve analysis

[4] Δ: refers to difference between the total mass of the specimen before sieving and the total mass of the specimen retained in each test sieve and sieve bottom.

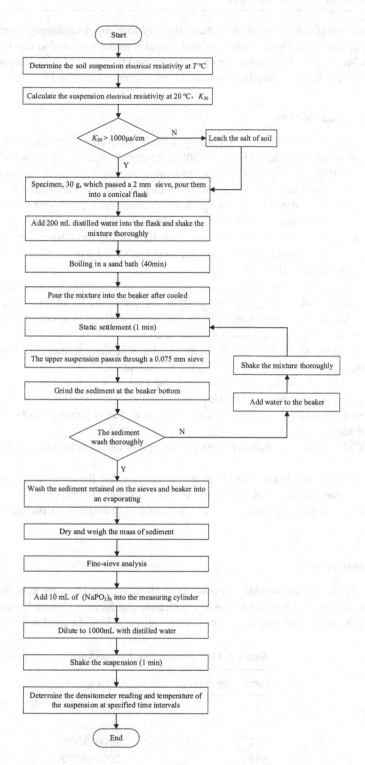

Figure 6.6 Determination of particle size distribution of soil (hydrometer method)

mass is added to sodium hexametaphosphate, thereby forming a uniformly distributed suspension of soil particles. The diameters of soil particles in the suspension can be calculated according to Stokes law. The total mass of specimen and the mass of specimen smaller than a specified particle diameter need be measured in densitometer method.

(1) Sieving analysis method

Step 1: Specimen shall be taken according to mass requirement for specimen (Table 6.2) If the mass of specimen is less than 500 g, determine and record its mass to the nearest 0.1 g. If the mass exceeds 500 g, determine and record its mass to the nearest 1 g.

Step 2: Sieve the specimen on a 2 mm test sieve. Weigh and record the mass of specimen retained on the sieve. If the specimen retained on a 2 mm test sieve is less than 10% of the total mass, only fine-sieve analysis should be used. If the specimen passing though the 2 mm test sieve is less than 10% of the total mass, only coarse-sieve analysis shall be used. Otherwise, both analyses shall be used. Specimen in this test is suitable for fine-sieve analysis.

Step 3: For fine-sieve analysis, arrange the sieves successively by their sizes. Pour the specimen passing through the test sieve in step 2 into the set of fine-sieves, cover the lid on the top of the sieves and start the fine-sieve analysis. For coarse-sieve analysis, follow the same procedure as fine-sieve analysis.

Step 4: Place the set of finer sieves on the mechanical sieve shaker. Turn on the shaker and start sieving. A 10 to 15 min shaking is adequate.

Step 5: Remove the sieves one by one from the shaker, weigh and record the masses of specimen retained on each test sieve and receiver to the nearest 0.1 g. The difference between the total mass of the specimen before and after sieving shall not exceed 1% of the total mass.

Step 6: Calculate the cumulative mass percentage of particles passing each sieve.

Take the particle size as abscissa on the logarithmic scale and the cumulative percentage as ordinate on the linear scale, to plot the particle size distribution curve.

If required, calculate the coefficient of uniformity (C_u) and the coefficient of curvature (C_c).

(2) Densimeter method

Step 1: Determine the soluble salt content of air-dried specimen by using conductivity method following the instruction of the electric conductivity meter. Measure the conductivity of the soil suspension (soil-water ratio is 1:5) and calculate the conductivity

Table 6.2 Mass requirement for specimen

Particle size (mm)	Mass (g)
<2	100–300
<10	300–1000
<20	1000–2000
<40	2000–4000
<60	>4000

of soil suspension at 20 °C. If the value is larger than 1000 µs/cm, the salt shall be leached from the soil according to formula 6.5.

$$K_{20} = \frac{K_T}{1+0.02(T-20)} \tag{6.5}$$

Where:

K_{20} – electrical resistivity of suspension in 20 °C (µs/cm),

K_T – electrical resistivity of suspension in T °C (µs/cm), and

T – temperature of suspension when testing (°C).

Salt-leaching method: the mass of air-dried specimen should be first calculated according to Formula 6.6 and 6.7. If the soluble salt content is less than 1%, Formula 6.6 is used, otherwise, Formula 6.7 is employed. Then, add 30 g of air-dried specimen and 200 mL distilled water into 500 mL conical flask. Filter the soil suspension by filter papers. Wash and filter the soil suspension until its electrical resistivity K_{20} is less than 1000 µs/cm (or until there is no white precipitates when the filtered soil suspension reacts with 5% acid silver nitrate solution and 5% acid barium chloride solution):

When soluble salt content is less than 1%,

$$m_0 = 30 \times (1+0.01\omega_0) \tag{6.6}$$

When soluble salt content is more than or equal to 1%,

$$m_0 = \frac{30 \times (1+0.01\omega_0)}{1-W} \tag{6.7}$$

Where: W – soluble salt content (%).

Step 2: Weigh 30 g of specimen that passes through the 2 mm test sieve and pour them into a 500 mL conical flask. Add 200 mL of distilled water into the flask and shake the mixture thoroughly. Place the conical flask in a sand bath and boil the flask for 40 min.

Step 3: Shake the mixture after cooling. Pour the mixture that has passed through a 0.075 mm sieve into a measuring cylinder. Then thoroughly wash the sediment retained in the conical flask with distilled water and pour the mixture into the cylinder. Grind the sediment on the sieve by using rubber pestle and rinse them with distilled water into the measuring cylinder.

Pour all the sediments retained on the sieves into an evaporating dish and decant the clear water, then dry and weigh the mass of the sediments. Conduct the fine-sieve analysis with the reference of the sieving method. Calculate the percentage of the mass of particles smaller than 0.075 mm in the total mass of the specimen.

Step 4: Add 10 mL of sodium hexametaphosphate ($NaPO_3$)$_6$ solution into the measuring cylinder, then dilute to 1000 mL with distilled water.

Step 5: Insert a stirrer in the measuring cylinder and stir the suspension up and down to make all soil aggregations break-up. Take out the stirrer and start the stopwatch immediately. Insert the hydrometer in the suspension and record the reading at specified time intervals. The hydrometer shall be inserted about 10 to 20 s before each recording. During the recording, the hydrometer bubble shall be in the center of the measuring cylinder and should not touch the inner wall. Record the reading at the top of the meniscus.

After reading, take out the hydrometer immediately and put it in a measuring cylinder filled with distilled water. Measure and record the temperature of the suspension in measuring cylinder. Carefully insert and remove the hydrometer (before and after taking each reading) without disturbing the suspension.

Step 6: Determine the water content of specimen with the reference of the water content test and determine the specific gravity of soil particles with the reference of the specific gravity of soil solids test.

Calculate the mass of air-dried specimen whose dry mass is 30 g according to the formula. The specific gravity shall be obtained from the correction table for specific gravity. Corrected temperature shall be obtained from the correction table for suspension temperature.

6.2.5 Atterberg limits

The consistency state of cohesive soil according to its water content can be divided into solid, semi-solid, plastic and liquid state. Atterberg limits of cohesive soil named Atterberg limits which includes shrinkage limit, plastic limit and liquid limit. Methods applicable for determining the liquid limit of soil are: cone penetrometer method and Casagrande method. The methods applicable for determining the plastic limit of soil are: rolling method and cone penetrometer method. Cone penetrometer method (Figure 6.7) is used for soil with particles smaller than 5 mm and organic matter content less than 5%, where the cone penetration and water content should be measured. This rolling method (Figure 6.8) is used for soil with particles smaller than 0.5 mm, and the parameters to be measured are the diameter and water content of soil thread.

(1) Cone penetrometer method

Procedures:

Step 1: If the specimen is of air-dried state, sieve the specimen with a 0.5 mm sieve and take 200 g specimen which is passed through the 0.5 mm sieve. If the specimen is of natural state, take 250 g. Determine the water content of remained specimen. Mix the specimen thoroughly with distilled water until the mass becomes a thick homogeneous paste. Allow the paste to stand for at least 24 h or for long enough to enable the water to permeate through the soil.

Thoroughly kneading shall be conducted for a dryer specimen. Push a portion of the mixed soil into the cup with a palette knife taking care not to trap air. Strike off excess soil with the straightedge to give a smooth level surface.

Step 2: Levelling the cone penetrometer. Coat a thin layer of petroleum jelly to the surface of the cone. Turn on the tester to lock the cone by electromagnet and set the dial gauge to zero.

Step 3: Put the sample cup on the seat of the tester. Raise the cup so that the tip of the cone just touches the surface of the specimen. Release the cone and let the cone fall freely into the test specimen, then keep 5 s and record the penetration reading.

Remove the specimen container and lift out the cone. Remove the portion of the specimen with petroleum jelly. Take a water content sample of about 10 g from the area penetrated by the cone and determine the water content.

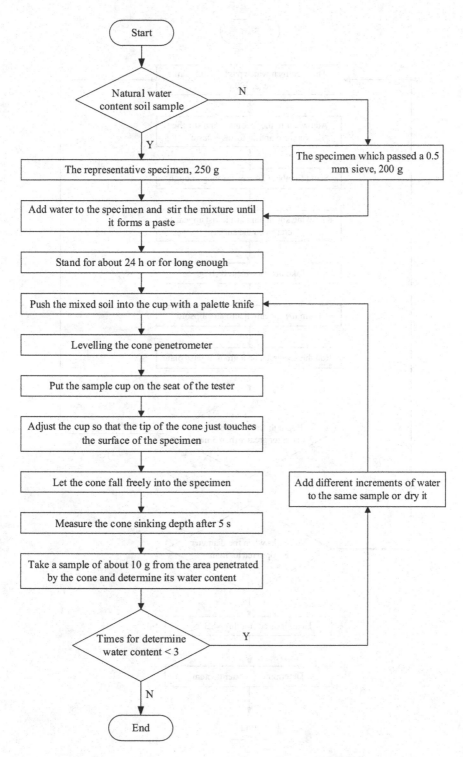

Figure 6.7 Determination of Atterberg limits of soil (cone penetrometer method)

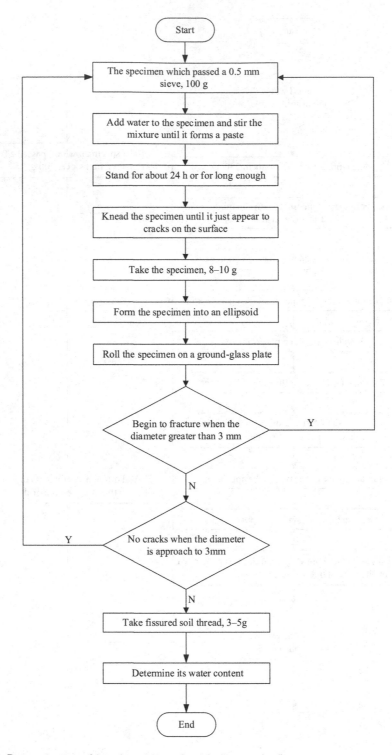

Figure 6.8 Determination of Atterberg limits of soil (rolling method)

Step 4: Adding different increments of water to the same sample or drying it. Then mix them thoroughly and repeat steps 1–3. The preparation of at least three specimens with different water contents is needed. These range of penetration values shall be distributed at 3–4 mm, 7–9 mm and 15–17 mm.

Plot the relationship between cone penetration and water content on a double-logarithmic coordinate with cone penetration as ordinate against water content as abscissa.

Draw the best straight line fitting the plotted points. Those three points shall be on a straight line as line "A". Otherwise, connect the point with the bigger water content and the remaining two points into two straight lines respectively. Two water contents at 2 mm penetration shall be obtained. If the difference between two smaller water contents is less than 2% then connect the midpoint of them and the point of high water content into a straight line as line B. If the difference between two smaller water contents is greater than 2%, repeat the test.

Report the water content corresponding to a cone penetration of 17 mm as the 17 mm liquid limit, a penetration of 10 mm as the 10 mm liquid limit and a penetration of 2 mm as the plastic limit. The values shall be expressed as the percentage with an accuracy of 0.1%.

Calculate the plasticity index and the liquidity index. The liquidity index shall reach its accuracy of 0.01.

(2) Rolling method

Procedures:

Step 1: Take 100 g of representative specimen passing through the 0.5 mm sieve and mix it with distilled water thoroughly and make the paste. Allow the paste to stand for at least 24 h or for long enough to enable the water to permeate through the soil.

Step 2: Knead the prepared specimen with hand until it is not sticky then pinch it to flat. The water content of the specimen is close to the plastic limit when the slight cracks appear after pinching it.

Step 3: Take about 8 to 10 g of the specimen near to plastic limit. Form the specimen into an ellipsoid and roll it on a ground-glass plate. It is important to maintain a uniform and sufficient rolling pressure. The thread shall not be hollowed and the length of the thread shall be no longer than the width of the palm.

Step 4: If fissures appear and the thread begins to fracture as the thread diameter approaches 3 mm, it indicates that water content of the specimen reaches the plastic limit water content. If no fissure appears when the thread diameter is 3 mm or thread fractures when the thread diameter is greater than 3 mm, it indicates that water content of the specimen is higher or lower than the plastic limit then repeat the test.

Step 5: Take 3 g to 5 g fissured soil thread with a diameter of 3 mm and determine its water content.

Another parallel test is necessary and the difference of parallel results should satisfy the requirements of water content test. Take the average of the results and record it.

It should be noted that: When the thread diameter is 3 mm it may be caused by uneven rolling pressure. And the fissure should be appeared in a shape of whorl. With soil that are marginally plastic it is often difficult to obtain the correct crumbling condition.

6.2.6 Relative density of cohesionless soils

Relative density is an index that reflects the tightness of the soil, is the ratio of the difference between the void ratio of a cohesionless soil in the loosest state and any given void ratio, to the difference between the void ratios in the loosest and in the densest states. The relative density of cohesionless soil is the index which reveals whether the soil is loose or dense. The determination of the relative density of cohesionless soil is composed in the minimum index density and the maximum index density. The minimum index density (Figure 6.9) can be determined by pouring the specimen through funnel in a graduated cylinder and inverting it if necessary. The maximum index density (Figure 6.10) can be determined by vibrating hammer method. These tests are applicable to soil with particle size less than 5 mm and which may contain up to 15% of particles having size greater than 2 mm. For minimum index density, the parameters to be measured are volume of soil after depositing in a cylinder and volume of soil after inverting the cylinder. For maximum index density, the parameters to be measured are the total mass of the mold and specimen after vibrating.

(1) Minimum dry density

Procedures:

Step 1: Insert the conical stopper from the bottom of the long stem funnel and lift it up to cover the end of the funnel tube with the cone. Put the funnel with the stopper in a 1000 mL measuring cylinder.

Step 2: Take oven-dried representative specimen 700 g, pour it into the funnel slowly and raise the funnel and conical stopper simultaneously. Leave the cone slightly out of the spout to maintain continuous flow of specimen without spout contacting the already deposited soil. Make sure all the specimen deposit in the cylinder as loosely as possible. If the specimen doesn't contain particles with diameter larger than 2 mm, then 400 g specimen shall be taken and the 500 mL measuring cylinder shall be used.

Step 3: Remove the funnel and conical stopper, flatten the surface of the sample. Record the volume of the specimen in the funnel to the nearest 5 mL.

Step 4: Stop the cylinder with a palm or a rubber stopper. Tip the cylinder upside down, and then quickly tilt it back to the original vertical position. Record the volume that the soil occupies in the graduated cylinder. Repeat this Step several times and record the maximum volume of the measuring cylinder to the nearest 5 mL.

Step 5: Take the larger value from the volumes measured after depositing the specimen and inverting the cylinder. Calculate the minimum index density. At least one parallel test is necessary. The difference of two parallel results shall not be greater than 0.03 g/cm^3 and take the mean of the two measured values.

(2) Maximum dry density

Procedures:

Step 1: Take representative specimen 2000 g and mix it thoroughly. Add a portion of the specimen to the mold and the volume of specimen shall be one-third of the mold. After the specimen is added into the mold, hit the sides of the mold with a vibration

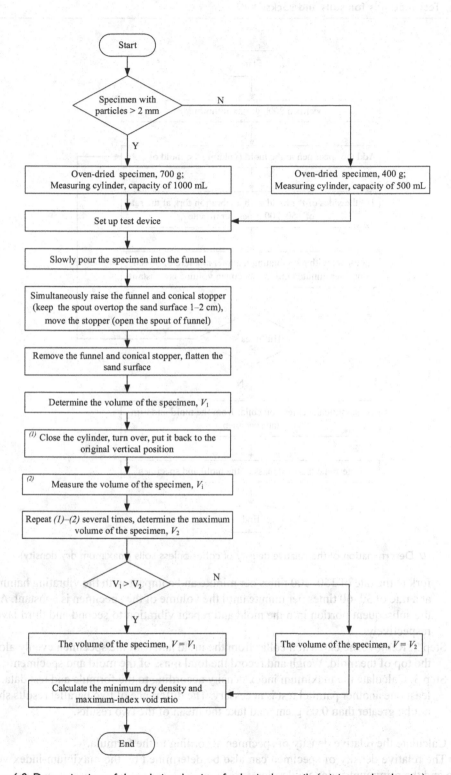

Figure 6.9 Determination of the relative density of cohesionless soils (minimum dry density)

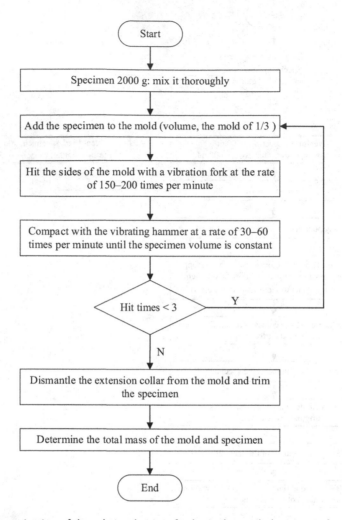

Figure 6.10 Determination of the relative density of cohesionless soils (maximum dry density)

fork at the rate of 150–200 times per minute and compact with the vibrating hammer at a rate of 30–60 times per minute until the volume of the specimen is constant. Add the subsequent portion into the mold and repeat vibration to second and third layers respectively.

Step 2: Remove the extension collar from the mold and trim the specimen evenly along the top of the mold. Weigh and record the total mass of the mold and specimen.

Step 3: Calculate the maximum index density according to the formula and test data. At least one another parallel test is necessary. The difference of two parallel results shall not be greater than 0.03 g/cm³ and take the mean of the two results.

Calculate the relative density of specimen according to the formula.

The relative density of specimen can also be determined by the maximum-index void ratio and the minimum-index void ratio.

6.3 Tests for mechanical properties

6.3.1 California Bearing Ratio

California Bearing Ratio (*CBR*) is mainly used to evaluate the bearing capacity of soil foundation and pavement materials. The test steps of which are composed of specimen preparation, immersion expansion and penetration test. As shown in Figure 6.11, the soil sample is prepared according to requirements of the maximum dry density and the optimum water content in the standard compaction test. The specimen is immersed for 96 h prior to the loading to simulate the material properties under the most severe conditions of the actual engineering. The standard penetration rod was pressed into the specimen at a speed of 1 to 1.25 mm/min during penetration. When the rod penetrates at the sample depth of 2.5 mm, the corresponding ratio of the unit pressure to the standard pressure is the California bearing ratio to be measured.

California Bearing Ratio is expressed as the ratio of the required pressure to penetrate 2.5 mm for the test soil to the required pressure to achieve equal penetration on standard soil. It is the strength index for subgrade and base materials, and is also one of the main parameters of flexible pavement design. This test is applicable for disturbed soil with maximum particle size not exceeding 20 mm (maximum particle size shall not exceed 40 mm for three layers compaction). Parameters to be measured include the swell during soaking, axial pressure and corresponding penetration. California bearing ratio test consists of several steps including specimen preparation, immersion and penetration. The specimen is prepared according to requirements of the maximum dry density and the optimum water content in the standard compaction test, immersed for 96 h before loading to learn the material properties under the actual conditions of engineering. The standard penetration rod is pressed into the specimen at the speed of 1 to 1.25 mm/min during penetration. When the rod reaches at 2.5 mm depth of the specimen, the corresponding ratio of the unit pressure to the standard pressure is determined as the California bearing ratio.

Procedures:

Step 1: specimen preparation.

Conduct a modified compaction test (with a bearer in the bottom of the mold) to determine the maximum index density and optimum water content of specimen in accordance with *Determination of Compaction Characteristics of Soil*. The procedure of preparing the specimen to maximum index density is similar.

Remove the collar, level off the soil surface along the mold with a scraper or a steel straightedge and fill the uneven surface with fine materials. Remove the bearer, weigh and record the mass of mold and specimen. Determine the water content of the compacted specimen. Each dry density needs three specimens, and the difference of them shall be less than 0.03 g/cm³.

Step 2: Immersion expansion.

Place a filter paper on the surface of specimen followed by a perforated baseplate. Invert the mold. Put a filter paper on the other end of the specimen, followed by the perforated swell plate with an adjustable stem. Fit the tie rod. Fit four annular surcharge weights around

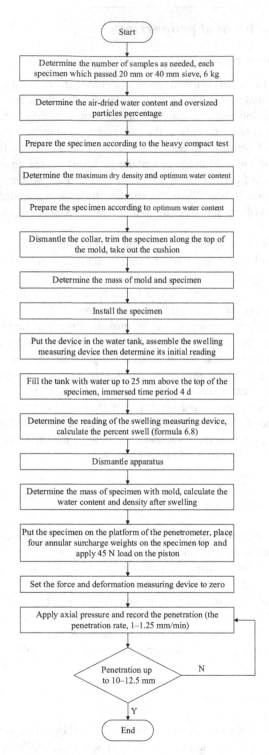

Figure 6.11 Determination of California bearing ratio of laboratory-compacted soil

the stem on the perforated plate. Place the top collar. Fix the collar, the mold and the perfo-rated baseplate by a tie rod.

Put the whole device in the water tank, assemble the swelling measuring device then read and record its initial reading. Fill the immersion tank with water up to 25 mm above the top of the specimen. After 4-days immersion, measure and record the height of the specimen.

Remove the specimen mold from the water tank. Absorb the water from the top of the specimen and take off the swell measurement device. Allow the specimen to drain down-ward for at least 15 min. Remove all the surcharge weights on specimen. Remove the per-forated swell plate, upper filter paper and top collar. Reverse the mold and remove the filter paper. Weigh the mass of specimen with mold. Record the reading.

Calculate the percent swell according to the formula 6.8 and calculate the water content and density of specimen after swelling.

$$\delta_w = \frac{\Delta h_w}{h_0} \times 100 \tag{6.8}$$

Where:
δ_w – the swell of the specimen (percent),
Δh_w – the height difference of the specimen before and after swelling (mm), and
h_0 – the initial height of the specimen.

Step 3: penetration.

Put the soaked specimen with mold on the platform of the penetrometer. Place the top sleeve on the mold. Put the penetration piston on the center of specimen. Clamp the dial indicator with the piston. Adjust the height of the platform to maintain sufficient contact between the load-transmitting members. Place four annular surcharge weights on the top of the specimen and apply 45N load on the piston. Set the force and deformation measuring device to zero.

Start the motor and apply axial pressure so that the piston penetrates the specimen at a rate of 1 to 1.25 mm/min. Record the penetration when the readings of the force indicator are integrated (for example 20, 40, 60 min) the force indicator is at regular reading. Make sure that there are more than 5 readings when penetration reaches 2.5 mm. Stop the penetration when it reaches 10 to 12. 5 mm.

Adjust the height of the platform. Take off the deformation measuring device. Remove all the surcharge weights on specimen. Remove top collar. Remove the specimen mold.

6.3.2 Modulus of resilience

Lever pressure apparatus method

The modulus of resilience is defined to be the ratio of the stress generated by pavement and road materials under load to the corresponding rebound strain. This reflects the strength and stiffness of the soil or base materials. In this test (Figure 6.12), the specimen is loaded/unloaded under pressures in stages to measure the resilient deformations of the soil.

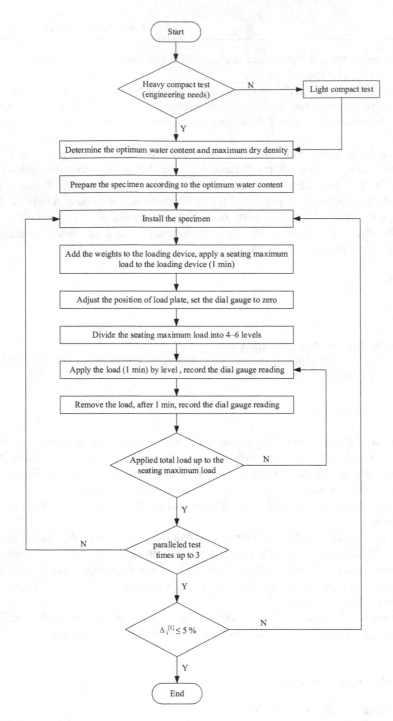

Figure 6.12 Determine of modulus of resilience of soil

Source:

[1]: $\Delta_i = E_{ei} - \overline{E}_e$

Where:

E_{ei} – modulus of resilience in the ith time (i = 1, 2, 3), and

$\overline{E}_e$ – average of parallel tests in modulus of resilience.

Procedures:

Step 1: Obtain the optimum water content and the maximum dry density by light or heavy compaction method according to the engineering requirements. Then prepare the compacted sample according to these data.

Step 2: Install the specimen. Place the cylinder with the specimen on the chassis of the lever pressure apparatus, where the bearing plate is placed at the center of the top surface of the specimen. Fix the dial gauge to the column and place the dial gauge probe on the bracket of the bearing plate.

Step 3: Add the weights on the loading device and apply a seating maximum load to the loading device for 1 min. Take a sample with water content less than the plastic limit as an example and apply the pressure of 100 to 200 kPa. If the water content of the sample is greater than the plastic limit, the preload pressure should be adjusted to 50 to 100 kPa. The preload should be carried out 1–2 times and each preload should be kept for 1 min. After preloading, adjust the position of load plate and set the dial gauge to zero.

Step 4: Divide the seating maximum pressure into 4–6 levels. Apply load by levels and record the dial gauge readings. Remove each load and after one minute record the corresponding dial gauge reading. The applied maximum pressure can be slightly greater than the seating maximum pressure.

Step 5: Three parallel tests should be conducted. The difference between resilience modulus of each parallel test and the average resilience modulus of all parallel tests should not exceed 5%. Otherwise, the test should be repeated.

6.3.3 Penetration test

The permeability of soil refers to property of water permeating in the pores of soil. According to Darcy's law, when the flow velocity of water in the soil is relatively low, the water permeating velocity of water is proportional to the hydraulic gradient, and the ratio of them is called the coefficient of permeability of soil. The coefficient of permeability is very useful in design of foundation pit protection, selection of dam body fill material, and calculation of consolidation settlement.

The penetration test is divided into constant-head method and variable-head method. The constant-head method is generally used for coarse-grained soil with high permeability ($k>10^{-3}$) and its procedure is shown in Figure 6.13. The constant head permeation device is used to measure amount of penetration water, temperatures of water, and the heights of water heads at different positions within a certain period of time, thereby the coefficient of permeability of the constant head is calculated. However, variable-head method is generally used for fine-grained soil with small coefficient of permeability ($k<10^{-3}$) and its test procedure is shown in Figure 6.14. The water seeps through the soil under variable head pressures and the coefficient of permeability of soil is calculated on basis of the relationship between declining height of water head and time.

The permeability of a soil is the capacity to allow the flow of water through the pore spaces between solid particles. According to Darcy's Law, when the velocity of flow in soil is low the permeability velocity is proportional to the hydraulic pressure gradient and its ratio is equal to the coefficient of permeability. The coefficient of permeability can be used in the design for deep excavation bracing, the selection of filled material for embank, the

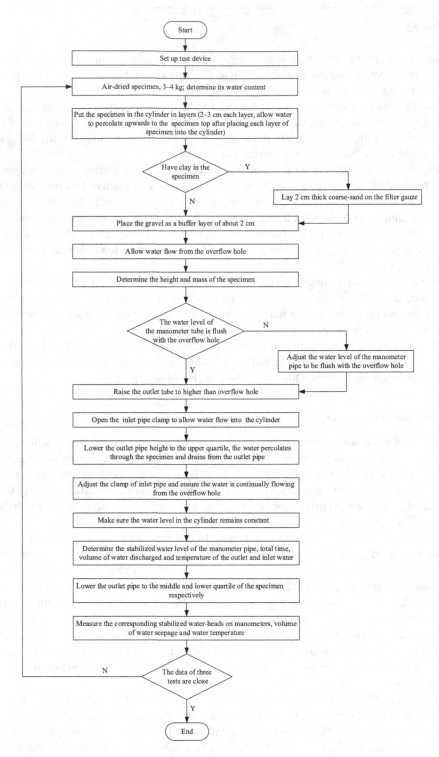

Figure 6.13 Determination of permeability of soil (constant-head method)

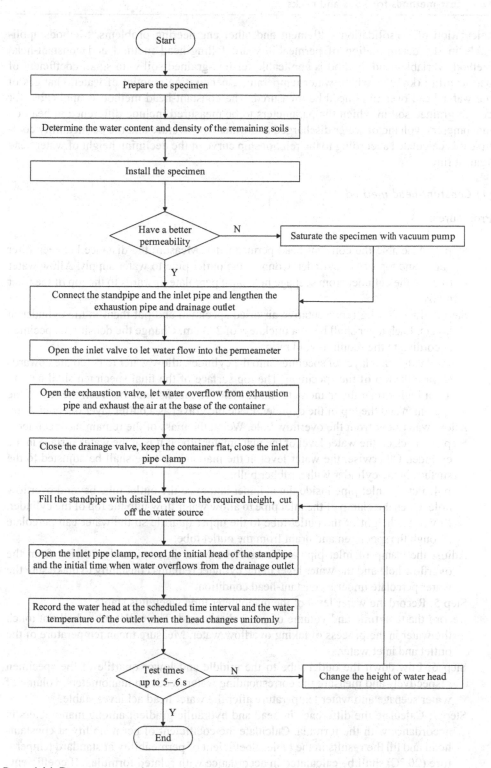

Figure 6.14 Determination of permeability of soil (variable-head method)

calculation of consolidation settlement and other engineering problems. Methods applicable in the determination of permeability are falling-head method and constant-head method. Variable-head method is applicable for fine-grained soil with small coefficient of permeability ($k<10^{-3}$), where water temperature, penetrating amount of water, changes of the water head over time need be measured. The constant-head method is applicable for coarse-grained soil in which the parameters to be measured include difference in head on manometers, volume of water discharged and water temperature and the permeability coefficient is calculated according to the relationship curve of the declining height of water head against time.

(1) Constant-head method

Procedures:

Step 1: Assemble the constant-head permeameter. Measure the distance between filter gauze and top of the cylinder. Connect the outlet pipe to water supply. Allow water to enter the cylinder from seepage hole until percolate upwards to the top of the filter gauze.

Step 2: Take 3–4 kg representative air-dried specimen and put them in the cylinder in layers. Each layer shall be of a thickness of 2–3 cm. Change the density of specimen according to the required void ratio.

After placing each layer of specimen into the cylinder, allow water to percolate upwards toward the top of the specimen. The top surface of the final specimen shall be 3 to 4 cm higher than the manometer hole. Measure the distance between the top of the specimen and the top of the cylinder. Place the gravel as a buffer layer of about 2 cm.

Allow water flow from the overflow hole. Weigh the mass of the remaining specimen.

Step 3: Check on the water level of the manometer tube if it is at the water surface in the cylinder. Otherwise, the water level in the manometer tube shall be adjusted to the surface in the cylinder with a rubber bulb.

Step 4: Put the inlet pipe inside the cylinder and raise the outlet tube beyond overflow hole. Open the clamp of the inlet pipe to allow water flow into the top of the cylinder. Lower the height of the outlet tube to the upper quartile so the water can percolate through the specimen and drain from the outlet tube.

Adjust the clamp of inlet pipe and ensure the water is continually flowing from the overflow hole and the water level in the cylinder remains constantly to make sure the water percolate under a constant-head condition.

Step 5: Record the water level of the manometer tube when it is stable.

Record the total time and volume of water discharged. The outlet tube shall not touch the water in the process of taking overflow water. Measure mean temperature of the outlet and inlet water.

Step 6: Low down the outlet tube to the middle and lower quartile of the specimen respectively and measure the corresponding water-heads on manometers, volume of water seepage and water temperature after the water head achieves stable.

Step 7: Calculate the difference in head and hydraulic gradient among manometers in accordance with the formula. Calculate the coefficient of permeability at constant-head and fill the results in the table. Coefficient of permeability at standard temperature (20 °C) shall be calculated in accordance with related formulas. If coefficients of permeability at different hydraulic gradient are within the allowable error, average of results shall be obtained. The maximum deviation shall not be greater than 2×10^{-n}.

It should be noted that: if there are large amount of clay in the specimen, a 2 cm thick coarse-sand shall be laid on the filter gauze as the filter layer to prevent the loss of finer grain.

(2) Variable-head method

Procedures:

Step 1: Apply a thin layer of petroleum jelly to the inner wall of the cutting ring. Cut the soil specimen along the outer wall of the ring. The specimen can also be prepared from disturbed soil. Measure the water content and density of soil from the remaining specimen.

Step 2: Coat a thin layer of petroleum jelly to the inner wall of the permeameter. Insert the cutting-ring with specimen into the permeameter. Place a porous stone on the base then place a waterproof gasket, the permeameter, a porous stone and the top plate. Tighten the screw to make sure all parts of permeameter are in proper contact. The leakage of water and air should not be allowed.

Saturate the specimen under the water head when a better permeability or under full vacuum when a worse permeability.

Step 3: Connect the standpipe and the inlet pipe of the permeameter and lengthen the exhaustion pipe and drainage outlet. Open the inlet valve to let water flow into the permeameter.

Open the exhaustion valve, let water overflow from exhaustion pipe and exhaust the air at the base of the container until no bubble exists in the overflowed water, then close the exhaustion valve.

Step 4: Fill the standpipe with distilled water to the required height then close the inlet valve. The height of water head is determined by porosity of the specimen, but generally it shall be lower than 2 m.

Step 5: Open the inlet valve. When water overflows from the drainage outlet, record the initial head of the standpipe and the initial time. When the head changes uniformly, record the water head at the scheduled time interval and the water temperature of the outlet.

Step 6: Adjust the head in the standpipe with distilled water to the different height than repeat the step 5 for 5–6 times.

Step 7: Calculate the permeability coefficient of variable head of soil according to the formula and record the results. The permeability coefficient at standard temperature shall be calculated in accordance with related formulas. The permeability coefficient at the void ratio shall be obtained from the average of 3 or 4 results within the allowable deviation ($\leq 2 \times 10^{-n}$ cm/s)

It should be noted that: When water flows too fast or the water through outlet is turbid, an immediate check is needed. If the leakage of the container or concentrated seepage in the specimen occurs, a new specimen shall be prepared and tested again.

6.3.4 Consolidation

Standard consolidation

The consolidation of soil is a comparatively sudden reduction in volume of a soil mass under an applied load principally because of an increase in compressive stress and compression of gas in the soil voids. The test is applicable for saturated clay. It is also applicable unsaturated soil if the compressibility is determined only. Parameters to be measured are heights of specimen during consolidation. The standard consolidation test is based on Terzaghi's unidirectional consolidation theory. In this test (Figure 6.15), the specimen in a container is vertically compressed by the load device under the confined conditions.

Procedures:

Step 1: Apply a thin layer of petroleum jelly in the inner wall of the cutting ring. Cut the soil specimen with the cutting ring. Trim both sides of the specimen and wipe the outside of the cutting ring. Determine the water content and density of specimen. Select a specimen from the remaining soil sample and determinate the specific gravity. When the specimen needs to be saturated, vacuum pump shall be used.

Assemble specimen ring, porous stones and filter paper in the oedometer. Insert the cutting-ring with the specimen. And place the guide ring, filter paper, porous stones and load plate successively on the specimen. The water of filter paper and porous stones shall be close to the natural water of the specimen.

Step 2: Fit the oedometer at the center of the loading device and align the load plate with the loading bar, then install the dial indicator.

Apply a seating load of 1 kPa to contact oedometer snugly with upper and lower parts of the loading apparatus. Set the dial indicator to zero or record the initial reading.

Step 3: Apply load of the first level which shall be based on the hardness of the soil and 12.5, 25 or 50 kPa are recommended. Add water to the water bath immediately to immerse the specimen. If the specimen is unsaturated, a wet cotton yarn is needed around the loading plate. The standard loading schedule shall consist of a load increment ratio (*LIR*) of one which is obtained by doubling the pressure on the soil to obtain values of approximately 12.5, 25, 50, 100, 200, 400, 800,1600, 3200 kPa. The final load shall be greater than the sum of the self-weight stress and additional stress. The maximum load is not less than 400 kPa if the compression coefficient is required only.

When estimating the preconsolidation stress, the load increment ratio shall be less than 1, 0.5 or 0.25. The load should ensure that the lower segment of the measured e-log p curve is a straight line. For over consolidated soil the secondary compression should be evaluated by unloading and reloading.

Step 4: Determine the settlement rate and coefficient of consolidation. After each load increment is applied, record the deformation of the specimen at the time intervals of the record form. When the deformation of each hour is not greater than 0.01 mm, record the deformation and apply the next stress level.

If the determination of the settlement rate is not needed, the standard load increment duration shall be approximately 24 h. If the coefficient of compression is required only, after applying each load increment it can be considered as stable when the deformation reaches 0.01 mm per hour.

Rebound the specimen back to the required load after consolidation at a specific load, if needed. Record deformation of specimen after 24 h.

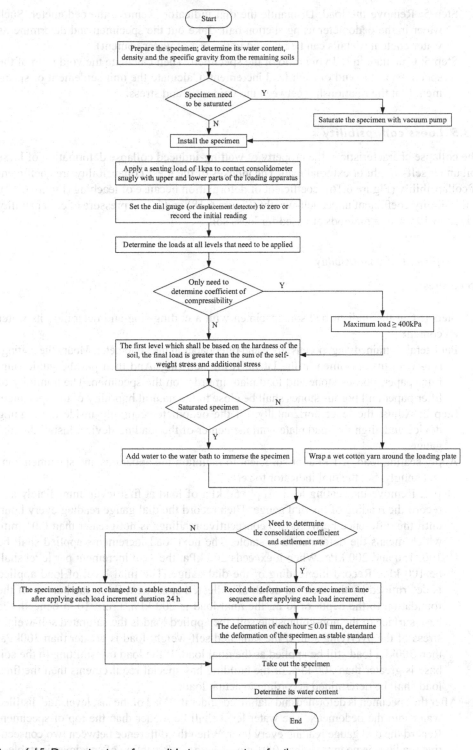

Start

Prepare the specimen; determine its water content, density and the specific gravity from the remaining soils

Specimen need to be saturated

Y → Saturate the specimen with vacuum pump

N

Install the specimen

Apply a seating load of 1kpa to contact consolidometer snugly with upper and lower parts of the loading apparatus

Set the dial gauge (or displacement detector) to zero or record the initial reading

Determine the loads at all levels that need to be applied

Only need to determine coefficient of compressibility

Y → Maximum load ≥ 400kPa

N

The first level which shall be based on the hardness of the soil, the final load is greater than the sum of the self-weight stress and additional stress

Saturated specimen

N → Wrap a wet cotton yarn around the loading plate

Y

Add water to the water bath to immerse the specimen

Need to determine the consolidation coefficient and settlement rate

N

The specimen height is not changed to a stable standard after applying each load increment duration 24 h

Y

Record the deformation of the specimen in time sequence after applying each load increment

The deformation of each hour ≤ 0.01 mm, determine the deformation of the specimen as stable standard

Take out the specimen

Determine its water content

End

Figure 6.15 Determination of consolidation properties of soil

Step 5: Remove the load. Dismantle the dial indicator. Remove the oedometer. Suck water in the oedometer using suction ball. Take out the specimen and determine its water content (details can be referred to the test of water content).

Step 6: Calculate initial void ratio of the specimen. Then calculate the void ratio of the specimen at the end of each load increment. Calculate the unit settlement of specimen. Plot the relationship between the void ratio and stress.

6.3.5 Loess collapsibility

The collapse characteristic is the property of wetting-induced collapse deformation of loess soil under self-weight or external load. The indexes of the loess collapsibility are coefficient of collapsibility (Figure 6.16), coefficient of deformation because of leaching (Figure 6.17), collapsibility coefficient under self-weight (Figure 6.18) and initial pressure of collapsibility (Figure 6.19). These methods are used for loess soil.

(1) Coefficient of collapsibility

Procedures:

Step 1: Cut the undisturbed soil specimen with a cutting-ring and determine its water content.

Put lateral restrained ring, porous stone and filter paper in the oedometer. Mount the cutting-ring with the specimen in the lateral restrained ring. And then put the guide ring, filter paper, porous stone and load plate in order on the specimen. The humidity of filter paper and porous stones shall be close to the natural humidity of the specimen.

Step 2: Adjust the lever horizontally. Fit the oedometer centrally inside the loading device and align the load plate with the center of the loading device. Install the dial gauge.

Apply seating load of 1 kPa, or sufficient to maintain the apparatus and specimen contact snugly. Set the dial indicator to zero.

Step 3: Remove the seating load. Apply 50 kPa of load at first level immediately and record the reading of the dial gauge. Then record the dial gauge reading every hour, until the difference between two consecutive readings is not greater than 0.01 mm, which means the deformation is stable. The next load increments applied shall be 100, 150 and 200 kPa. When it exceeds 200 kPa, the load increment per level shall be 100 kPa. Record the reading of the dial gauge. The final level of load applied is determined by the location depth of collected specimen. From the bottom of the foundation to the depth of 10 m, the final load is 200 kPa. From 10 m beneath the base surface to the non-collapsible soil, the applied load is the saturated self-weight stress of the overlying soil. If the saturated self-weight load is greater than 300kPa, then 300kPa load still be applied as the final load. If the load transmitting to the soil base is greater than 300 kPa, or the building has special requirements then the final load shall be determined according to actual load.

After the specimen is deformed and stabilized under the load of the last level, add distilled water into the oedometer. The water level shall be higher than the top of specimen. Record the dial gauge reading every hour. When the difference between two consecutive readings is no greater than 0.01 mm then the deformation of specimen is stable.

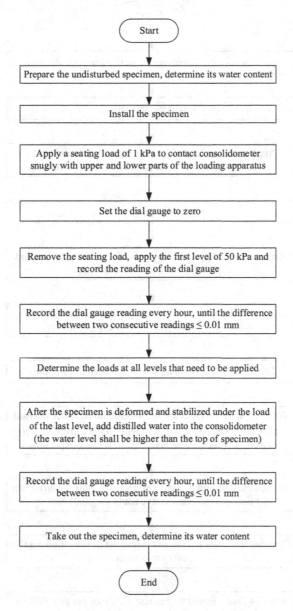

Figure 6.16 Collapse characteristics of loess (coefficient of collapsibility)

Step 4: Release the load and take down the oedometer. Use the suction ball to suck water in the oedometer. Dismantle all parts of the apparatus quickly and take out the whole specimen. Then determine the water content of the specimen (refer to the test of the water content).

Step 5: Calculate the coefficient of collapsibility according to the formula and data in the table and then fill in the form.

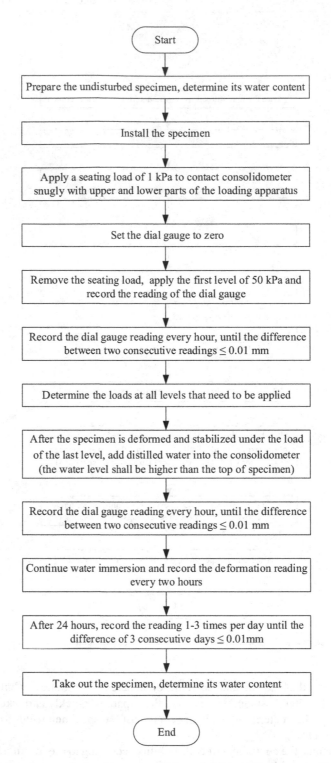

Figure 6.17 Collapse characteristics of loess (coefficient of deformation because of leaching)

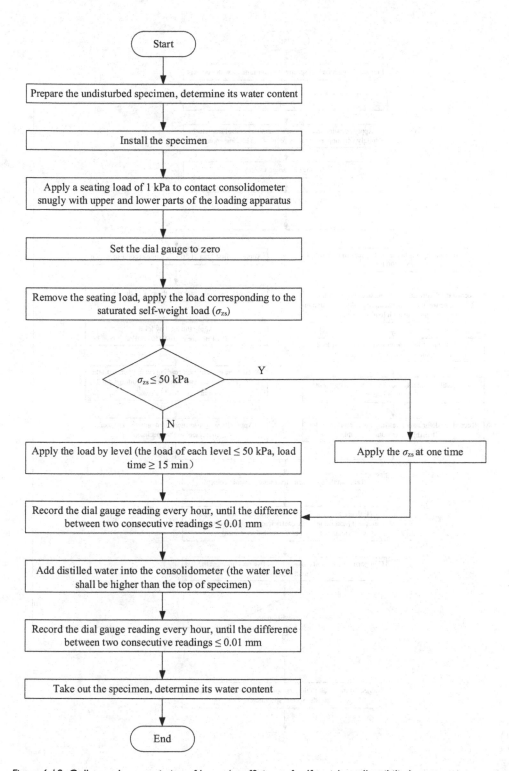

Figure 6.18 Collapse characteristics of loess (coefficient of self-weight collapsibility)

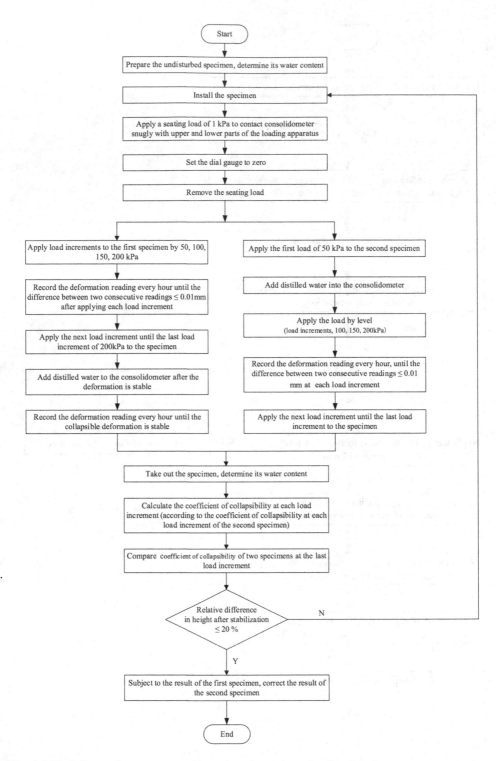

Figure 6.19 Collapse characteristics of loess (initial pressure of collapsibility)

(2) Coefficient of deformation because of leaching

The leached deformation of loess is caused by salt leach and continuous compression to the pores in the soil under the long-term effects of load and water immersion. Coefficient of deformation because of leaching is the ratio of the leached deformation to the initial height of the specimen. Parameter to be measured is the change of height of the specimen over time.

Procedures:

Step 1: After the collapsible deformation of specimen is stable in the test for collapsibility coefficient, continue water immersion and record the deformation reading every 2 h. After 24 h record the reading 1–3 times per day until the difference of 3 consecutive days is no greater than 0.01 mm, which means that the deformation is stable.

Step 2: After the lixiviation deformation is stable, release the load and take down the oedometer. Using suction ball to suck water in the oedometer. Dismantle all parts of the apparatus quickly and take out the whole specimen. Then determine the water content of the specimen.

Step 3: Calculate the coefficient of lixiviation deformation according to the formula.

(3) Coefficient of self-weight collapsibility

Coefficient of self-weight collapsibility is expressed as the percentage of collapsible amount of soil when the soil sample is applied by the saturated self-weight load. Coefficient of self-weight collapsibility can be used to distinguish the self-weight collapsible loess and non-self-weight collapsible loess. Parameter to be measured is the change of the specimen height over time.

Procedures:

Step 1: Insert the assembled oedometer in the loading device. Apply seating load of 1kPa or sufficient to maintain the apparatus and specimen contact snugly. Set the dial gauge to zero.

Step 2: Release the seating load. Apply the load corresponding to the saturated self-weight load. If the saturated self-weight load is less than or equal to 50 kPa, it can be applied at one time. If the saturated self-weight load is greater than 50 kPa, the load shall be applied by level. The load of each level shall be no greater than 50 kPa and the load time shall not be less than 15 min.

After the final load increment is applied, record the dial gauge reading every hour until the difference between two consecutive readings is no greater than 0.01 mm, which means the deformation is stable.

Step 3: Add distilled water into the oedometer and the water level shall be higher than top of the specimen.

Record the dial gauge reading every hour until the difference between two consecutive readings is no greater than 0.01 mm, which means the wetting-induced deformation is stable.

Step 4: Calculate the collapsibility coefficient under self-weight according to the formula.

(4) Initial pressure of collapsibility (double line method)

The initial pressure of collapsibility is the minimum load of the wetting-induced collapse deformation under load-induced strains after wetting-induced collapse deformation has occurred (when loess is soaked), which can be obtained by the relation curve of coefficient of collapsibility and the pressure. Parameter to be measured is the change of height of the specimen over time under load of a certain level.

Procedures:

Step 1: Prepare two specimens. Install the assembled oedometer centrally inside the loading device. Apply seating load of 1 kPa, or sufficient to maintain the apparatus and specimen contact snugly. Set the dial gauge to zero.

Step 2: Release the seating load. Apply load increments to the first specimen by 50, 100, 150, 200kPa under natural humidity. At each load increment, record the deformation reading every hour until the difference between two consecutive readings is no greater than 0.01 mm. Then apply the next load increment. After applying the last load increment of 200 kPa to the specimen, add distilled water to the oedometer after the deformation is stable. Record the deformation reading every hour until the collapsible deformation is stable.

After applying the first load increment of 50 kPa to the second specimen, add distilled water into the oedometer and apply the load by level. If the applied load is within 150kPa, each load increment shall be 25 to 50 kPa. If the load is greater than 150kPa, each load increment shall be 50 to 100 kPa. At each load increment, record the deformation reading every hour. When the difference between two consecutive readings is no greater than 0.01 mm, the deformation of specimen is stable.

Step 3: Release the load and remove the oedometer. Use suction ball to suck water in the oedometer. Dismantle all parts of the apparatus quickly and take out the whole specimen. Then determine the water content of specimen (refer to the test of water content)

Step 4: According to the formula and final height of specimen, calculate the coefficient of collapsibility at each load increment.

According to the collapsibility coefficient at each load increment of the second specimen, plot the curve of collapsibility coefficient and load. The initial pressure of collapsibility is the load when the collapsibility coefficient is 0.015.

6.3.6 Triaxial compression

Triaxial compression test covers the determination of strength and stress-strain relationships of a cylindrical specimen of either an intact or reconstituted soil. For different engineering requirements, the condition of compression test may be chosen from consolidated-undrained, unconsolidated-undrained and consolidated-drained. At least three specimens are required for tests at different chamber pressures. If it is difficult to obtain enough specimens for a group test, the triaxial compression test with multistage chamber pressures should be conducted. These methods are applicable for fine-grained soil and coarse-grained soil finer than 20 mm. Parameters to be measured are the maximum principal stress and axial deformation.

(1) Specimen preparation and saturation

1   The minimum and maximum diameters of the specimen used in triaxial compression test is 35 mm and 101 mm, respectively, and the height of the specimen should be 2–2.5 times the diameter of the specimen. The allowable maximum particle size of the specimen should meet the requirements in Table 6.3. For the specimens with cracks, weak surfaces and structural surfaces, the diameter of the specimen should be greater than 60 mm.

2   Cut undisturbed soil samples in cylindrical shape according to the specified size in Table 6.3.

   1)   For soft soil specimen, cut the specimen in column with wire saw or soil cutter, the size of the specimen is slightly larger than the specified size. Then, place the soil column between upper and lower cutting plates. Cut down and turn the disc while cutting. The specimen should be prevented from being disturbed during cutting. In case there are holes on the surface of the cut specimen, the holes should be filled with the remaining soil.

   2)   For hard soil samples, the rough column sample is placed on the soil-cutting stand for polishing.

   3)   Take out the specimen, flatten both ends to gain the specified height, and weigh it. Take the remaining soil for the determination of water content of the specimen.

   4)   The disturbed soil sample should be prepared on basis of the predetermined dry density and water content. After the sample is prepared, it is layered and compacted in the sampler (3–5 layers for silts, 5–8 layers for clay) The amount of soil in each layer should be the same, and the contact surfaces between layers should be shaved to gain a roughness. After the last layer is compacted, both ends of the specimen in the sampler should be flattened. Finally, the weight and size (diameter and height) of the prepared specimen should be measured. The average diameter of the specimen should be calculated according to formula 6. 9.

$$D_0 = \frac{D_1 + 2D_2 + D_3}{4} \qquad\qquad (6.9)$$

   Where, D_1, D_2 and D_3 refer to the diameters at the upper, middle and lower third of the specimen, respectively.

   5)   Preparation of specimen for cohesionless soil. Impervious plates, rubber membrane and split round mold should be sequentially placed on the base of the pressure chamber. According to the dry density and volume of the specimen, weigh the required specimen and divide it into three equal parts. Then, fill each specimen into the rubber membrane to the required height of the layer. Subsequently, the second and the third layer are poured into the rubber membrane until the membrane is full. For preparation of saturated specimen, porous disc, rubber membrane and

Table 6.3 Allowable maximum particle size of specimen (mm)

Specimen diameter	allowable particle size
<100	1/10 of the specimen diameter
>100	1/5 of the specimen diameter

split round mold are placed in order on the base of the pressure chamber, and then inject distilled water into the mold to one-third of the specimen height. Similar to unsaturated cohesionless soil, the specimen is divided into three equal parts to boil in water. Then cool the specimen down and fill the specimen of each layer in rubber membrane according to a prespecified dry density. When the required dry density is relatively large, the split needs to gently tap round mold during filling process. Level the sample surface, place the impervious plate or porous disc, the specimen cap and then tighten the rubber membrane. Apply a negative pressure of 50kPa is the inside of the specimen for a certain time. Finally, Dismantle the split round mold.

3   Sample saturation

1)   Vacuum saturation. Successively insert porous disk, filter paper, mold with a speci-men, top filter paper, top porous disk and top plate into the saturator and tighten the assembled device. Put the saturator with specimen in a vacuum container. Coat a thin layer of petroleum jelly on the contact area between the container and its lid. Install the split funnel and connect the vacuum container to the pump. Start the pump. When the vacuum pressure is closed to the atmospheric pressure, add water in the split funnel and open the clamp to allow water to flow into the vacuum container slowly. The vacuum pressure shall be constant during the procedure. Stop the pump until the saturator is submerged. Open the clamp to let air go into the cylinder. Stand for a while to enable the specimen fully saturated (Stand 10 h for fine-grained soil) Open the vacuum container, take out the saturator and remove the specimen with the mold. Weigh the total mass of the specimen and mold. Calculate the degree of saturation in accordance with the formula. If degree of saturation is lower than 95%, the pumping process shall be continued.

2)   Water-head saturation. Install the specimen in the pressure chamber according to the step 1 of the consolidation undrained test. In water-head saturation, filter paper does not required around the specimen and apply ambient pressure of 20 kPa to specimen. Adjust the water levels of the measuring tube at the bottom and the top of the specimen. The water head at the bottom of specimen shall be higher than the top about 1 m. Allow the water to percolate from the bottom to the top of the specimen. Then, open the pore fluid pressure valve, the measuring tube valve and the drain valve to allow the distilled water to overflow from the top of the specimen. When the saturation degree of the specimen needs to be increased, dioxide gas with pres-sure of 5–10kPa could be used to replace the air in sample, and then the specimen is saturated by water-head method.

3)   Back pressure saturation. The back-pressure method is applied to the specimen that is required to be completely saturated. The back-pressure system is similar to the surrounding pressure system. However, it should be noted that the variable tube of double layer is used in back-pressure system, but the water drainage tube in the surrounding pressure system. After the specimen is completely installed, adjust the pore water pressure to atmospheric pressure. Then close pore water pressure valve, back-pressure valve and body variable tube valve. Record the body variable tube reading. Next, open the ambient pressure valve, apply a pressure of 20kPa to the specimen, and open the pore water pressure valve in order. Record the reading until pore water pressure is constant. Finally, close the pore water pressure valve. Back

pressure should be applied by levels to reduce disruption for the specimen. Preferred increment of ambient pressure and back-pressure should be 30kPa. Open the variable pressure valve and the back-pressure valve and simultaneously apply the ambient pressure and back-pressure, and then open the pore water pressure valve slowly to record the pore water pressure and the volumetric tube reading until the pore water pressure is stable. Then apply the next level of ambient pressure and the pore water pressure to the specimen and record the pore water pressure and the volumetric tube reading. As such, the pore water pressure increment caused by each ambient pressure increment can be determined. When the ratio of the pore water pressure increment to the ambient pressure increment is greater than 0.98, the specimen is considered to be saturated.

(2) Unconsolidated-undrained shear test

As show in Figure 6.20, triaxial compression test under UU conditions describes specimen compressed at unconsolidated and undrained conditions. As no consolidation and drainage, the porous plates usually should be replaced to watertight plates, except that the initial pore water pressure coefficient is required or the back-pressure are needed. The strain rate of compression should be 0.5–1%/min.

Procedures:

Step 1: Specimen preparation and saturation.

For undisturbed specimens, cut a specimen with a diameter slightly larger than the required specimen by a wire saw or steel straightedge, then put the specimen on a soil lathe, and trim the specimen vertically to the final diameter with a wire saw or steel straightedge. Remove the specimen. Cut the specimen to the final height. Where the voids appeared, carefully fill the voids with the trimmings. Weigh the final specimen. Determine the water content of sample obtained from the trimmings. Emery wire cutting machine may be used to trim the original specimen to the required diameter.

For disturbed specimens, add water to produce the desired water content. The specimen shall be reconstituted to the desired density.

Specimen may be saturated by vacuum pump, water head and back-pressure method in accordance with the type of soil and desired saturation degree.

Step 2: Exhaustion of apparatus.

The air in chamber system shall be forced out completely by flushing with distilled water or applying pressure to dissolve the air in water and then overflowing from the chamber pedestal.

Step 3: Mounting specimen.

Sheathe the rubber membrane around the specimen with the membrane expander. Put porous disk, specimen, and porous disk on the pressure cell pedestal successively. Roll the membrane on the base and seal it with elastic band or O-ring and remove the expander. Roll the membrane on the cap and seal it.

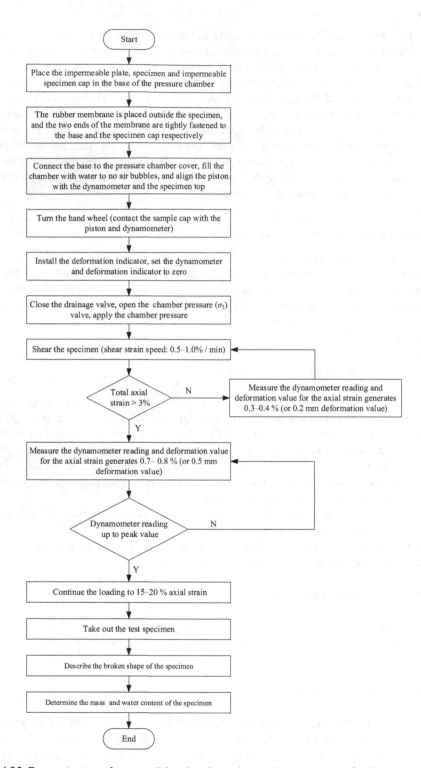

Figure 6.20 Determination of unconsolidated-undrained triaxial compression of soil

Assemble the triaxial chamber and bring the load piston into contact with the specimen cap. Align the piston and the specimen cap. Tighten the bolts on the base of the chamber. Rise the base of the chamber until the reading of the load indicator changes slightly. Open the vent screw on the chamber and the chamber pressure valve. Fill the chamber with water, ensuring that water fills the chamber and overflows from the bleed plug. Close the chamber pressure valve and tighten the bleed plug.

Step 4: Specimen shearing.

Select UU test, and input test parameters. Apply chamber pressure to the specimen to desired value, the chamber pressure should be determined according to the actual load of the engineering project (chamber pressures of 100, 200, 300 and 400 kPa can also be used)

Until the chamber pressure stabilizes to the set value. Apply axial load. The strain rate should be 0.5–1%/min. Axial load should be stopped when the axial strain is 15–20%.

Step 5: Unload chamber pressure. Open the chamber pressure valve, and loosen the vent screw, then drain the water in the sample chamber. Lower down the sample chamber to allow the load cell to leave away from the load piston. Remove the cover, take out the specimen, describe the failure of the specimen and determine the water content of total specimen.

(3) Consolidated undrained shear test

As shown in Figure 6.21, after the specimen is drained and consolidated under a certain confining pressure, axial pressure is applied under undrained conditions until the specimen is damaged.

Procedures:

Step 1: Specimen preparation and saturation.

The details of this procedure can be referenced to unconsolidated-undrained shear test.

Step 2: Exhaustion of apparatus.

The air in the system shall be forced out completely by flushing with distilled water or applying pressure to dissolve the air in water and then overflowing from the pedestal.

Step 3: Mounting specimen.

Put porous disk, soaked filter paper, specimen, soaked filter paper and porous disk on the chamber pedestal successively. Cover the specimen periphery with 7–9 filter-paper trips. Sheathe the rubber membrane around the specimen with the membrane expander. Roll the membrane on the base and seal it with elastic band or O-ring and remove the expander.

Open the pore-pressure valve. Allow the water to flow through the specimen from the bottom. Remove the air pockets from between membrane and the specimen by light stroking upwards. Close the pore pressure valve.

Open the drainage valve to moisten the top cap. Fit the cap on to the porous disk. Roll the membrane on the cap and seal it with elastic band. Remove surplus water between the cap and the membrane, and close the drainage valve.

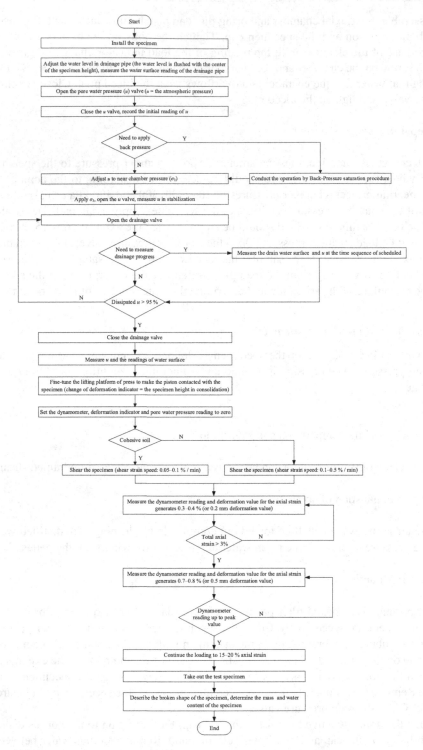

Figure 6.21 Determination of consolidated undrained triaxial compression of soil

Assemble the triaxial chamber and bring the load piston into contact with the specimen cap. Align the piston and the specimen cap. Tighten the bolts on the base of the chamber. Rise the base of the chamber until the reading of the load indicator changes slightly. Open the vent screw on the chamber and the chamber pressure valve. Fill the chamber with water, ensuring that water fills the chamber and overflows from the bleed plug. Close the chamber pressure valve and tighten the bleed plug.

Step 4: Consolidation.

Select CU test, and input the parameters.
Adjust the drainage burette to make the water level align with the center of the specimen, and record the reading of water level in the drainage burette before the specimen is consolidated.
Apply chamber pressure to the specimen to desired value, the chamber pressure should be determined according to the actual stress of the engineering (chamber pressures of 100, 200, 300 and 400 kPa may also be used)
Record the reading of pore pressure. When the pore pressure rises to the chamber pressure, open the drainage valve and start the process of consolidation. Draw a plot of pore pressure versus square root of time and a plot of pore pressure versus log time. Allow consolidation to continue until the excess pore-pressure dissipates at least 95%. Close the drainage valve after the consolidation is finished. Record the reading of the drainage burette. The amount of consolidated drainage is the change of amount of water in the drainage burette before and after consolidation.

Step 5: Specimen Shearing.

Adjust the pedestal of the sample chamber slightly to allow the piston to contact with the specimen cap to permit proper seating load. Apply axial load, input the amount of drained water during consolidation, and start shearing. The shear strain rate of clay should be 0.05–0.1% per minute for clay and 0.1–0.5% per minute for silt. Observe and record the change of curve and specimen during shear process. Continue the loading to 15–20% strain.

Step 6: Unload chamber pressure. Open the chamber pressure valve, and loosen the vent screw, then drain the water in the sample chamber. Low down the chamber to allow the load cell separate from the load piston. Remove the chamber and specimen quickly, describe the failure specimen, determine the water content of total specimen.

(4) Consolidated-drained shear test

The strength parameters in terms of effective stress may be determined by this test method. A strain rate should be slow enough to allow excess pore pressure to dissipate during compression. The installation, consolidation and shearing of the specimen are the same as the procedures in consolidated undrained (CU) shear test. Compared with the test under CU conditions, the drainage valve shall be opened and the strain rate shall be 0.003–0.012%/min during compression under consolidated-drained CD conditions.

(5) Multi-level loading test

This test is applicable for soil difficult to obtain enough specimens and for undisturbed soil with a low sensitivity. Multistage test is to apply the first confining pressure, and then to steadily increase the axial load until the sample deforms plastically. The next stage confining pressure is then increased.

Compared with conventional test under UU conditions, the compression in multistage UU method (Figure 6.22a) shall be conducted until the reading of the dynameter is constant at the first stage of chamber pressure (i.e. the specimen deforms plastically) Record the reading of the dynameter and the axial deformation. Unload the axial force then. Apply the second stage of chamber pressure and continue to shear the specimen. Then shear the specimen under the third and fourth stage of chamber pressure. The cumulative axial strain should not be greater than 20%. The axial strain of the specimen shall be calculated according to the accumulated deformation. Other details should follow the procedures of test under UU conditions.

Compared with conventional test under CU conditions, the first stage of chamber pressure should be 50 kPa during consolidation of multistage CU test (Figure 6.22b). Start shearing after the consolidation is finished. The strain rate should be 0.05–0.1%/min for clay or be 0.1–0.5%/min for silt. Shear until the reading of dynameter is constant, then stop shearing and record the readings. Unload the axial pressure then. After the pore pressure is stable, apply the second stage of chamber pressure and continue to shear the specimen. The secondary and subsequent chamber pressures shall not be less than that of the major principal stress under the previous stage of chamber pressure. Then continue the shearing under the third stage of chamber pressure in the same way. The cumulative axial strain shall not be greater than 20%. In the calculation of the axial deformation, the stable height at the stage of chamber pressure after previous unload shall be taken as the initial height of specimen at the next stage. Calculate the axial strain under the chamber pressure at every stage. Other details should follow the procedures of test under CU conditions.

It should be noted that:

1) The test shall be carried out under a constant temperature.
2) When determining the stress-strain relation of soil, two circular membranes with silicone grease inside shall be placed between the specimen and the porous plate, and a hole with diameter of 1 cm shall be left in the middle of the membrane to drain.

6.3.7 Unconfined compression

The unconfined compression test (Figure 6.23) is used for the determination of the unconfined compressive strength of cohesive soil in the intact, remolded, or reconstituted condition, using strain-controlled application of the axial load. This test method is applicable for saturated clay, where the axial stress and axial displacement need be measured.

Procedures:

Step 1: For the preparation and saturation of undisturbed specimen, the details can be referred to the triaxial compression test. The specimen shall have a diameter of 35–50 mm and a height-to-diameter ratio of 2.0–2.5.

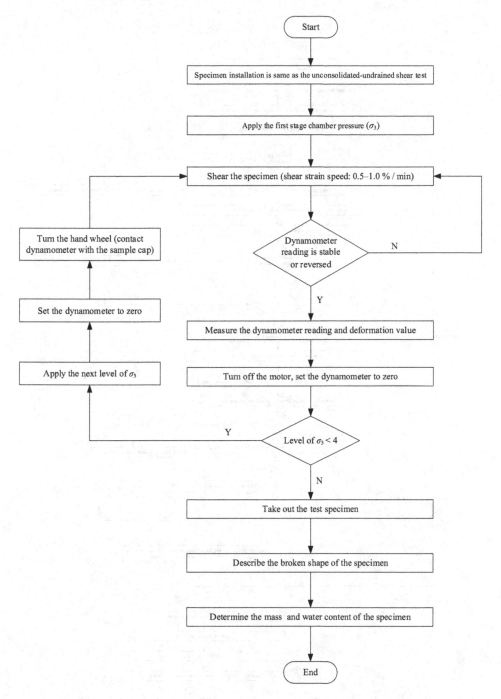

a) Unconsolidated-undrained triaxial compression

Figure 6.22 Determination of multi-stage loading of soil

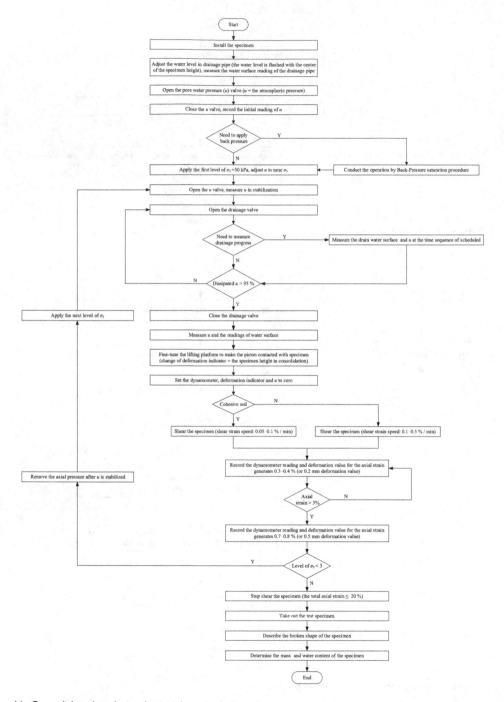

b) Consolidated undrained triaxial compression

Figure 6.22 Continued

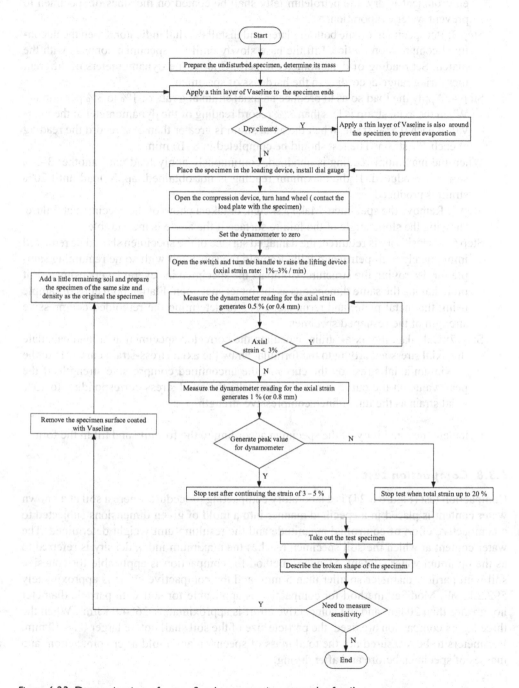

The flowchart contains the following elements:

Start

Prepare the undisturbed specimen, determine its mass

Apply a thin layer of Vaseline to the specimen ends

Dry climate — Y → Apply a thin layer of Vaseline is also around the specimen to prevent evaporation

N

Place the specimen in the loading device, install dial gauge

Open the compression device, turn hand wheel (contact the load plate with the specimen)

Set the dynamometer to zero

Open the switch and turn the handle to raise the lifting device (axial strain rate: 1%–3% / min)

Measure the dynamometer reading for the axial strain generates 0.5 % (or 0.4 mm)

Axial strain < 3% — Y

N

Measure the dynamometer reading for the axial strain generates 1 % (or 0.8 mm)

Generate peak value for dynamometer — N → Stop test when total strain up to 20 %

Y

Stop test after continuing the strain of 3 – 5 %

Take out the test specimen

Describe the broken shape of the specimen

Need to measure sensitivity — Y

N

End

Remove the specimen surface coated with Vaseline

Add a little remaining soil and prepare the specimen of the same size and density as the original the specimen

Figure 6.23 Determination of unconfined compressive strength of soil

Step 2: Coat a thin layer of petroleum jelly on both ends of the specimen. When the environment is dry, the petroleum jelly shall be coated on the sides of specimen to prevent water evaporation.

Step 3: Put specimen on the bottom platen and install the dial indicator. Open the unconfined compression device. Lift the base slowly until the specimen contacts with the platen. Set reading of dynamometer to zero. Choose the dynamometers of different measuring range according to the hardness of specimen.

Step 4: Apply the load so as to produce an axial strain at a rate of 1% to 3% per minute. When the axial strain is less than 3%, record reading of the dynamometer at the interval of each 0.5% strain. When the axial strain is greater than 3%, record the reading at each 1% strain. The test should be completed in 8–10 min.

When the maximum reading is obtained, continuously apply load until another 3–5% strain is produced. If the maximum reading is not obtained, apply load until 20% strain is produced.

Step 5: Remove the specimen. Make a sketch, or take a photo of the specimen at failure, showing the slope angle of the failure surface if the angle is measurable.

Step 6: If sensitivity is required, the damaged surface of the specimen should be removed immediately with petroleum jelly. Remold the specimen with some remaining sample for destroying the structure. Compact the specimen in several layers using a split mold having the same dimensions as the intact specimen. Flatten both ends of sample using the metal plate, and rerun steps 1–5 to determine the remolded compressive strength of the reshaped specimen.

Step 7: Calculate the axial strain, calculate the corrected specimen area, and calculate the axial stress according to the formula. Draw the axial stress-strain curve. Take the maximum axial stress on the curve as the unconfined compressive strength. If the peak value on the curve is not obvious, take the axial stress corresponding to 15% axial strain as the unconfined compressive strength.

Calculate the sensitivity of the specimen according to the formula and fill in the form.

6.3.8 Compaction test

Compaction test (Figure 6.24) is a laboratory compacting procedure where a soil at a known water content is placed in a specified manner into a mold of given dimensions subjected to a compactive effort of controlled magnitude and the resulting unit weight determined. The water content at which the soil specimen reaches the maximum index density is referred to as the optimum water content. Standard method for compaction is applicable for cohesive soil with particle diameter smaller than 5 mm, and the compactive effort is approximately 592.2 kJ/m³. Modified method for compaction is applicable for soil with particle diameter not greater than 20 mm and the compactive effort is approximately 2684.9 kJ/m³. When the three layers compaction are used, the particle size of the soil shall not be larger than 40 mm. Parameters to be measured are the total mass of specimen and mold after compaction, and masses of specimen before and after drying.

Specimen preparation for light compaction test

Dry process: take 20 kg of air-dried soil sample (50 kg for heavy compaction test) to grind and sieve with a 5 mm standard sieve (20 mm or 40 mm sieve for heavy compaction test)

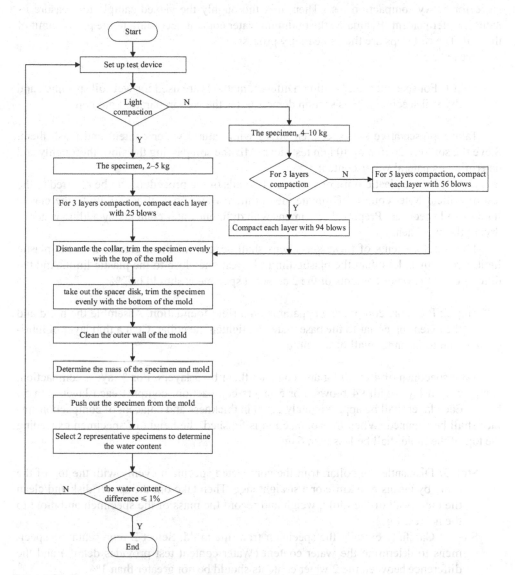

Figure 6.24 Determination of compaction characteristics of soil

Then, mix thoroughly the sieved sample to measure its water content. Evaluate the optimum water content according to the plastic limit of the soil. Weigh the air-dried and sieved soil sample and then place it in an enamel pan to spray water evenly on the soil sample. The mixture is stirred thoroughly, then transferred into a soil container and allowed to stand for 24 h. At least five specimens with different water contents should be prepared. Two of them are prepared with water contents greater than the plastic limit (*PL*), the other two with water contents less than *PL*, and the rest one with water content comparable to *PL*. In addition, the difference in water content among specimens in the same group should be less than 2%.

Wet process: take 20 kg of air-dried soil sample with natural water content (50 kg for heavy compaction test) to grind and sieve with a 5 mm standard sieve (20 mm or 40 mm sieve for heavy compaction test) Then, mix thoroughly the sieved sample to measure its natural water content. Evaluated the optimum water content according to the plastic limit of the soil. The rest steps are the same to dry process.

Procedures:

Step 1: For specimen preparation. Different methods are used for wet soil specimen and dry soil specimen. This section demonstrates the wet preparation method.

Take representative soil specimen 50 kg with natural water content and mash them. Sieve the soil on a 20 mm or 40 mm test sieve. Mix the soil passing the sieve thoroughly and determine its natural water content.

Determine the plastic limit of specimen (Details of the procedure can be referred to the test of critical water content) Estimate the optimum water content according to the plastic limit of soil specimen. Prepare 5 specimens with different water contents by adding water or drying the specimen.

The water contents of these specimens shall be, two of them greater than the plastic limit, two of them less than the plastic limit, the rest one close to the plastic limit. And the difference of the water contents of the 2 adjacent specimens should be 2%.

Step 2: Place the compacting apparatus on a rigid foundation. Assemble the mold and the extension collar to the base plate and tighten the collar. Coat a thin layer of lubricant to the inner wall of the mold.

Take specimen of 4 to 10 kg and compact them by 3 layers. For 3 layers compaction, compact each layer with 94 blows. For 5 layers compaction, compact each layer with 56 blows. Each layer shall be approximately equal in thickness and adjacent of compaction surface shall be trimmed. When the compaction is finished, the height of specimen exceeding the top of the mold shall be less than 6 mm.

Step 3: Dismantle the collar, trim the compacted specimen evenly with the top of the mold by means of a knife or a straightedge. Then take out the spacer disk and clean the outer wall of the mold, weigh and record the mass of the specimen and mold to the nearest 1 g.

Step 4: Carefully extrude the specimen from the mold. Select 2 representative specimens to determine the water content (Water content test provides details) and the difference between the 2 water contents should be not greater than 1%.

Compact and weigh the masses of other specimens with different water content and determine the water contents of the specimens after compaction.

Calculate the wet density and dry density of each specimen.

6.3.9 Direct shear test under consolidated-drained conditions

This test method (Figure 6.25) covers the determination of the consolidated-drained shear strength of a soil material under direct shear boundary conditions. Test data enable the effective shear strength parameters of cohesion (c) and internal friction angle (φ) of soil to be derived.

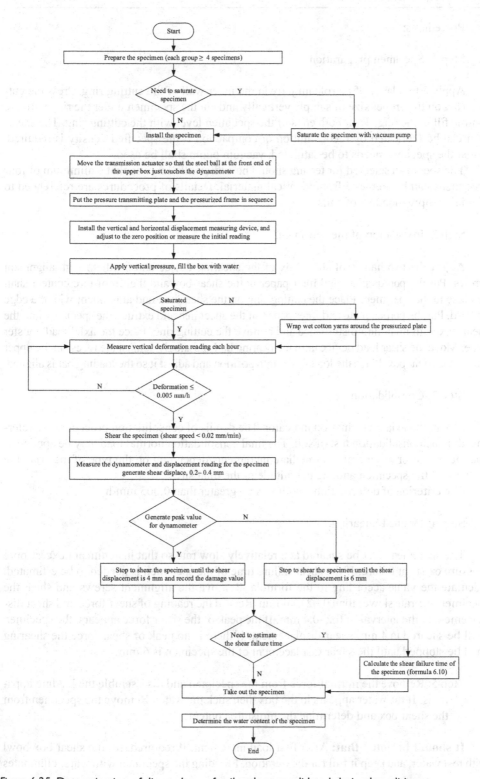

Figure 6.25 Determination of direct shear of soil under consolidated-drained conditions

Procedures:

Step 1: Specimen preparation.

Apply a thin layer of petroleum jelly in the inner wall of the cutting ring. Press the cutting ring on the proper size of sample vertically and cut the specimen under the ring until the sample fill in the ring. Trim both ends of the specimen level with the cutting ring. The specimen can be fabricated by reconstitution or compaction if the specified density is required. When the specimen needs to be saturated, vacuum pump shall be used.

The specimen selected for testing should be sufficiently large so that a minimum of four specimens can be prepared from identical material. Details of procedures are referenced to triaxial compression test of soils.

Step 2: Installation of the specimen.

Adjust the two halves of shear box of the shearing device and fix them with alignment screws. Put the porous plate and filter paper in the shear box and their moisture content shall be close to the specimen. Place the cutting ring on the shear box and align them with the edge upward. Put the porous plate and filter paper on the specimen and extrude the specimen into the shear box with minimum disturbance and remove the cutting ring. Place the axial load transfer plate. Move the shear load device until the bearing ball sits snugly against the recess in the upper half of the shear box. Place the load frame into position and adjust it so the loading bar is aligned.

Step 3: Consolidation.

Mount the axial deformation indicator. The details of consolidation procedure are referenced in the consolidation test of soil. The final consolidation normal load may be applied in one increment or in several intermediate increments depending on the type of material, the stiffness of the specimen and the magnitude of the final stress.

The criterion of deformation stability is no greater than 0.005 mm/h.

Step 4: Drained Shearing.

The specimen must be sheared at a relatively slow rate so that insignificant excess pore pressure exists at failure. If the shear failure time of the specimen is needed to be estimated, calculate the value according to the formula. Remove the alignment screws and shear the specimen at a rate slower than 0.02 mm/min. Record the reading of shear force and shear displacement at the interval of 0.2–0.4 mm. If the peak of the shear force appears, the specimen shall be sheared to 4 mm shear displacement. If there is no peak of shear force the shearing shall be stopped until the shear displacement of the specimen is 6 mm.

Step 5: Remove the normal force from the specimen and disassemble the loading apparatus. If the water appears in the box then suck the water. Remove the specimen from the shear box and determine the moisture content.

It should be noted that: After first load increment, if required, fill the shear box bowl with test water, and keep it full for the duration. Flooding the specimen with water eliminates negative pore pressure because of surface tension and also prevents evaporative drying during the test.

6.3.10 Reversal direct shear test

The reversal direct shear test (Figure 6.26) is used to determine the residual shear strength of the soil under the drainage condition. In this method, the soil sample is repeatedly sheared by the direct shear apparatus until the shearing strength is stable.

Procedures:

Step 1: Specimen preparation.

1) Flatten the two ends of the undisturbed specimen and cut off the soft surface downwards to the half-height (10 mm) of the original specimen. For the specimen with the relative low density, the specimen so prepared should be slightly higher than the half-height (10 mm) of the original specimen. For the undisturbed soil sample without soft surface, the specimen should be prepared according to the steps of undisturbed soil sample.

2) For the soil on the interlayer or landside level, it can be scraped and prepared to be the 10 mm soil paste in the liquid limit state. Then it is filled into the cutting-ring by layer without air bubbles. The density difference of the specimen in the same group should be less than 0.03 g/cm^3.

3) Saturate the specimen by the vacuum pump method.

Step 2: The installation and consolidation drainage of the specimen can be referenced to those in the slow shear test.

Step 3: Unplug the fixed pin, turn on forward switch of the motor, and shear the specimen at the speed of 0.02 mm/min (0.06 mm/min for silt) Record the reading of dynamometer and displacement when the shear displacement is 0.2–0.4 mm. After the shear stress reaches its peak value, record the dynamometer and displacement at an interval of shear displacement of 0.5 mm. Stop shearing until the displacement reaches 8–10 mm.

Step 4: When the first shearing test is accomplished, turn on the backward switch, return the shear box to its original position and insert the fixing pin. The reversing speed should be less than 0.6 mm/min.

Step 5: Wait for 30 min and then repeat step 3 and step 4 to conduct the second shearing test. The specimen is sheared repeatedly until the difference between two consecutive dynamometer readings is quite close. For silty clay, the shearing test needs to be conducted for 5–6 times and the total shear displacement should be 30–40 mm. For the clay soil, it needs to be conducted for 3–4 times and the total shear displacement should be 30–40 mm.

Step 6: Remove water in the box. Take out the specimen on the shear surface and determine its water content.

6.3.11 Tensile strength

Tensile strength is the load per unit area at which an unconfined cylindrical specimen will fail in a simple tension (pull) test. This test (Figure 6.27) is applicable for all kinds of soil

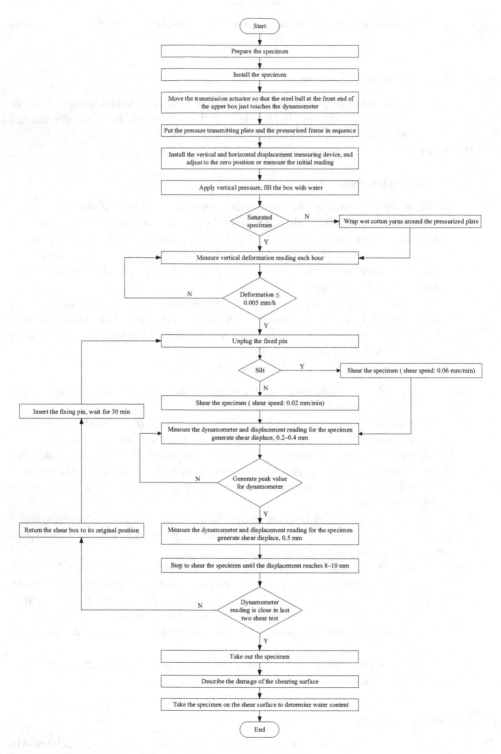

Figure 6.26 Determination of residual shear strength of soil

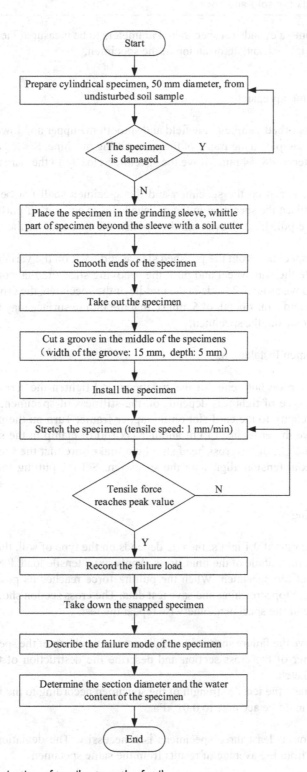

```
                    ┌─────────────┐
                    │    Start    │
                    └─────────────┘
                           │
          ┌────────────────▼─────────────────┐
          │ Prepare cylindrical specimen, 50 mm diameter, from │◄──┐
          │      undisturbed soil sample      │                  │
          └────────────────┬─────────────────┘                  │
                           │                                     │
                    ╱──────▼──────╲                              │
                   ╱  The specimen  ╲         Y                  │
                  ╲   is damaged    ╱─────────────────────────────┘
                   ╲──────┬──────╱
                         │ N
          ┌──────────────▼───────────────────┐
          │ Place the specimen in the grinding sleeve, whittle │
          │ part of specimen beyond the sleeve with a soil cutter │
          └──────────────┬───────────────────┘
                         │
          ┌──────────────▼───────────────────┐
          │    Smooth ends of the specimen    │
          └──────────────┬───────────────────┘
                         │
          ┌──────────────▼───────────────────┐
          │      Take out the specimen        │
          └──────────────┬───────────────────┘
                         │
          ┌──────────────▼───────────────────┐
          │ Cut a groove in the middle of the specimens │
          │ (width of the groove: 15 mm, depth: 5 mm)   │
          └──────────────┬───────────────────┘
                         │
          ┌──────────────▼───────────────────┐
          │       Install the specimen        │
          └──────────────┬───────────────────┘
                         │
          ┌──────────────▼───────────────────┐
          │ Stretch the specimen (tensile speed: 1 mm/min) │◄──┐
          └──────────────┬───────────────────┘                │
                         │                                     │
                  ╱──────▼──────╲                              │
                 ╱  Tensile force  ╲        N                  │
                ╲ reaches peak value ╱───────────────────────────┘
                 ╲──────┬──────╱
                       │ Y
          ┌────────────▼─────────────────────┐
          │       Record the failure load     │
          └────────────┬─────────────────────┘
                       │
          ┌────────────▼─────────────────────┐
          │   Take down the snapped specimen  │
          └────────────┬─────────────────────┘
                       │
          ┌────────────▼─────────────────────┐
          │ Describe the failure mode of the specimen │
          └────────────┬─────────────────────┘
                       │
          ┌────────────▼─────────────────────┐
          │ Determine the section diameter and the water │
          │        content of the specimen    │
          └────────────┬─────────────────────┘
                       │
                ┌──────▼──────┐
                │     End     │
                └─────────────┘
```

Figure 6.27 Determination of tensile strength of soil

that can be made into a cylindrical specimen. Parameters to be measured include diameter of failure surface, axial load and deformation of the specimen.

Procedures:

Step 1: Specimen preparation.

Take the undisturbed sample in the field and smooth the upper and lower surface of the sample. Place the sample on the frame of the wire cutting machine. Set the cutting path as a circle with diameter of 48–54 mm. Move the specimen closely to the wire and start cutting the soil specimen.

After cutting, check on the specimen and the specimen shall not be used if incomplete or cracked. Place the specimen in the polished sleeve and cut off portion of the specimen exceeding the polishing sleeve, and then polish the two ends of the specimen with a sandpaper.

Take out the specimen from the polished sleeve and fix it on the variable diameter cutting device. Rotate the handwheel and push the grooving slice simultaneously and slowly. Cut an annular groove around the side in the middle of the specimen, the groove should have a width of 15 mm and a max depth of 5 mm so that the failure surface may be very likely in the groove. Then take out the specimen.

Step 2: Specimen Installation.

Wrap the clamp on both ends of the specimen and tighten the screws to clamp the specimen. The degree of tightness depends on the stiffness of specimen. Insert one side of the lax rod of clamp to the card slot on the upper frame. Turn on the universal testing machine and move the cross head to fit another lax rod of clamp in the slot on the cross head. Adjust the height of the cross head slowly to make sure that the specimen is pulled snugly and the axial tension align with the specimen. Set the pulling force and the displacement to zero.

Step 3: Loading.

Set the tensile rate of 0.1 mm/s, the rate depends on the type of soil, the stiffness of the specimen, and the magnitude of the final stress. Apply axial tensile load. Record the tension and deformation of the specimen. When the pulling force reaches its peak, the failure of specimen happens. Stop stretching and save test data. The cross section should be located in the middle groove of the specimen, otherwise repeat the test.

Step 4: Remove the failure specimen and dismantle the clamp of the specimen. Measure the diameter of the cross section and describe the destruction of the specimen or make its sketch.

Step 5: Calculate the tensile strength of the specimen according to the formula, the calculation should be accurate to 0.01 kPa.

Parallel test on at least three specimens is a necessity. The deviation should not be greater than 20% from the average of results from the same specimen.

6.3.12 Shrinkage test

The water evaporation leads to its volume shrinkage of the soil. In this test (Figure 6.28), the contraction tester is used to measure the shrinkage deformation of the specimen under the air-dried condition. The shrinkage indexes include the linear shrinkage rate, the volume shrinkage rate and the coefficient of shrinkage of the soil sample.

Step 1: Specimen preparation can be referenced to undisturbed soil or disturbed soil preparation.

Push the specimen out of the cutting-ring and placed on a perforated plate. Weigh the total mass of the specimen and the perforated plate with an accuracy of 0.1 g. Then, put the perforated plate and the specimen on the block of contraction tester. Set the dial gauge to zero or record its initial reading.

Step 2: Conduct the shrinkage test at room temperature (less than 30 °C) According to water content and shrinkage rate of the specimen, record the dial gauge readings every 1–4 h. Weight the total mass of the equipment and the specimen with accuracy of 0.1 g. After 48 h, record the dial gauge readings every 6–24 h and weight the total mass of the equipment and the specimen until the two consecutive readings of dial gauge are very comparable. It is noted that the dial gauge reading should be kept constant when determining the total mass of the equipment and the specimen.

Step 3: Take out the specimen and dry it at 105–110 °C. When the mass of dried specimen is constant, cool the specimen down to room temperature and determine the total mass of dry soil and perforated plate with an accuracy of 0.1 g.

Step 4: Determine the volume of the specimen according to the wax sealing method, which can refer to the density test of soil sample.

6.3.13 Free swelling of clay

The free swelling test of soil (Figure 6.29) is generally used to determine the potential expansive capacity of loose soil particles in water, and to evaluate the initial swell-shrink characteristics of clay. In the test, the loose soil of known mass is dissolved in NaCl solution to determine the volume increment when the expansion of the specimen is stable.

Procedures:

Step 1: A certain amount of air-dried soil are ground to be fine particles, and pass with a 0.5 mm sieve. Thoroughly mix the soil under the sieve. Dry the specimen in an oven at 105–110 °C and cool down to room temperature.

Step 2: Place the neckless funnel on the stand, align the bottom of the funnel with the center of the soil cup and keep their distance being10 mm.

Step 3: Take a proper amount of the specimen with a soil spoon. Pour the sample into the funnel and slightly stir the poured specimen with a thin wire. When the soil sample is full of soil cups, remove the funnel, scrape off the excess soil from the cup, transfer the specimen into the soil spoon, and place the soil cup under the funnel again. Move all soil samples into the funnel, scrape off the excess soil above the cup, and determine the total mass of soil cup and the sample. Repeat the step 1–3 and the difference in the total mass of soil cup and the sample should not less than 0.1 g.

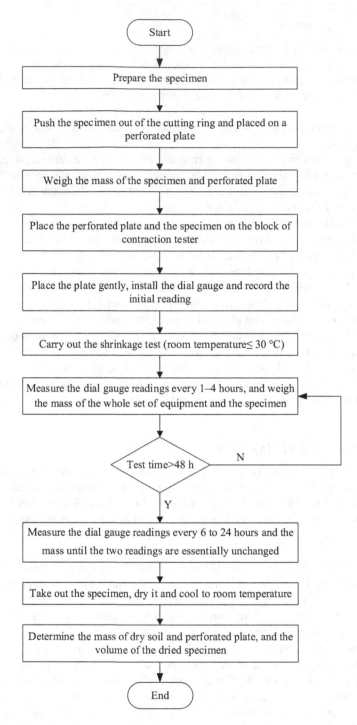

Figure 6.28 Determination of soil shrinkage

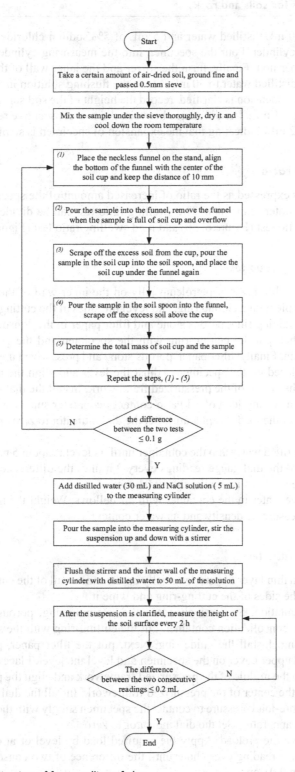

The flowchart contains the following text boxes:

Start

Take a certain amount of air-dried soil, ground fine and passed 0.5mm sieve

Mix the sample under the sieve thoroughly, dry it and cool down the room temperature

(1) Place the neckless funnel on the stand, align the bottom of the funnel with the center of the soil cup and keep the distance of 10 mm

(2) Pour the sample into the funnel, remove the funnel when the sample is full of soil cup and overflow

(3) Scrape off the excess soil from the cup, pour the sample in the soil cup into the soil spoon, and place the soil cup under the funnel again

(4) Pour the sample in the soil spoon into the funnel, scrape off the excess soil above the cup

(5) Determine the total mass of soil cup and the sample

Repeat the steps, (1) - (5)

the difference between the two tests ≤ 0.1 g   N

Y

Add distilled water (30 mL) and NaCl solution (5 mL) to the measuring cylinder

Pour the sample into the measuring cylinder, stir the suspension up and down with a stirrer

Flush the stirrer and the inner wall of the measuring cylinder with distilled water to 50 mL of the solution

After the suspension is clarified, measure the height of the soil surface every 2 h

The difference between the two consecutive readings ≤ 0.2 mL   N

Y

End

Figure 6.29 Determination of free swelling of clay

Step 4: Add 30 mL distilled water and 5 mL of 5% sodium chloride solution into the measuring cylinder. Pour the specimen into the measuring cylinder and thoroughly stir the suspension. Finally, flush the stirrer and the inner wall of the measuring cylinder with distilled water to 50 mL. Move the flushing solution in the suspension.

Step 5: As the suspension is clarified, record the height of the soil surface every 2 h with accuracy of 0.1 mL. The difference between the two consecutive readings should be less than 0.2 mL, indicating that the expansion of specimen is stable.

6.3.14 Swelling ratio

The swell of soil is expressed as the ratio of increased amount of the specimen height to its initial height after water immersion. The swelling ratio test can be divided into two types: no-load swelling ratio test (Figure 6.30) and load swelling ratio test (Figure 6.31).

(1) No-load swelling ratio test

Step 1: Apply a thin layer of petroleum jelly on the inner wall of the cutting ring, cut the soil sample with a cutting ring, and level the sides of the cutting-ring and wipe it. Place the retaining ring, porous stone and filter paper in the consolidation container in order. Then put the cutting-ring with the specimen and the guide ring into the retaining ring. Finally, filter paper, porous stone and pressurized upper cover are successively placed on the specimen. Adjust the lever and align the pressurized upper cover and the center of the pressurized framework. Install the dial indicator.

Step 2: Apply a seating load of 1 kPa to contact oedometer snugly with the upper and lower parts of the loading apparatus. Set the dial indicator to zero or record the initial reading.

Step 3: Add distilled water into the container until its level is about 5 mm over the specimen. Record the dial gauge readings every 2 h until the difference between the two consecutive readings is smaller than 0.01 mm.

Step 4: Remove water in the container with air bellows. Weight the mass of the specimen and measure its density and its water content.

(2) Load swelling ratio test

Step 1: Apply a thin layer of petroleum jelly on the inner wall of the cutting ring. Cut the soil, level the sides of the cutting-ring and wipe it.

Specimen installation. Successively place the retaining ring, porous stone and filter paper in the consolidation container. Put the cutting-ring with the specimen into the retaining ring. Install the guide ring. Next, put the filter paper, porous stone and pressurized upper cover on the specimen and level the lever. Place the consolidation container in the middle of the pressurized framework and align the pressurized upper cover with the center of the pressurized framework. Install the dial indicator.

Step 2: Apply pre-load pressure to contact the specimen snugly with the upper and lower parts of the apparatus. Set the dial indicator to zero.

Step 3: Remove the preload. Apply the required load by level or at one time. Record the dial gauge reading every hour until the difference of two consecutive reading is smaller than 0.01 mm.

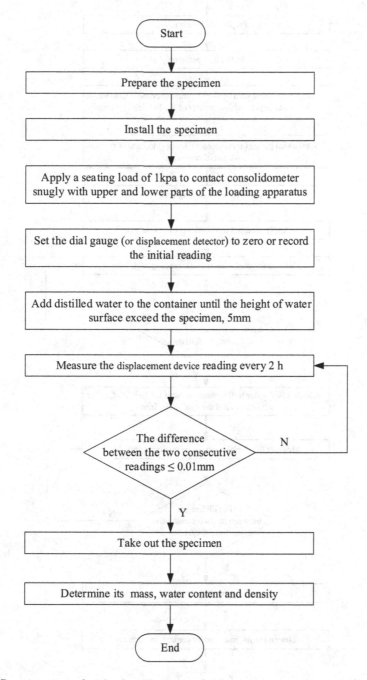

Figure 6.30 Determination of no-load swelling ratio of soil

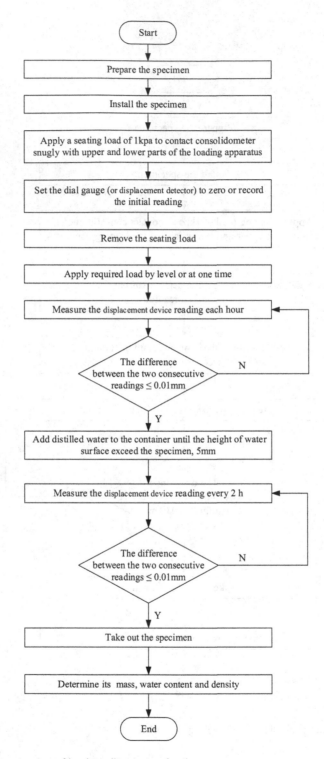

Figure 6.31 Determination of load swelling ratio of soil

Step 4: Inject distilled water into the container until its level is 5 mm over the specimen. Record the dial gauge readings every 2 h after water injection until the difference between the two consecutive readings does not exceed 0.01 mm.

Step 5: Remove the load. Remove water in the container. Take out the specimen, weigh its mass and measure its density and water content.

6.3.15 Swelling pressure

The swelling pressure refers to the internal stress generated by the expansion of clay in water. It features the strength of the cohesive soil. In this method (Figure 6.32), the vertical maximum internal stress of the specimen can be determined by the balanced loading method without any lateral deformation and vertical strain.

Step 1: Specimen preparation can be referenced to undisturbed soil or disturbed soil preparation.

Step 2: Specimen installation can be referenced to consolidation test. Apply a seating load of 1kPa to contact oedometer snugly with the upper and lower parts of the loading apparatus. Set the dial gauge or displacement detector to zero. Add the distilled water to the container until the water level is over the specimen.

Step 3: Immediately the appropriately balanced load to return the dial indicator to the original position once the dial indicator starts to rotate clockwise.

Step 4: Anticlockwise rotate the dial to a value which equals the deformation of apparatus when the apparatus is sufficiently deformed.

Step 5: Measure the dial gauge reading every 2 h until the difference between the two consecutive readings is smaller than 0.01 mm. This reveals that the specimen is stable under the certain level load. As such, the total balanced load is recorded.

Step 6: Remove the loads. Take out the specimen and determine its mass and water content.

6.4 Tests for chemical properties

6.4.1 pH value

The degree of acidity or alkalinity in soil materials is measured by pH value. It mainly depends on the concentration of hydrogen ions in the soil suspension. Measurements of pH of soil can be conducted either with the colorimetric method or the electrometric method. Principle of colorimetric method (Figure 6.33) is to add the indicator to the sample tube, and compare its color with standard colorimetric tube, then the pH value of sample can be obtained. Comparing with colorimetric method, the electrometric method is more accurate and convenient. Electrometric method can be used for all types of soil. Parameter to be measured is the pH value of soil suspension.

Procedures:

Step 1: Preparation of reagents.

The pH value of buffer solutions should be stable at 20 °C.
Prepare standard buffer solution with pH value of 4.01:

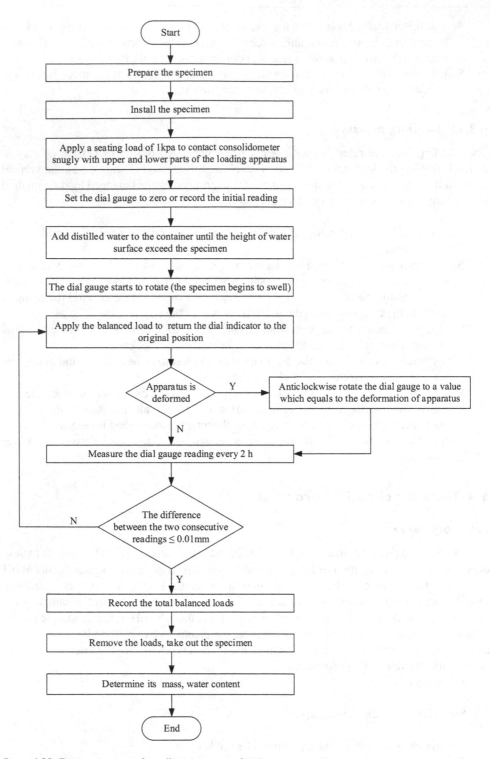

Figure 6.32 Determination of swelling pressure of soil

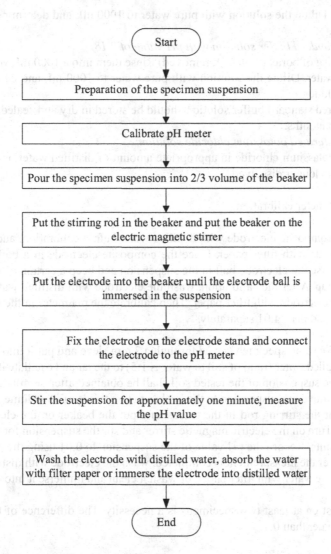

Figure 6.33 Determination of the pH value of soil

Weigh 10.21 g of potassium hydrogen phthalate (KHC$_8$H$_4$O$_4$) Place the funnel over the volumetric flask. Pour the potassium hydrogen phthalate (KHC$_8$H$_4$O$_4$) into the funnel slowly, and rinse it into the 1000 mL volumetric flask with distilled water. Dilute the solution with pure water to the position of the calibration line, and use a dropper to reach 1000 mL. Plug the cork and reverse the volumetric flask repeatedly. After the solution is mixed uniformly, pour part of solution into the jar. Wash the head of the pH meter with distilled water and put it into the jar to determine the pH value of the solution.

Prepare standard buffer solution with pH value of 6.87:

Weigh 3.53 g of dried sodium phosphate dibasic (Na$_2$HPO4) and 3.39 g of dried potassium dihydrogen phosphate (KH$_2$PO$_4$) Rinse them into a 1000 mL volumetric flask with distilled water. Dilute the solution with pure water to 1000 mL and determine the pH value of the solution.

Prepare standard buffer solution with pH value of 9.18:

Weigh 3.80 g of borax (with CO_2 removed) Rinse them into a 1000 mL volumetric flask with distilled water. Dilute the solution with pure water to 1000 mL and determine the pH value of the solution.

The prepared standard buffer solution should be stored in dry and sealed plastic bottles for at most two months.

Prepare saturated potassium chloride solution:

Dissolve potassium chloride in appropriate amount of distilled water and stir the solution until it is no longer dissolved.

Step 2: pH meter calibration.

Wash the composite electrode with distilled water before calibration, and absorb water from the electrode with filter paper. Place the composite electrode in a buffer solution of pH=6.87, immerse the electrode ball in the suspension and keep a certain distance from the bottom of the cup. After the calibration, wash the electrode with distilled water, and absorb water from the electrode with filter paper. Then calibrate the pH meter in the buffer solution with pH =9.18 and pH =4.01 separately.

Step 3: Take some specimen sieved through a 2 mm sieve and put it into the jar, add 50 mL distilled water (ratio of soil to water is 1:5) to the jar and oscillate it for three min. Then the suspension of the tested soil shall be obtained after 30 minute's standing.

Step 4: Remove the suspension to the beaker to two thirds of its volume.

Step 5: Put the stirring rod in the beaker and put the beaker on the electric magnetic stirrer. Turn on the electric magnetic stirrer and stir the suspension for approximately one minute. Record the pH value of the suspension to 0.01 using the pH meter.

Step 6: After the measurement is completed, wash the electrode with distilled water and absorb the water with filter paper, or wash it clean and immerse it into distilled water.

Parallel test on at least two specimens is a necessity. The difference of two pH values shall not be greater than 0.1.

6.4.2 Content of soluble salt

The soluble salts in soil exhibits either solid state or liquid state, which easily cause the collapsing and swelling hazard of the roadbed. In general, the fillers containing high content of soluble salts are not suitable in the engineering. The determination of soluble salt in soil is composed of leaching solution preparation and ion content measurement. Various ions be measured to are the anions (CO_3^{2-}, HCO_3^-, Cl^-, SO_4^{2-}) and the cations (Ca^{2+}, Mg^{2+}, Na^+, K^+).

(1) Leaching solution preparation

As shown in Figure 6.34, completely dissolve the soluble salt of soil in water to obtain the transparent filtrate.

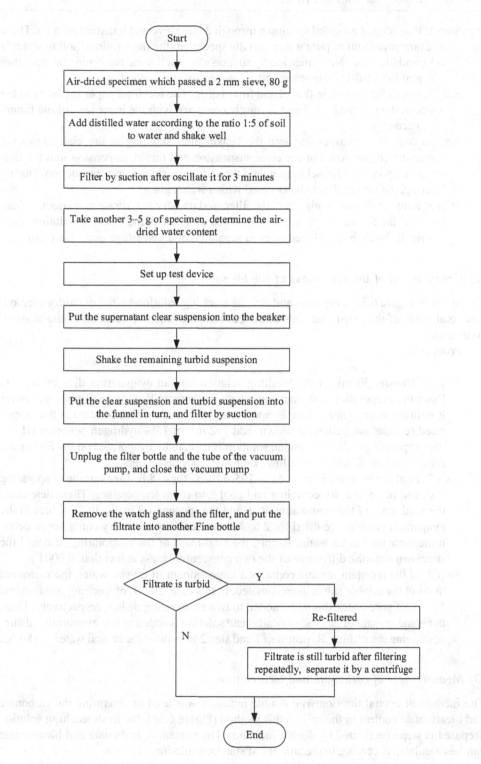

Start

Air-dried specimen which passed a 2 mm sieve, 80 g

Add distilled water according to the ratio 1:5 of soil to water and shake well

Filter by suction after oscillate it for 3 minutes

Take another 3–5 g of specimen, determine the air-dried water content

Set up test device

Put the supernatant clear suspension into the beaker

Shake the remaining turbid suspension

Put the clear suspension and turbid suspension into the funnel in turn, and filter by suction

Unplug the filter bottle and the tube of the vacuum pump, and close the vacuum pump

Remove the watch glass and the filter, and put the filtrate into another Fine bottle

Filtrate is turbid

Y

N

Re-filtered

Filtrate is still turbid after filtering repeatedly, separate it by a centrifuge

End

Figure 6.34 Preparation of leaching solution

Step 1: Pass 80 g of air-dried specimen through 2 mm sieve and transfer into a jar. Then, add proper amount of pure water into the specimen (the mass ratio of soil to water is 1:5), and filter the soil suspension by suction after oscillating for 3 min. Take another 3–5 g of the air-dried specimen and measure its water content.

Step 2: Install flat porcelain funnel and filter bottle. The wet filter paper is placed to the inside bottom of the funnel and is snugly contacted with the inner wall of the funnel by a vacuum pump.

Put the supernatant suspension into the beaker after standing in the vial. Shake the remaining suspension. Put the clear suspension and turbid suspension into the flat-bottomed porcelain funnel in turn, and then filter the suspensions by suction. During filtering, the funnel should be covered with a watch glass.

Step 3: Remove the watch glass and the filter, and transfer the filtrate into another clean bottle. If the filtrate is not turbid, it is stored in a vial as the leaching solution. Otherwise, it should be re-filtered or even separated by a centrifuge until it is clarified.

(2) Measurement of the total mass of soluble salt

As shown in Figure 6.35, evaporate and dry the leaching solution by the oven-dry method. The total mass of the dried materials obtained from the leaching solution is the mass of soluble salt.

Procedures:

Step 1: Transfer 50 mL of the leaching solution into an evaporating dish by pipette. Place the evaporating dish on a water bath kettle or an E-purpose furnace and cover it with the watch glass. Then, Evaporate and dry the leaching solution If the evaporated residues are yellowish brown, add 1–2 mL of 15% hydrogen peroxide (H_2O_2) the evaporating dish. The residue solution is repeatedly evaporated on the E-purpose furnace at 100 °C until the yellowish-brown color disappears.

Step 2: Heat the residues in an oven at 105–110 °C for 4–8 h. Take out the evaporating dish and place in a dry container and cool it to room temperature. Then, determine the total mass of the evaporating dish and the specimen. Finally, the residues in the evaporating dish are re-dried for 2 to 4 h and cool down in a dry container to determine their total mass again. Record the total mass of the evaporating dish and the specimen until the difference of the two consecutive mass is less than 0.0001 g.

Step 3: If the resultant residue contains a large amount of crystal water, the measured mass of the soluble salt is overestimated. In this case, 50 mL of leaching solution and 50 mL of pure water should be added to two evaporating dishes, respectively. Then, the equal amount of 2% sodium carbonate solution is added to two evaporating dishes containing the residue. Repeat step 1 and step 2 to remove the crystal water at 180 °C.

(3) Measurement of carbonate and bicarbonate

The method of neutral titration with double indicator was used to determine the carbonate and bicarbonate content in the soil. In this method (Figure 6.36), the fresh leaching solution prepared is stepwise titrated by double indicator. The content of carbonate and bicarbonate can be calculated according to the amount of standard solution.

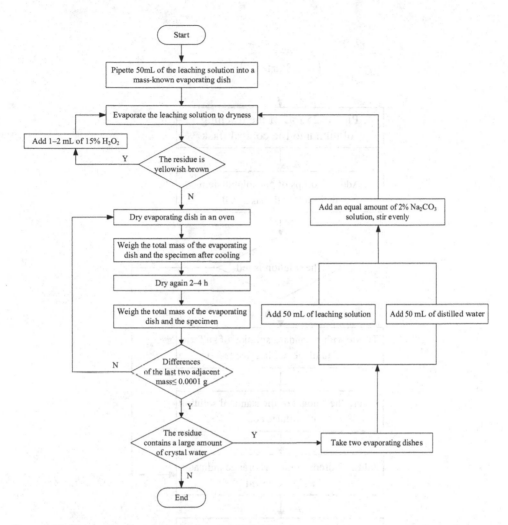

Figure 6.35 Determination of the total mass of soluble salt

Step 1: Immediately inject 25 mL of the fresh leaching solution into the Erlenmeyer flask. Then add 2 to 3 drops of the phenolphthalein indicator into the leaching solution. If the solution is not red, it indicates that there is no carbonate in the leaching solution. Otherwise, it should be titrated with a standard solution of sulfuric acid until the red color fully disappears. Record the amount of the standard solution of sulfuric acid with accuracy of 0.05 mL.

Step 2: Add 1–2 drops of methyl orange indicator in the test solution. Next, titrate the well mixed solution with a standard solution of sulfuric acid until its color changes from yellow to orange. Record the amount of the standard solution of sulfuric acid with accuracy of 0.05 mL.

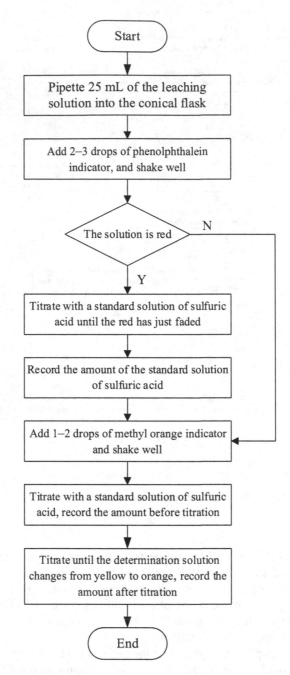

Figure 6.36 Determination of CO_3^{2-} and HCO_3^-

(4) Measurement of chloride ion

Several methods were used to determine the chloride ion content of the soil, including silver nitrate volumetric method, mercury nitrate titration method, and thiocyanate spectrophotometry and ion chromatography. Among these methods, silver nitrate volumetric method (Figure 6.37) are easy to handle, especially for the specimen with low chloride concentration. Potassium chromate is used as an indicator and the neutral leaching solution is titrated with the silver nitrate standard solution to form a brick-red precipitate in silver nitrate volumetric method. The content of chloride ion can be obtained according to the consumption of silver nitrate standard solution.

Step 1: Transfer 25 mL of the leaching solution into the conical flask by a pipette and then add 1–2 drops of methyl orange indicator into the leaching solution. Next, sodium bicarbonate is dropwise added to leaching solution until the solution is yellow (pH=7) Finally, add 5–6 drops of potassium chromate indicator to the leaching

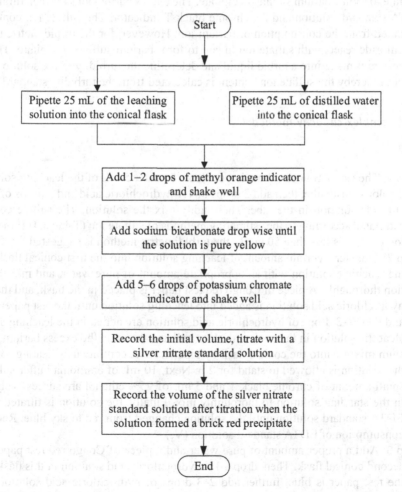

Figure 6.37 Determination of Cl⁻

solution and shake them well. Record the initial volume and titrate the solution with a silver nitrate standard solution until a brick red precipitate is formed in the solution. Record the volume of the silver nitrate standard solution and calculate the consumption amount of silver nitrate standard solution with accuracy of 0.05 mL.

Step 2: Take 25 mL of pure water as a blank specimen and repeat the step 1 to record the consumption amount of the silver nitrate standard solution when the brick red precipitate is formed in the solution.

(5) Measurement of sulfate ion

As shown in Figure 6.38, the methods for the determination of sulfate ion content in the soil highly depend on the concentration of sulfate ions. The EDTA complex capacity method is suggested to measure the sulfate ion content when the sulfate content is greater than 0.025%, otherwise, the turbidimetric method is suggested.

For the EDTA complex capacity method, an excess of barium magnesium mixture reacts with sulfate to form a barium sulfate precipitate. The excess barium ions are then titrated with the EDTA standard solution and the chrome black T indicator. The sulfate ion content can be calculated from the consumption of barium ion. However, for the turbidimetric method, barium chloride reacts with sulfate ion in soil to form barium sulfate precipitate. Then, the barium sulfate is made into a turbid liquid and determine the turbidity of the solution by the colorimeter, thereby the sulfate ion content is calculated from the turbidity standard curve.

I) EDTA complex capacity method

Procedures:

Step 1: The estimation of sulfate ion content. Transfer 5 mL of the leaching solution in a colorimetric tube, then add 2 drops of 1:1 hydrochloric acid and 5 drops of barium chloride solution in the tube. Thoroughly mix the solution. The sulfate content is estimated according to the estimating methods of sulfate ion (Table 6.4) If the sulfate ion content is less than 50 mg/L, the turbidimetric method is suggested.

Step 2: Transfer a certain amount of leaching solution into the first conical flask, dilute the leaching solution with an appropriate amount of pure water, and mix the solution thoroughly. A piece of Congo red test paper is placed in the flask, and then a 1:4 hydrochloric acid solution is added into leaching solution until the test paper is blue, and then 2–3 drops of hydrochloric acid solution are added in the leaching solution. Heat the solution in the conical flask to boil and then add the excess barium magnesium mixture into the conical flask by pipette. After continuously heating for 5 min, the solution is allowed to stand for 2 h. Next, 10 mL of ammonia buffer solution, a small amount of chrome black T and 5 mL of 95% ethanol are successively added in the standing solution to thoroughly mix. Finally, the solution is titrated with an EDTA standard solution until the solution changed from red to sky blue. Record the consumption of EDTA standard solution (V_1)

Step 3: Add a proper amount of pure water and a piece of Congo red test paper in the second conical flask. Then, drop a 1:4 hydrochloric acid solution in this flask. When the test paper is blue, further add 2–3 drops of hydrochloric acid solution in this flask. Next, the same amount of barium magnesium mixture as in step 2, 10 mL of

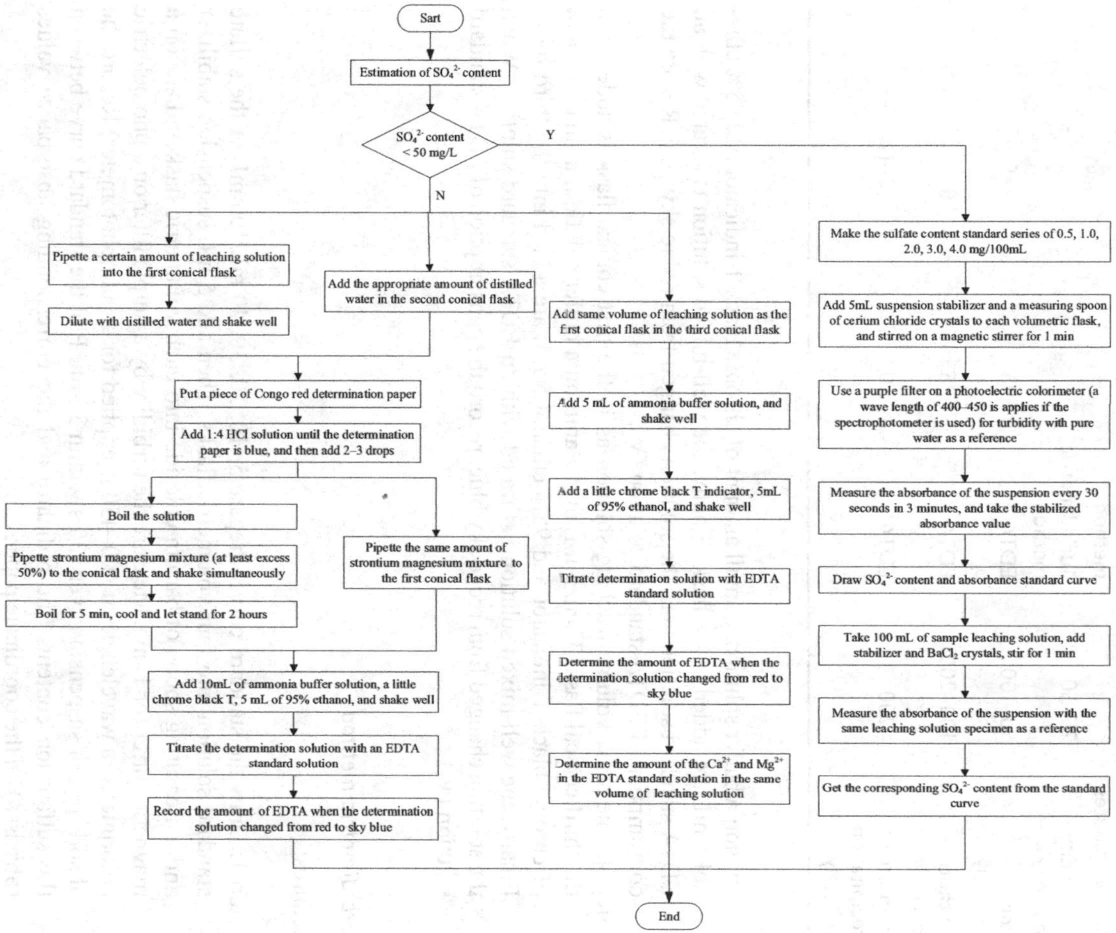

Figure 6.38 Determination of $SO_4{}^{2-}$

Start

Estimation of $SO_4{}^{2-}$ content

$SO_4{}^{2-}$ content < 50 mg/L

Y branch:

Make the sulfate content standard series of 0.5, 1.0, 2.0, 3.0, 4.0 mg/100mL

Add 5mL suspension stabilizer and a measuring spoon of cerium chloride crystals to each volumetric flask, and stirred on a magnetic stirrer for 1 min

Use a purple filter on a photoelectric colorimeter (a wave length of 400-450 is applies if the spectrophotometer is used) for turbidity with pure water as a reference

Measure the absorbance of the suspension every 30 seconds in 3 minutes, and take the stabilized absorbance value

Draw $SO_4{}^{2-}$ content and absorbance standard curve

Take 100 mL of sample leaching solution, add stabilizer and $BaCl_3$ crystals, stir for 1 min

Measure the absorbance of the suspension with the same leaching solution specimen as a reference

Get the corresponding $SO_4{}^{2-}$ content from the standard curve

End

N branch:

Add the appropriate amount of distilled water in the second conical flask

Add same volume of leaching solution as the first conical flask in the third conical flask

Add 5 mL of ammonia buffer solution, and shake well

Add a little chrome black T indicator, 5mL of 95% ethanol, and shake well

Titrate determination solution with EDTA standard solution

Determine the amount of EDTA when the determination solution changed from red to sky blue

Determine the amount of the Ca^{2+} and Mg^{2+} in the EDTA standard solution in the same volume of leaching solution

Pipette a certain amount of leaching solution into the first conical flask

Dilute with distilled water and shake well

Put a piece of Congo red determination paper

Add 1:4 HCl solution until the determination paper is blue, and then add 2-3 drops

Boil the solution

Pipette strontium magnesium mixture (at least excess 50%) to the conical flask and shake simultaneously

Boil for 5 min, cool and let stand for 2 hours

Pipette the same amount of strontium magnesium mixture to the first conical flask

Add 10mL of ammonia buffer solution, a little chrome black T, 5 mL of 95% ethanol, and shake well

Titrate the determination solution with an EDTA standard solution

Record the amount of EDTA when the determination solution changed from red to sky blue

Table 6.4 the estimating methods of SO_4^{2-}

Turbidity of solution after adding $BaCl_2$	SO_4^{2-} content (mg/L)	Methods	The amount of leaching solution (mL)	The amount of strontium magnesium mixture (mL)
Slight turbidity after a few minutes	<10	Turbidimetric method	–	–
Slight turbidity immediately	25–50	Turbidimetric method	–	–
Turbidity immediately	50–100	EDTA	25	4–5
Precipitation immediately	100–200	EDTA	25	8
Large amount of precipitation immediately	>200	EDTA	10	10–12

ammonia buffer solution, a small amount of chrome black T indicator and 95% ethanol 5 mL are added to the flask. Finally, the well-mixed solution is titrated with an EDTA standard solution until the solution changed from red to sky blue. Record the consumption of EDTA standard solution (V_2)

Step 4: The same volume leaching solution as in the first conical flask is added into the third conical flask. Then, 5 mL of the ammonia buffer solution, a small amount of chrome black T indicator and 95% ethanol 5 mL are added into the third flask. Finally, the well-mixed solution is titrated with an EDTA standard solution. When the solution changed from red to sky blue, record the consumption of EDTA standard solution (V_3)

2) Turbidimetric method

Procedures:

Step 1: Draw the standard curve. Prepare 5, 10, 20, 30, 40 mL/100 mL of the sulfate standard solution into four volumetric flasks. Then add 5 mL suspension stabilizer and a measuring spoon of barium chloride into each volumetric flask and stir on a magnetic stirrer for 1 min. Measure the turbidity by a purple filter on a photoelectric colorimeter (a wavelength of 400–450 is applied for spectrophotometer) Record the absorbance of suspension every 30 s within 3 min. Plot the standard curve between the sulfate ion contents (the ordinate) and the corresponding absorbance values (abscissa) on the coordinate paper.

Step 2: Measurement of sulfate ion content. Transfer 100 mL of the specimen leaching solution into a beaker by a pipette. Measure absorbance of the suspension as in step 1. Compare the standard curve to obtain the corresponding sulfuric acid content (mg/100 mL) Parallel tests need to be conducted for the reference.

(6) Measurement of calcium and magnesium ions

The EDTA volumetric method is commonly used to determine calcium and magnesium ions in the soil, which is convenient and has the simple equipment. In this method (Figure 6.39), Mg^{2+} ion in the leaching solution is first precipitated by the sodium hydroxide solution, and then the Ca^{2+} in the leaching solution is determined by the titration method using the EDTA standard solution and the chrome black T.

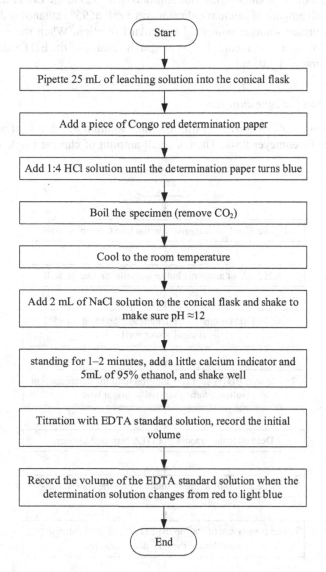

Figure 6.39 Determination of Ca^{2+}

Procedures:

1) Measurement of calcium ions

Step 1: Put 25 mL of leaching solution and a piece of Congo red test paper into the Erlenmeyer flask, and then slowly drop 1:4 hydrochloric acid solution into the leaching solution until the test paper is blue. Next, heat the solution to remove carbon dioxide and then cool down to room temperature.

Step 2: Add 2 mL of sodium hydroxide solution (pH ≈12) to the Erlenmeyer flask. Then add a small amount of calcium indicator and 5 mL of 95% ethanol in the flask. Titrate the well-mixed solution with EDTA standard solution. When the solution changes from red to light blue, record the consumption volume of the EDTA standard solution with accuracy of 0.05 mL.

2) Measurement of magnesium ions

As shown in Figure 6.40, Transfer 25 mL leaching solution and 5 mL of ammonia buffer solution into the Erlenmeyer flask. Then, a small amount of chrome black T indicator and

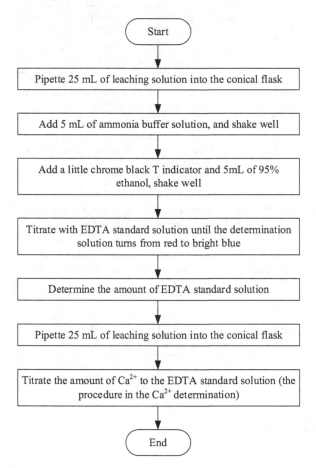

Figure 6.40 Determination of Mg^{2+}

5 mL of 95% ethanol are added in the flask. Titrate the well-mixed solution with EDTA standard solution until it changes from red to bright blue, record the consumption of EDTA standard solution.

(7) Measurement of sodium and potassium ions

The subtraction method and flame photometer method are mainly used to determine sodium and potassium ions. The subtraction method can only provide the total amount of sodium and potassium ions. However, the flame photometer method is more sensitive and convenient. The sodium and potassium ions can be separately measured by the flame photometer method (Figure 6.41).

Step 1: Successively turn on the power supply, the main engine, the air compressor, and the gas valve. Perform the ignition. Adjust the flame to be blue.

Step 2: Add 0, 1, 5, 10, 15, 25 ml of Na^+ and K^+ standard solution to six of volumetric flasks (50 mL), and dilute to 50 mL solutions with pure water, respectively,

Step 3: Adjust the Na^+ and K^+ ranges according to the concentrations of Na^+ and K^+ solutions. Calibrate Na^+ and K^+ with distilled water.

Step 4: Place the specimen tubes in standard solution bottles. And then measure the Na^+ and K^+ absorbance from low concentration to high concentration. It is noted that the specimen tube should be cleaned after each test.

Step 5: Plot the standard curves of Na^+ and K^+, respectively, or establish the regression equation via the least square method.

Step 6: According to absorbance of Na^+ and K^+ in the specimen, the corresponding sodium and potassium ion concentrations is obtained from the standard curve or by the regression equation.

6.4.3 Organic matter

Several methods, e.g., mass method, volumetric method, potassium dichromate volumetric method, colorimetric method, and hydrogen peroxide oxidation method, etc., can be used for determining the organic matter. Compared to the other methods, the potassium dichromate volumetric method is much convenient and accurate. It is also can be used for measurement of a large amount of carbonate. In this method (Figure 6.42), the organic matters completely react with an excess of potassium dichromate, and then the organic matter is determined according to the consumption of potassium dichromate.

Procedures:

Step 1: Specimen preparation. If the organic carbon content in the specimen is less than 8 mg, remove the plant roots in the air-dried specimen. Take 0.1–0.5 g of the specimen sieved with a 0.15 mm sieve. Place the specimen into the dried conical flask. Then slowly drop 10 mL of potassium dichromate standard solution with a burette into the dried conical flask and mix thoroughly. Take the same amount of pure sand for the blank test.

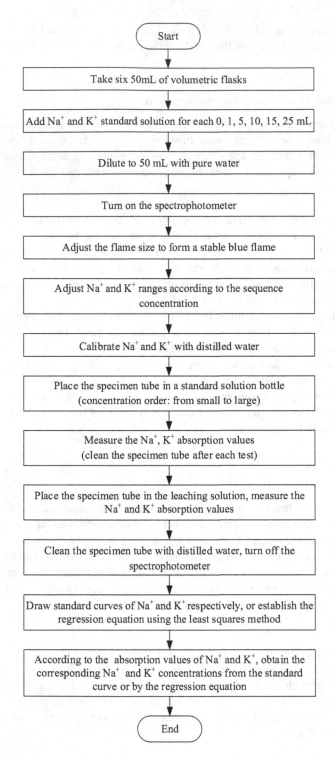

Figure 6.41 Determination of Na^+ and K^+

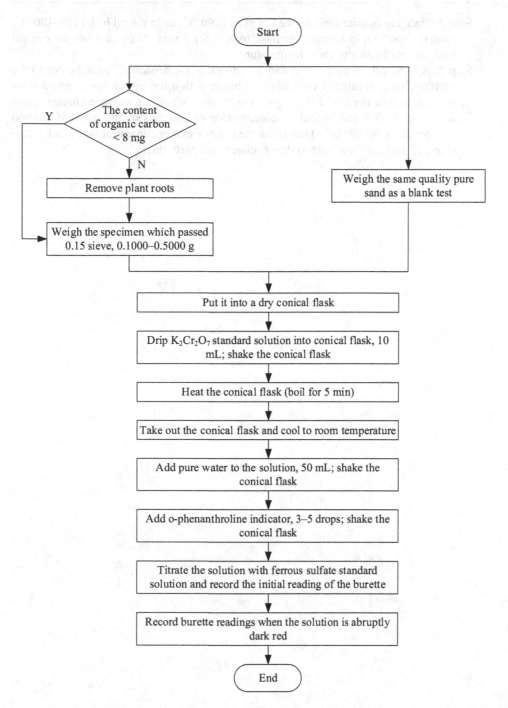

Figure 6.42 Determination of the organic matter of soil

Step 2: Place the conical flask on the hot plate (190 °C) or in the oil bath (170–180 °C) Heat the solution in Erlenmeyer flask to boil for 5 min. Then take out the conical flask and cool down to room temperature.

Step 3: Add 50 mL of pure water and 3–5 drops of o-phenanthroline indicator to the solution. Then, titrate the well-mixed solution with a ferrous sulfate standard solution and record the initial reading of the burette. When the solution changes from dark green to dark red, record the consumption of a ferrous sulfate standard solution with accuracy of 0.05 mL. Determine the organic matter content of pure sand. Compare and analyze the results of the specimen and pure sand.

Chapter 7

Procedures of tests on rocks

In this chapter, the procedures of rock tests are described according to *the Standard for Engineering Rock Mass Determination Methods GB/T 50266-2013*. Combining with the industrial standards, e.g., highways and water conservancy, we have made revisions for some key steps of rock tests. The preparation and specific requirements of rock specimens are unified and standardized in Sections 7.1 and 7.2. The overviews of the test procedures are provided in the form of flowcharts.

7.1 Requirements of specimen

7.1.1 Water content

(1) Take the specimen on sites and maintain its natural moisture content during handling, transporting, storing and preparing process. The blasting method is used for this test.
(2) The specimen should be more than 10 times the diameter of the largest mineral particles of the rock, and the mass of each specimen should be in the range from 40 to 200 grams. At least five specimens are required in each test group.
(3) Determination of the water content of the structural surface filling should comply with the National Standard of Determination Methods for Geotechnical Engineering (GB/T 50123)

7.1.2 Determination of grain density

(1) Crush the rock into powder by a shredding machine and then pass through a mesh with a hole diameter of 0.25 mm, and then remove iron scraps with magnet.
(2) The rock containing the magnetic minerals should be crushed with a porcelain mortar or agate mortar and then passed through a mesh with a hole diameter of 0.25 mm.

7.1.3 Density

(1) Method A-Direct measurement

1) The specimen should be 10 times larger than the diameter of the largest mineral particle of the rock. The minimum size of the specimen should not be less than 50 mm.
2) The cylinder, a square cylinder or a cube specimen are needed in the test.
3) The deviation in diameter along the height of the specimen shall be within 0.3 mm;

4) The non-parallelism deviation at both end faces of specimen shall be within 0.05 mm.
5) Both end faces of the specimen shall be perpendicular to the axis of the specimen and the deviation shall not exceed 0.25 degrees.
6) Adjacent sides of the square cylinder or cube specimen should be perpendicular to each other and the deviation shall not exceed 0.25 degrees.

(2) Method B-Water displacement

The side length of the wax seal rock specimen should be in the range of 40~60 mm.

7.1.4 Water absorption

(1) Regular specimen should comply with the specimen requirements of the volume method in section 7.1.3.
(2) Irregular specimen in a round rock should have a side length of 40–60 mm.
(3) Three samples in each test group are required.

7.1.5 Slake durability

(1) Take specimens with natural water content by on-the-spot sampling method and seal them well.
(2) Mass of each specimen in round shape should be in the range from 40 to 60 g.
(3) Ten samples in each test group are required.

7.1.6 Swelling properties

(1) Take the specimen by on-the-spot sampling method. The specimen should be kept in a natural water-containing state and processed by dry method.
(2) For the free swelling test of the cylinder or square specimen, the diameter of specimen is in the range of 48–65 mm, the height should be equal to the diameter, and the top and bottom surfaces should be parallel. The side length of the specimen should be in the range of 48–65 mm. Three samples in each test group are required.
(3) For the lateral restraint swelling test and the swell pressure test under the condition of constant volume, the height of specimen should not be less than 20 mm, or smaller than 10 times the maximum mineral particle size of the constituent rock, and the top and bottom surfaces should be parallel to each other. The specimen should have a diameter of 50–65 mm and be less than the inner diameter of the metal collar. Three samples in each test group are required.

7.1.7 Uniaxial compressive strength

(1) Diameter of the cylindrical specimen should be in the range from 48 to 54 mm.
(2) Diameter of the specimen should be greater than 10 times the diameter of the largest particle in the rock.
(3) The ratio of the height to the diameter for the specimen piece should be in the range from 2.0 to 2.5.

(4) Top and bottom surfaces of the specimen should be parallel.
(5) Deviation in diameter along the height of the specimen shall be within 0.7 mm.
(6) Top and bottom surfaces of specimen should be perpendicular to the axis of the specimen and with the deviation shall be within 0.25°.

Specimen requirements for durability under freezing and thawing conditions and deformability in uniaxial compression should comply with the requirements 1–6.

7.1.8 Strength in triaxial compression

(1) Diameter of the cylinder specimen should be in the range from 0.96 to 1.00 times the diameter of the bearing plate of the test machine.
(2) Five specimens having the same water content are required.
(3) Other requirements should comply with 7.1.7.

7.1.9 Indirect tensile strength by Brazilian test

(1) Diameter of the cylindrical specimen should be 48–54 mm. The thickness of the specimen is 0.5–1.0 times the diameter of specimen and greater than 10 times the maximum particle diameter of the rock.
(2) Other requirements should comply with 7.1.7.

7.1.10 Shear strength of rock joints

(1) Side length of the cube specimen is greater than 50 mm;
(2) Structural surface should be parallel to the applied shear stress and located in the middle of the specimen;
(3) At least 5 specimens are required in each test group. They are used to determine the shear strength under different normal loads.

7.1.11 Point load strength

(1) For the specimen of rock core, the ratio of length to diameter of the specimen should be more than 1.0. However, for those used for axial test, the ratio of length to diameter of the specimen should be 0.3–1.0.
(2) The size of block specimen should be 15–85 mm, and the ratio of the distance between the two loading points to the average width of specimen at the loading points should be in the range from 0.3 to 1.0.

7.1.12 Wave velocity by ultrasonic pulse transmission technique

(1) Diameter of the cylindrical specimen should be in the range from 48 to 54 mm.
(2) Diameter of the specimen should be greater than 10 times the diameter of the largest particle in the rock.
(3) The ratio of the height to the diameter for the specimen piece should be in the range from 2.0 to 2.5.

(4) Top and bottom surfaces of the specimen should be parallel.
(5) Deviation in diameter along the height of the specimen shall be within 0.3 mm.
(6) Top and bottom surfaces of specimen should be perpendicular to the axis of the specimen and with the deviation shall be within 0.25°.

7.2 Description of specimens

7.2.1 Water content

(1) Rock type, color, mineral constituents, structure, texture, weathering degree and cementing properties.
(2) Moisture conditions and the methods of saturation and drying.

7.2.2 Grain density

(1) Rock type, color, mineral constituents, structure, texture, weathering degree and cementing properties.
(2) Moisture conditions and the methods of saturation and drying.

7.2.3 Density and water absorption

(1) Rock type, color, mineral constituents, structure, texture, weathering degree and cementing properties.
(2) Moisture conditions and the methods of saturation and drying.
(3) Development degree and distribution of joint fissures.
(4) Shape of the specimen.

7.2.4 Slake durability

Rock type, color, mineral constituents, structure, texture, weathering degree and cementing properties.

7.2.5 Swelling properties

(1) Rock type, color, mineral constituents, structure, texture, weathering degree and cementing properties.
(2) Relation among loading orientation and stratifications, joints and fissures of the specimen.
(3) Phenomena in specimen processing

7.2.6 Uniaxial compressive strength

(1) Rock type, color, mineral constituents, structure, texture, weathering degree and cementing properties.
(2) Relation among loading orientation and stratifications, joints and fissures of the specimen.
(3) Moisture conditions and the methods of saturation and drying.
(4) Describe phenomena such as crack, peel or split in the processing specimen.

7.2.7 Shear strength of rock joints

(1) Rock type, color, mineral constituents, structure, texture, weathering degree and cementing properties.
(2) Development degree of the layers, the texture, and the joint fissures.
(3) Relationship between the shear direction and the layer (texture or joint fissure)
(4) Filling properties and filling degree of the structural surface.
(5) Methods to take the specimen and disturbance degree in the preparation process.

7.2.8 Point load strength

(1) Rock type, color, mineral constituents, structure, texture, weathering degree and cementing properties.
(2) Specimen shape and methods of specimen preparation.
(3) Relationship between loading direction and bedding, phylogeny or joints.
(4) Water content of specimen and Moisture conditions.

7.2.9 Wave velocity by ultrasonic pulse transmission

(1) Rock type, color, mineral constituents, structure, texture, weathering degree and cementing properties.
(2) The relationship between loading direction and rock specimen bedding, joints and fissures.
(3) The state of water and the method used to prepare the specimens.
(4) The phenomenon that occurs during the processing of the specimens.

7.3 Physical tests

7.3.1 Water content

Water content of rock is expressed as the ratio of the lost water mass to rock solid particle mass when the rock is dried to have a constant mass at 105–110 °C. Water content of rock indirectly reflects its voids and the compactness of the rock. The test process is shown in Figure 7.1.

Procedures:

Step 1: Determine the mass of the container. Place the specimen in the container and determine the total mass of the specimen and container with accuracy of 0.01 g.

Step 2: Place the container containing the specimen in an oven and dry at 105–110 °C for 24 hours. For rocks having crystallization water and volatile minerals, the specimen is usually dried at 55–65 °C or by vacuum evacuation at room temperature.

Step 3: Cool down to room temperature and determine the total mass of the specimen and the container with accuracy of 0.01 g.

Step 4: Calculate the water content according to the formula with accuracy of 0.01. Take the average of five specimens as water content in rock.

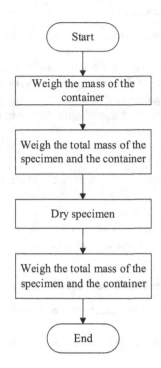

Figure 7.1 Flowchart of determining water content

7.3.2 Grain density

The grain density of rock is the ratio of the mass of rock to its volume when the rock is baked to have a constant mass at 105–110 °C. Pycnometer method and weighing method in water are usually used to determine the rock grain density. In this section, the pycnometer method is present for determining the grain density of various rocks. The test process is shown in Figure 7.2.

Procedures:

Step 1: Take two samples of rock powder by the quadruple method. Each sample of rock powder is 15 g.

Step 2: Put the rock powder into the dried pycnometer. Add the test solution to half of the volume of the pycnometer and mix the rock powder and water thoroughly. Plug the stopper.

Step 3: Remove the gas in the solution by boiling method or vacuum pumping method when water is used in the test. When kerosene is used, only the vacuum can be used to remove the gas in the solution. Place the pycnometer in a sand bath and heat the solution to boil. Then, heat the solution at a relatively low temperature and keep at a slow boil for about 1 hour. In vacuum pumping method, the vacuum gauge reading should be the same as the local atmospheric pressure. Continuously pump the water solution for more than 1 h after bubbles are completely removed.

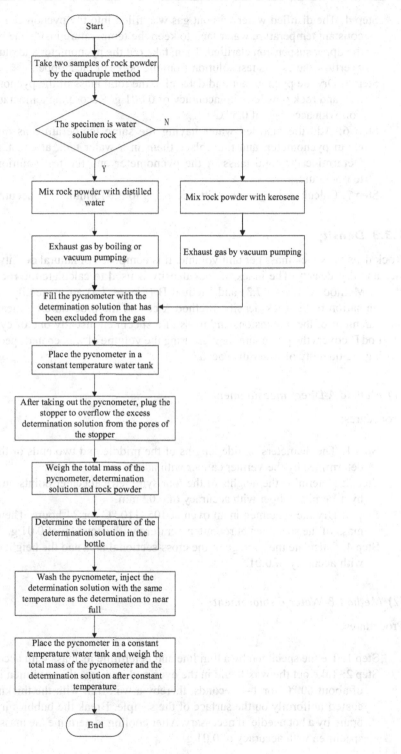

Figure 7.2 Flowchart of determining grain density

Step 4: The distilled water without gas was filled into the pycnometer and placed in a constant temperature water tank to keep the temperature inside the bottle stable and the upper suspension clarified. Then, take out the pycnometer and plug the stopper to overflow the excess test solution from the pores of the stopper.

Step 5: Dry the pycnometer and determine the total mass of the pycnometer, test solution and rock powder with accuracy of 0.001 g. Record the temperature of the solution with accuracy of 0.5 °C.

Step 6: Add the distilled water having the same temperature as the solution to a clean pycnometer, and then place them in a water tank at constant temperature. Determine the total mass of the pycnometer and the test solution at a constant temperature.

Step 7: Calculate the grain density according to the formula with accuracy of 0.01.

7.3.3 Density

Rock density is rock mass per unit volume. It is composed of natural density, saturated density and dry density. The index of rock density is used to calculate the rock's own weight stress. Method A (Figure 7.3) and Method B (Figure 7.4) are generally employed in the determination of the rock density. Method A covers the procedure by means of the direct measurement of the dimensions and mass of a specimen, usually one of cylindrical shape. Method B covers the procedure for measuring the volume of wax coated specimens by determining the quantity of water displaced.

(1) Method A-Direct measurement

Procedures:

Step 1: The diameters or side lengths of the middle and two ends of the specimen are determined by the Vernier caliper with accuracy of 0.02 mm.

Step 2: Determine the heights of the four symmetrical center points on the end surface by a Vernier caliper with accuracy of 0.02 mm.

Step 3: Dry the specimen in an oven at 105–110 °C for 24 hours. Then, determine the mass of the specimen at room temperature with accuracy of 0.01 g.

Step 4: Calculate the average of the cross-sectional area and the height of the specimen with accuracy of 0.01.

(2) Method B-Water displacement

Procedures:

Step 1: Tie the specimen by a thin line and determine its mass with accuracy of 0.01 g.

Step 2: Take out the wax liquid in the oven and immerse the specimen in a molten wax of about 60 °C for 1–2 seconds, thereby a wax film with the thickness of 1 mm is coated uniformly on the surface of the sample. Break the bubbles in the wax membrane by a hot needle, if necessary. After cooling, determine the mass of the wax seal specimen with accuracy of 0.01 g.

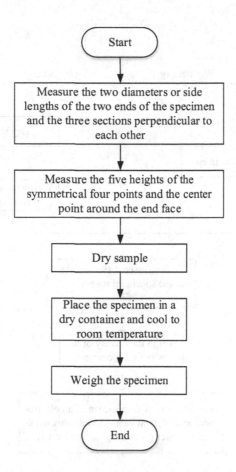

Figure 7.3 Flowchart of determining rock density (direct measurement)

Step 3: Hang the wax seal specimen in an electro-optical analytical balance and immerse the specimen in a beaker containing distilled water without touching the beaker wall. Adjust the balance and record the readings.

Step 4: Take out the specimen, wipe the water on the wax surface, and then determine the mass of the wax seal specimen with accuracy of 0.01 g. If the mass of the specimen increases after immersion, the wax film of specimen should be stripped and the test should be re-conducted.

Step 5: Determine the water content of the rock.

Step 6: calculate the wet density of the rock according to the formula with accuracy of 0.01. For dry density of the rock, the specimen should be dried at 105–110 °C for 24 hours. Store the specimen in a dry container. Determine the mass of dried specimen at room temperature. Re-conduct step 1–4 and calculate the wet density of the specimen with accuracy of 0.01.

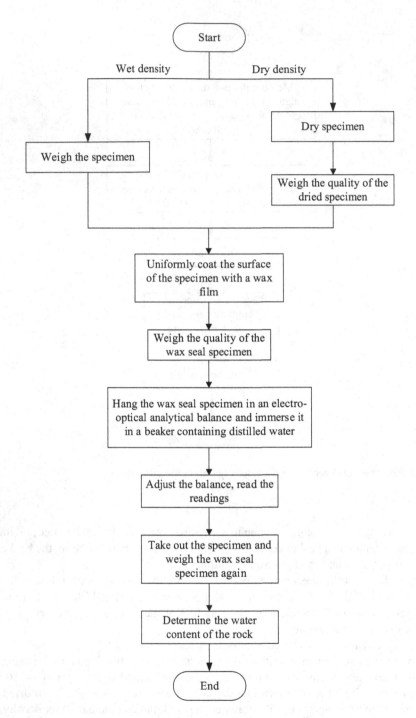

Figure 7.4 Flowchart of determining rock density (water displacement)

7.3.4 Water absorption

Rock absorption includes rock water absorption test and rock saturated water absorption test. The water absorption of rock is the ratio of the mass of water being absorbed by rock at atmospheric pressure and room temperature to the mass of rock solid particles. Natural water immersion method is applicable to the rock water absorption. The saturated water absorption of rock is expressed as the mass ratio of the maximum water absorption of rock under forced conditions to rock solid particles. Rock water absorption is determined by boiling method or vacuuming saturation method. The test process is shown in Figure 7.5.

Procedures:

Step 1: Dry the specimen in an oven at 105–110 °C for 24 hours. Transfer the dried specimen into a dry container and cool down to room temperature. Determine the mass of the specimen.

Step 2: Freely immerse the specimen in pure water. Put the specimen into the water tank and fill the water to the 1/4 height of the specimen. And then fill the water to the half-height of the specimen after 2 hours, and finally add water to the 3/4 height of

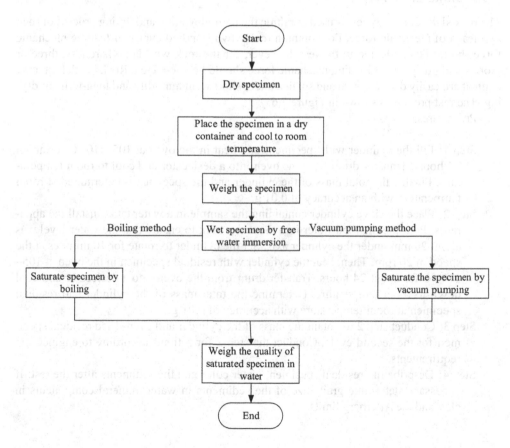

Figure 7.5 Flowchart of determining water absorption

the specimen after 4 hours. Fully immerse the specimen in the water after 6 hours. When the specimen is immersed in water for 48 hours, take out the specimen and remove water on the surface of specimen. Determine the mass of the specimen with accuracy of 0.01 g.

Step 3: In boiling method, the boiling water surface should always be higher than the height of specimen, and the boiling time should be more than 6 hours. After boiling, cool the specimen down to room temperature. Remove water on the surface of specimen and determine the mass of the specimen with accuracy of 0.01 g.

Step 4: In vacuum pumping method, keep the initial water surface in the saturated container to be higher than the height of specimen. Pump the saturated specimen about 4 hours until no bubbles are found. After the pumping is completely accomplished, the saturated specimen is allowed to stand for 4 hours at atmospheric pressure. Remove water on the surface of specimen and determine the mass of the specimen with accuracy of 0.01 g.

Step 5: Calculating rock water absorption, saturated water absorption, dry density and grain density according to the formula.

7.3.5 Slake durability

The rock slake durability test is used to mimic the natural wetting and drying process of rock by means of mechanic force. For common rocks, two standard cycles of loading mechanic force should be conducted in the test, however, for the rock with large hardness, three or more standard cycles of loading mechanic force should be conducted. Rocks with high clay content are easily disintegrated and spalled under short-term humidity and long-term air drying. The test process is shown in Figure 7.6.

Procedures:

Step 1: Fill the cylinder with specimens and heat in the oven at 105–110 °C for about 24 hours. Transfer drum from the oven into a desiccator and cool to room temperature. Finally, the total mass of the cylinder and the specimen is determined at room temperature with an accuracy of 0.01 g.

Step 2: Place the sieve cylinder containing the sample in a water tank. Install the apparatus. Pour the distilled water into the water tank to ensure that the water level was about 20 mm under the cylinder axis. Set the cylinder to rotate for 10 minutes at the speed of 20 rpm. Then, heat the cylinder with residual specimen in the oven at 105–110 °C for about 24 hours. Transfer drum from the oven into a desiccator and cool down to room temperature. Determine the total mass of the cylinder and residual specimen at room temperature with accuracy of 0.01 g.

Step 3: Conduct step 2 to obtain the mass of the cylinder and oven-dried residual specimen for the second cycle. Conduct the step 2 for 5 times according to engineering requirements.

Step 4: Describe the residual specimen, water color and the sediments after the test. If necessary, determine grain size of the sediments in water, mineral components in clay and the Atterberg limits.

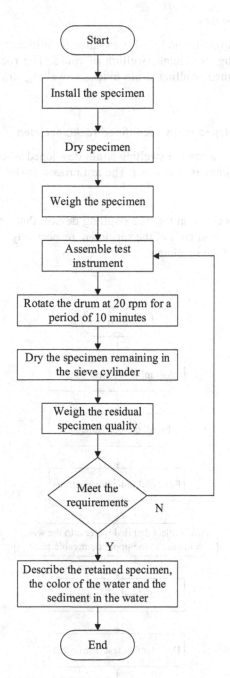

Figure 7.6 Flowchart of determining slake durability

7.4 Mechanical tests

7.4.1 Swelling properties

Rocks that contain hydrophilic and easily expandable minerals can absorb an amount of water, thereby leading to volume swelling of rocks. The rock swelling indexes are composed of an unconfined swelling strain index, a swelling strain index and a swelling pressure index.

(1) Swelling strain developed in an unconfined rock specimen

This test is intended to measure the swelling strain developed when an unconfined, undisturbed rock specimen is immersed in water. The test process is shown in Figure 7.7.
   Procedures:

Step 1: Place the specimen in the free swelling device. Put a water permeable plate is on the upper and lower ends of the specimen, respectively. Additionally, put a metal plate on the top of the specimen.

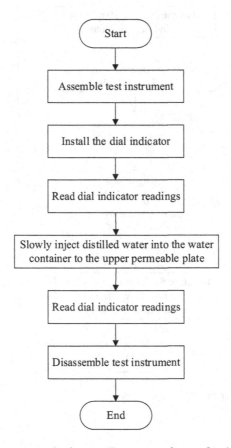

Figure 7.7 Flowchart of determining the free swelling strain of unconfined specimen

Step 2: Install the dial gauges on the center of the specimen and the symmetrical center of the four sides. Measure the axial deformation and radial deformation of the specimen. A piece of thin copper is required between the specimen and the dial gauges of four sides.

Step 3: Record the dial indicator readings every 10 minutes until three consecutive readings are very close.

Step 4: Slowly inject distilled water into the water container to the upper permeable plate and record the dial reading immediately. Immerse the sample should be immersed in water for about 48 hours. Record the dial gauge readings every 10 minutes at the first hour of the test; and then, Record the dial gauge readings every 1 hour until the difference between three consecutive readings is less than 0.001 mm, indicating that the sample swelling is stable.

If the specimen is disintegrated, cracked, dropped, softened or has surface muddy after the test, describe the specimen in detail.

Step 5: Calculate the axial and radial swelling strain of the rock according to the formula.

(2) Swelling strain index for a radially confined rock specimen with axial surcharge

This test is intended to measure the axial swelling strain developed against a constant axial pressure or surcharge, when a radially confined, undisturbed rock specimen is immersed in water. The test process is shown in Figure 7.8.

Procedures:

Step 1: Apply a thin layer of petroleum jelly to the inner wall of the metal collar and transfer the specimen of the known size into the metal collar, and then put the filter paper and the permeable plate on the upper and lower ends of the specimen, respectively.

Step 2: Put a metal load block with a pressure of 5 kPa on the top of the upper permeable plate. Install a dial gauge.

Step 3: Record the dial indicator reading every 10 minutes until three consecutive readings are very close.

Step 4: Slowly inject distilled water into the water container and record the dial indicator reading immediately. Immerse the specimen in water for less than 48 hours. Record the dial gauge reading every 10 minutes during the first hour of test; and then, record the reading every 1 hour until the difference of three consecutive readings was smaller than 0.001 mm. This indicates that the specimen swelling is stable. If the specimen has muddy surface or is softened after the test, describe the specimen in detail.

Step 5: Calculate the swelling strain index for a radially confined specimen with axial surcharge according to the formula.

(3) Swelling pressure index under conditions of zero volume change

This test is intended to measure the pressure is required to keep the original volume of the rock specimen after water immersion. The test process is shown in Figure 7.9.

Procedures:

Step 1: Apply a thin layer of petroleum jelly to the inner wall of the metal collar and install the specimen into the metal collar. Put the filter paper and metal permeable plate on the top of the specimen.

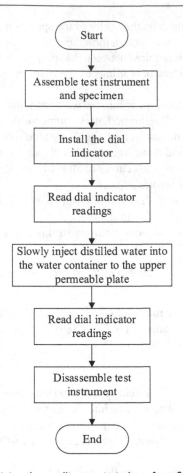

Figure 7.8 Flowchart of determining the swelling strain index of confined specimen

Step 2: The pressurization system and the dial gauge are installed. The instrument and the specimen should be installed on the same axis without eccentric load.

Step 3: Apply a load of 10 kPa to the specimen and record the dynamometer and dial gauge readings every 10 minutes until three consecutive readings are very close.

Step 4: Slowly inject pure water into the water container to the upper metal permeable plate. When the observed deformation is more than 0.001 mm, adjust the applied load to ensure that no swelling deformation or thickness change of the specimen take place. Record the dynamometer reading.

Immerse the specimen in water for less than 48 hours. Record the dial gauge reading every 10 minutes during the first hour of test; and then, record the reading every 1 hour until the difference of three consecutive readings was less than 0.001 mm. If the specimen has muddy surface or is softened after the test, describe the specimen in detail.

Step 5: Calculate the swelling pressure index under conditions of zero volume change according to the formula.

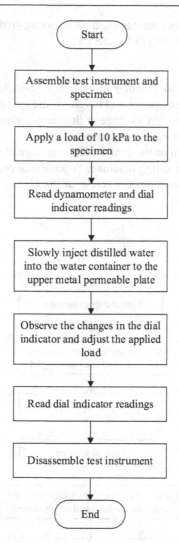

Figure 7.9 Flowchart for determining the swelling pressure

7.4.2 Uniaxial compressive strength

The uniaxial compressive strength of rock is the ultimate compressive stress when the rock sample breaks under unidirectional pressure, and is usually used for the strength grading and lithological description of the rock. This test method is applicable for rocks that can be made into regular specimens. Parameters to be measured are the dimensions of specimen and failure load. The specimens could be made into regular shape by drilling cores or rock blocks. Avoid making any cracks on the rock during the process of collection, transportation and preparation. Specimens with natural moisture state, dry state, saturated state and other moisture states are selected according to engineering requirements. In this method, the specimen is loaded by a microcomputer controlled electro-hydraulic servo pressure testing machine. The test process is shown in Figure 7.10.

Procedures:

Step 1: Turn on the microcomputer control electro-hydraulic servo pressure testing machine. Measure the diameter and height of the specimen at different positions by a Vernier caliper. Record the average of the diameter and height of the specimen and input these data to the computer system.

Step 2: Place the specimen on the center of the pressure plate of the testing machine and lower the screw of the testing machine. When the upper plate is 2–3 mm away from the top of the specimen, stop the screw of the testing machine and then manually

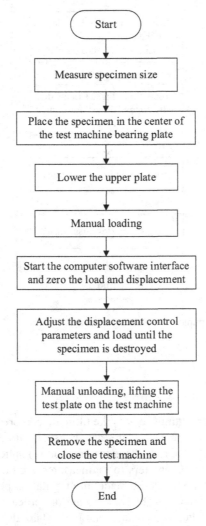

Figure 7.10 Flowchart for determining uniaxial compressive strength

apply the load to contact the two ends of the specimen snugly with the upper and lower plates of the testing machine.

Step 3: Start the computer and set the load and displacement to zero. Adjust the displacement control parameters and apply the load on the specimen until failure. Record the testing data, the phenomenon and the load until it is damaged.

Step 4: After test, describe the failure mode of the specimen.

Step 5: Calculate the uniaxial compressive strength and softening coefficient of the rock according to the data and formula.

7.4.3 Durability under freezing and thawing conditions

The ratio of the compressive strength of the rock specimen before and after freezing and thawing is defined as freeze-thaw coefficient. In this test, the rock is frozen and heated to dissolve for several times at ±25 °C. Rock freezing-thawing resistance is an index of the saturated rock to prevent its damage from freezing-thawing cycles. The test process is shown in Figure 7.11.

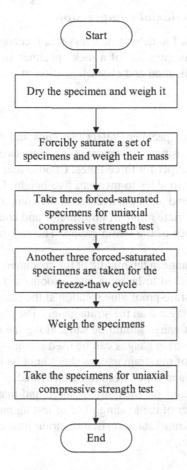

Figure 7.11 Flowchart for determining slake durability under freezing and thawing conditions

Procedures:

Step 1: Determine the mass of dried specimen with accuracy of 0.01 g. A group of specimens is saturated. The method of which can refer to rock water absorption test. Then determine the mass of saturated specimen with an accuracy of 0.01 g. Each group of tests should contain at least six samples. Three of them are used to the compressive strength test after saturation, and another three samples are used for the compressive strength test after freeze-thaw cycle.

Step 2: Place the specimen in a tin box and cool at −22 to −18 °C for 4 hours.

Remove the tin box and fill the box with water to immerse the specimen. Then, place the tin box in the test box and heat at 18–22 °C for 4 hours.

The freezing-thawing cycles can be conducted for 25, 50 or 100 times in term of the engineering environment. After each cycle, check the specimen if it has any blockage or cracks, etc. Record the test results after all freezing-thawing cycles.

Step 3: Take out the specimen, remove water on the surface of specimen, and determine its mass of specimen with accuracy of 0.01 g.

7.4.4 Deformability in uniaxial compression

This method of test is intended to determine stress-strain curves and Young's modulus and Poisson's ratio in uniaxial compression of a rock specimen of regular geometry. The test is mainly intended for classification and characterization of intact rock. The test process is shown in Figure 7.12.

Procedures:

Step 1 Take the prepared specimen, select two ends and the middle of the specimen, these three different positions to separately measure two diameters perpendicular to each other with the caliper for three times. Choose four symmetrical points and the center point around two sides to measure five height. Calculate the average of the diameter and height, and input the average in the computer.

Step 2: Select an appropriate range of multimeter and connect with the strain gauge. Measure the resistance of the strain gauge. The resistance value shall be around 120 Ω.

Step 3: Stick the strain gauge in the middle of the specimen. The place to stick the strain gauge should be polished to be even and smooth, and cleaned with cleaner fluid. Apply a layer of moisture-proof glue solution at the place, with a thickness less than 0.1 mm and an area bigger than the strain gauge. The place should also avoid cracks or patches. The strain gauge should be stuck along the radial and axial direction in relative faces. 2 or 4 strain gauges can be used along the radial and axial direction. Insulation resistance of the strain gauges should not be less than 200 MΩ. Mark the two lateral sides of A and B that need to be measured.

Step 4: Weld the conductor on each strain gauge and mark the conductor. Place the specimen on the center of the loading plate of testing machine and connect the conductor to the instrument of static resistance strain gauge. Set instrument parameters of resistance strain gauge.

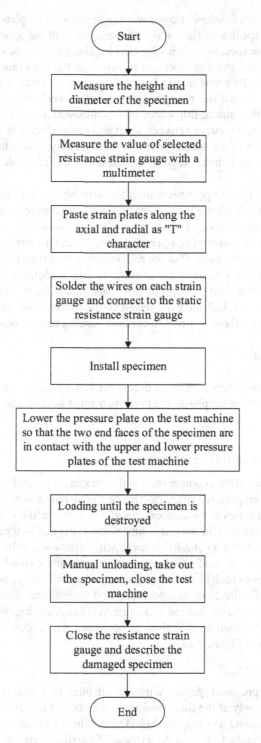

Figure 7.12 Flowchart for determining the deformability in uniaxial compression

Step 5: Start the test machine. Low down the upper loading plate until it is 2–3 mm to the top of the specimen. Then add load manually until the upper plate contact snugly to the top of the specimen, then adjust the loading plate to make load attribute evenly.

Step 6: Set the load and displacement to zero. Set the load rate of 0.002 mm/s (load rate of 0.5–1 MPa/s may be used according to the engineering requirement) Apply the load and record the phenomenon of crack, peel or split during the loading process, plot the relationship curve between load and displacement, and the deformation relationship curve between axial and radial direction. As the load increases the deformation of the specimen increases gradually. When the specimen is failure and stop the load then. Record the failure load, axial deformation and radial deformation.

Step 7: Unload. Lift the upper pressure plate, turn off the test machine and remove the conductor. Switch off the instrument of electrical resistance strain gauges. Remove the specimen and describe the destruction of specimen.

Step 8: Calculate the uniaxial compressive strength and the stress of every level according to the data in the table. Plot the relationship curve between stress and axial strain and radical strain. Calculate the average elastic modulus and Poisson's ratio. The elastic modulus should calculate to three significant digits. The Poisson's ratio shall be accurate to 0.01. Calculate the secant elastic modulus and the corresponding Poisson ratio. Repeat the step 1 to 8 to conduct parallel tests to other specimens.

It should be noted that:

(1) Protective measures should be taken during the test, such as wearing safety glasses.
(2) During the pressurization phase, the performer must keep a safe distance from the test machine.

7.4.5 Strength in triaxial compression

The strength of rock in triaxial compression is the maximum axial stress that a specimen can resist under triaxial compressive strength. This test is intended to measure the triaxial compression strength of a series of rock specimens at different confining pressure, and calculate the shear strength of the rock in triaxial compression. This test is applicable for all kinds of rocks that can be made into a cylindrical specimen. Parameters to be measured include the dimensions and the failure load of specimen. The specimens could be made into regular shape by drilling cores or rock blocks. Any cracks on the rock should not be made as possible during the process of collection, transportation and preparation. Specimens with different moisture conditions should be selected according to the engineering requirements. Under the same water content and loading direction, 5 specimens are needed in each test group. The test process is shown in Figure 7.13.

Procedures:

Step 1: Take the prepared specimen, measure diameters, which is perpendicular to each other, respectively at the three positions of ends and middle of specimen following the schematic. And five heights should be obtained from circumferential and middle points at both ends. Calculate the average of the diameters and heights, and input the average dimensions.

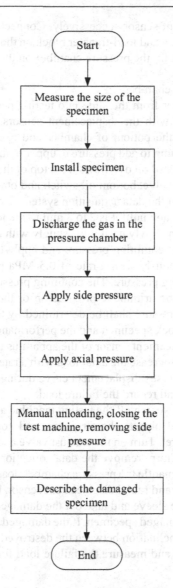

Figure 7.13 Flowchart for determining the rock strength in triaxial compression

Step 2: Assemble the specimen

Put the specimen between two platens and wind the insulating tape on the connection of the platen and the specimen. Oil proof measures shall be used. Cut 140 to 150 mm heat-shrinkable sleeve and wrap the specimen and the platens. Heat the sleeve thoroughly and repeatedly using a hot blower until no bubble exist between the sleeve and the specimen and then tighten both ends of the sleeve by iron wire.

Lift the confining pressure chamber. Insert the specimen into the confining base. Mount the confining base with specimen on base plate of the chamber. Install axial and radial displacement sensor successively. Connect the conductor of the sensor. Mount the confining cap and load-transfer block on the specimen. Place the pressure chamber carefully and fix the pressure chamber on the base plate of chamber using a clamping band.

Step 3: Exhaust air in the chamber.

Push the pressure chamber from the slide into testing position. Connect displacement data acquisition lines with the displacement sensors. Connect the oil barrel (i.e. chamber liquid) with the bottom of chamber and open the vent valve and the oil switch. Open the air pump to add pressure to upper of the oil surface. Transfer oil into the pressure chamber until oil overflows from top of the vent pipe. Stop the air pump and close the top vent valve, bottom oil switch and other valves connected to the oil line immediately. Open the data acquisition system. Turn on the testing machine and low down the upper platen until it is 2 to 3 mm to the loading piston of the chamber. Lower the upper platen slowly until it fits snugly with the loading piston.

Step 4: Open the valve of confining pressure and apply the axial load and the confining pressure simultaneously with a rate of 0.5 MPa per second until reaching the predetermined confining pressure. The confining pressure on each specimen shall be applied according to the arithmetic progression or the geometric progression. The maximum confining pressure shall be determined by the engineering requirements, the characteristics of rock specimen and the performance of triaxial testing machine. Set the load and displacement sensor of the apparatus to zero.

Step 5: Apply axial load. Increase the axial load with a rate of 0.5 to 1.0 MPa per second and observe the load versus displacement curve during loading until the specimen is failure. Stop loading and record the failure load.

Step 6: Unload manually, switch off the testing machine and unload the confining pressure. Connect the vent pipe to the air pump, and connect the bottom oil pipe of chamber to the oil barrel. Turn on the exhaust valve and air pump and start removing the oil. After the oil return, remove the data collection conductor and move out the pressure chamber. Dismantle the pressure chamber, load transfer block and confining cap. Remove the axial and radial displacement sensor. Dismantle the confining base, cut the heat-shrinkable sleeve and take out the damaged specimen.

Describe or sketch the damaged specimen. If the damaged specimen has an intact failure surface, measure the inclination between the destroyed plane and the axial direction.

Step 7: Repeat steps 1–6 and measure the failure load for other specimens at different confining pressures.

Calculate the maximum principal stress under different confining pressures according to the failure load.

Construct the Mohr's circles according to the calculated maximum axial loads and the corresponding confining pressures on the specimens.

According to the Mohr-Coulomb criterion, determine the shear strength parameters of rock, including the friction coefficient, f, and the cohesion, c, under triaxial stress.

7.4.6 Indirect tensile strength by Brazilian test

This test is intended to measure the uniaxial tensile strength of prepared rock specimens indirectly by the Brazilian test. This test is used for all kinds of rocks that can be made into regular shape. Parameters to be measured include the dimensions of specimen and the failure load. The specimens could be made into regular shape by drilling cores or rock blocks. It should be avoided to introduce any damage of samples during the process of collection, transportation and preparation. Specimens with different moisture conditions should be selected according to the engineering requirements. At least three specimens shall be tested to obtain a meaningful average value under the same water content and loading direction. The test process is shown in Figure 7.14.

Procedures:

Step 1: Take the prepared specimen and measure diameters at the three positions of ends and middle of specimen. And five heights should be obtained from circumferential and middle points at both ends. Calculate the average of the diameters and heights,

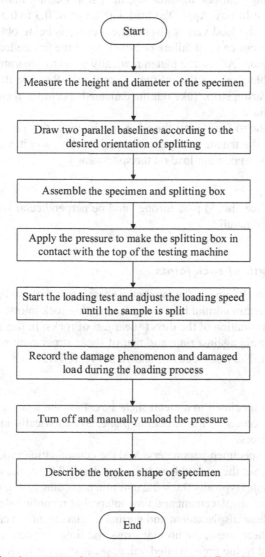

Figure 7.14 Flowchart for determining the indirect tensile strength (Brazilian test)

and input the average dimensions. According to the engineering requirements of the splitting direction, mark two parallel loading baselines passing the diametric ends of the specimen.

Step 2: Assemble split box and specimen. Put the split box at the center of the loading plate, then put the bottom bearing strip in the split box and place the specimen. Tighten the screws at both ends so as to fix the specimen. Put the top bearing strip on the specimen and make sure that both strips are fixed along the marked loading reference line on both sides of the specimen. Install the top bearing block. Ensure that bearing strips and the specimen are on the same loading axis.

Step 3: Open the universal testing machine. Add load manually to lift up the split box. When the top bearing strip is 2 to 3 mm above the upper plate of the testing machine, switch off the manual loading rotary switch. Then switch to computer control, when the load reading changes significantly, then stop loading immediately. Loosen the screws on the split box. Apply the load with a rate of 0.3 to 0.5 MPa per second. At the same time, the load versus displacement curve is being obtained and record the behavior of specimen until failure occurs. Record the failure load then.

Step 4: Stop loading. Adjust the platen manually so as to separate the platen from the load transfer block above the split box completely. Remove the load transfer block and the top bearing strip. Take out the damaged specimen, then record the type and location of failure.

Repeat the step one to four to test the other specimens in this group.

Step 5: Calculate the tensile strength of the rock to three significant digits according to dimensions and maximum load on the specimen.

It should be noted that:

The fracture surface should pass through and be perpendicular to the diameter, otherwise, the test should be invalid.

7.4.7 Shear strength of rock joints

The shear strength of rock joints is the maximum shear stress that can be resisted by the rock when subjected to a certain normal load along the existing rock joints. The flat push method is applicable for determination of the direct shear test of rocks. In this method, the relationship curve of shear stress against time and that of shear stress against horizontal displacement could be plotted. The test process is shown in Figure 7.15.

Procedures:

Step 1: Place the specimen in a metal shear box. Pass the loading shear stress through the geometric center of the structural plane. Symmetrically install the normal displacement sensors.

Step 2: Input the specimen parameters. Add the consolidation module and the standard shear module. Set the normal load to be less than 1.2 times the engineering pressure. Set the shear rate (typically 0.05–0.2 mm/min according to engineering practice), the maximum shear displacement and time interval of recording data t as well.

Step 3: Set the shear displacement and normal displacement to zero. The y1, y2, x axis is set as the shear stress, the normal stress and time, respectively. When the normal load is applied to the pre-specified value, shear stress is loaded to specimen until it

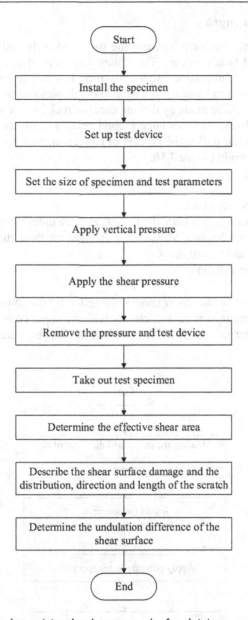

Figure 7.15 Flowchart for determining the shear strength of rock joints

features the maximum horizontal displacement. Plot the relationship curve of shear stress against time and that of shear stress against horizontal displacement.

Step 4: Dismantle the testing device and take out the specimen. Determine the effective shear surface area of the specimen and the damage degree of shear surface and describe the distribution, direction and length of the scratch. Determine the shear surface undulation to plot the relation curve of the shear surface height again the shear direction.

7.4.8 Point load strength

This test is to place the specimen between the two cone ends and apply an increasingly concentrated load until failure occurs. The failure load is used to calculate the point load strength index and the point load strength anisotropy index. This test is applicable for all kinds of rocks. Parameters to be measured include the dimensions and failure load of specimen. The specimens could be made by drilling cores or rock blocks from outcrops, exploration pits, tunnels, roadways or other underground chamber. Any cracks on the specimens should be avoided to occur during the process of collection, transportation and preparation. The test process is shown in Figure 7.16.

Procedures:

Step 1: Measure specimen size
Take prepared specimen. Measure the heights at three different positions with a Vernier caliper and record the data. Measure the diameters at three different positions with a Vernier caliper and record the data.
Step 2: specimen installation
(1)Axial Test
Place the specimen on the lower cone of the tester, lift the lower cone until the upper plate of the specimen is contact closely with the upper cone. The loading direction shall be perpendicular to the both ends of the specimen, the connection line

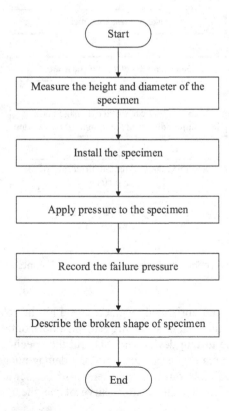

Figure 7.16 Flowchart for determining point load strength

between the upper and lower conical tip shall pass through the center of the cross section of the specimen.

(2)Radial Test

Insert a specimen in the truncated conical plates of the tester. The upper and lower cone end shall closely contact with the diameter of the core specimen. Make sure that the distance 'L', between the contact points and the nearest free end is at least half of the distance between the loading points.

(3)Block and Irregular Lump Tests:

The loading direction shall be along the direction with the smallest size of the specimen. The distance between the loading point and the width or average width of the minimum cross section through the two loading points shall be measured. The distance between the loading point and the free end of the specimen shall not be less than half of the loading point spacing.

Step 3: Steadily increase the load to make sure that failure occurs within 10 to 60 s, and record failure load. If condition permitted, measure the distance between two contact points at the moment specimen failure occurs. Describe the shape of the specimen after destruction. The fracture surface should pass through two loading point and throughout the specimen

Step 4: Calculate the value of equivalent core diameter of different specimens to three significant digits according to the formula.

7.4.9 Wave velocity by ultrasonic pulse transmission

Ultrasonic testing is intended to measure the spread time of the longitudinal and transverse acoustic waves in the specimen. Accordingly, the velocity of the ultrasonic in the rock could be calculated. The transmission of elastic waves is highly related to the compactness degree of the medium. In general, the longitudinal wave velocity in the intact rock block is faster than in the rock mass with various structural planes. The test process is shown in Figure 7.17.

Procedures:

Step 1: Apply the couplant (petroleum jelly or butter) to the surface of the transmitter and receiver. For the shear wave velocity test, the couplants should be solid materials such as aluminum foil, copper foil or salicylic acid phenolate.

Step 2: Determine the heights of the specimen at three different positions with accuracy of 1 mm. Take the average of the determined specimen heights as the distance between the two transducers. Apply couplant on the surface of the transducer.

Step 3: Place the specimen on the testing frame and the transducer on both ends of the specimen. Apply a certain amount of pressure to contact the transducer snugly with the rock mass.

Create a new project is created on the rock ultrasonic measuring device, including the project name, specimen number and transducer distance.

Step 4: Determine the ultrasonic of the longitudinal wave in the specimen is determine ten times and take the average of determined velocities. Under the same water content and the same loading direction, three specimens should be tested in each group.

The method of rock transverse wave velocity test is similar to that of longitudinal wave velocity test.

Step 5: Calculate elastic parameters of rocks according to the measured longitudinal and shear wave velocities.

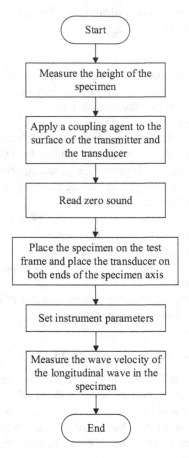

Figure 7.17 Flowchart for determining the wave velocity (ultrasonic pulse transmission)

7.4.10 Rebound hardness

Rebound distance of the elastic rod is defined as the rebound value when the rock surface is impacted by the elastic rod of the rebound apparatus which. Rebound distance reflects the surface hardness of the rock. The test process is shown in Figure 7.18.

Procedures:

Step 1: Measurement of the rock rebound hardness.

Support the surface of the specimen by the rebounding rod of the rebound apparatus, and slowly press the button of the rebound apparatus to extend rebounding main rod, as such, the rebound hammer is hung on the hook.

Hold the rebound tester and apply pressure to the specimen slowly and evenly. As the hammer is unhooked and impacted, the hammer moves the pointer backward. At this time, the rebound value on the scale is recorded. Change the measurement points and repeat the steps in Figure 7.18. Collect the test data of 16 points in each measuring area.

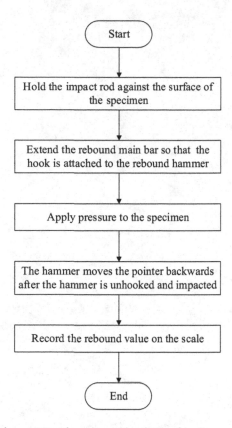

Figure 7.18 Flowchart for determining the rebound hardness of rocks

Step 2: Data processing

Among sixteen data of each measurement area, the three maximum values and three minimum values in the measurement area are discarded. The average of the remaining data could be used as the rebound value of rocks.

Chapter 8

Data analysis and documentation of geotechnical testing

This section summarizes the calculation formulas for laboratory tests of soil and rocks together with the basic requirements of each calculation into a set of tables. This section covers the most commonly used tests for soil, such as water content (Table 8.1), density (Table 8.2), specific gravity of soil solids (Table 8.3), particle size distribution (Tables 8.4 and 8.5), Atterberg limits (Table 8.8), relative density of sand (Table 8.9), California Bearing Ratio (CBR) (Table 8.10), modulus of resilience (Table 8.11), penetration (Table 8.12), consolidation (Table 8.13), collapse characteristics test of loess (Table 8.14), triaxial compression (Table 8.15), unconfined compressive strength (Table 8.16), compaction (Table 8.17), direct shear under consolidated-drained conditions (Table 8.18), reversal direct shear strength (Table 8.19), tensile strength (Table 8.20), Free swell test of clay (Table 8.21), swelling ratio (Table 8.22), swell pressure (Table 8.23), shrinkage (Table 8.24), soluble salt content (Tables 8.25–8.31) and organic matter content (Table 8.32), and those for rocks, such as water content (Table 8.33), grain density (Table 8.34), density (Tables 8.35–8.37), absorption (Table 8.38), slake durability (Table 8.39), swelling properties (Table 8.40), uniaxial compressive strength (Table 8.41), durability under freezing and thawing conditions (Table 8.42), the deformability in uniaxial compression (Table 8.43), the strength in triaxial compression (Table 8.44), the indirect tensile strength by Brazil test (Table 8.45), the shear strength of rock joints (Table 8.46), point load strength (Table 8.47), the sound velocity test by ultrasonic pulse transmission technique (Table 8.48) and rebound hardness test (Table 8.49).

8.1 Physical tests on soils

8.1.1 Water content

Table 8.1 Data analysis in test for water content

Calculation	Formula	Symbol
General formula	$\omega_0 = \left(\dfrac{m_0}{m_d} - 1 \right) \times 100$	m_d – dry specimen mass (g); m_0 – wet specimen mass (g).
Water content of layered and reticulated frozen soil	$\omega = \left[\dfrac{m_1}{m_2} (1 + 0.01\omega_h) - 1 \right] \times 100$	ω – water content (%); m_1 – frozen specimen mass (g); m_2 – pasty specimen mass (g); ω_h – water content of pasty specimen (%).

Note: The water content is expressed as a percentage and should be calculated to the nearest 0.1%. This test shall be performed in parallel and the results shall be taken as the average of the two measurements. When $\omega<40\%$, the parallel difference is allowed to be 1%; when $\omega\geq40\%$, the parallel difference is allowed to be 2%; for the layered and reticulated frozen soil, the parallel difference is allowed to be 3%. For the specimens with very small water content, the parallel difference might not comply with thee requirements. For instance, the parallel difference is allowed to be 0.3% for specimens with $\omega < 5\%$ in the road standard.

8.1.2 Density

Table 8.2 Data analysis in test for soil density

Calculation	Formula	Symbol
Wet density (cutting ring method)	$\rho_0 = \dfrac{m_0}{V}$	m_0 – mass of the specimen (g); V – volume of the cutting-ring (cm³).
Wet density (wax-sealing method)	$\rho_0 = \dfrac{m_0}{\dfrac{m_n - m_{nw}}{\rho_{wT}} - \dfrac{m_n - m_0}{\rho_n}}$	m_n – mass of specimen and wax coating (g); m_{nw} – apparent mass of specimen and wax coating when suspended in water (g); ρ_{wT} – density of pure water at T °C (g/cm³); ρ_n – density of paraffin wax (g/cm³).
Dry density	$\rho_d = \dfrac{\rho_0}{1 + 0.01\omega_0}$	ω_0 – water content of specimen.

Note: Parallel test on at least two specimens is a necessity. If the difference is not greater than 0.03 g/cm³, calculate the average of the two calculations as the dry density of the specimen. Otherwise, repeat the test.

8.1.3 Specific gravity of soil solids

Table 8.3 Data analysis in test for specific gravity of soil solids

Scope of application	Formula	Symbol
Average gravity	$G_{sm} = \dfrac{1}{\dfrac{P_1}{G_{s1}} + \dfrac{P_2}{G_{s2}}}$	G_{s1} – specific gravity of soil particles with particle size greater than or equal to 5 mm; G_{s2} – specific gravity of soil particles with a particle size less than 5 mm; P_1 – percentage of particle mass with particle size greater than, or equal to 5 mm in the total specimen mass of soil (%); P_2 – percentage of particle mass with particle size less than 5 mm in the total specimen mass of soil (%).
Pycnometer method	$G_s = \dfrac{m_d}{m_{bw} + m_d - m_{bws}} \cdot G_{iT}$	m_{bw} – mass of pycnometer and water (g); m_{bws} – mass of the pycnometer, water and specimen (g); G_{iT} – specific gravity of water or neutral liquid at T °C.

Note: In order to ensure the accuracy and reliability of the results measured by the pycnometer method, the weighing should be accurate to 0.001 g, and the calculation should be accurate to 0.01. The specific gravity of the water can be checked in the physics manual (the specific gravity of the neutral liquid should be measured).

A parallel test must be conducted. The difference of the two results shall not be greater than 0.02, otherwise repeat the test. Calculate the average of two calculations as specific gravity of solids.

8.1.4 Particle size distribution

Table 8.4 Data analysis in test for particle size distribution (sieving method)

Calculation	Formula	Symbol
The percentage of the mass of the specimen is smaller than a certain particle size in the total mass of the specimen (%)	$X = \dfrac{m_A}{m_B} \cdot d_X$	m_A – mass of the specimen smaller than a certain particle size (g); m_B – mass of the specimen taken in the fine sieve analysis; the total mass of the specimen in the coarse sieve analysis (g); d_X – percentage of mass of the specimen smaller than 2 mm in the total mass of the specimen (%).
Coefficient of non-uniformity	$C_u = {d_{60}}\big/{d_{10}}$	d_{60} – limited particle size, a certain particle size on the particle size distribution curve, and the soil content smaller than the particle size accounts for 60% of the total mass; d_{10} – effective particle size, a certain particle size on the particle size distribution curve, the soil content less than the particle size accounts for 10% of the total mass.
Coefficient of curvature	$C_c = \dfrac{d_{30}^2}{d_{10} \cdot d_{60}}$	d_{30} – a certain particle size on the particle size distribution curve, the soil content less than the particle size accounts for 30% of the total mass.

Plot the particle size distribution curve on the single logarithmic coordinate by using the mass cumulative percentages of the specimens that are smaller than a given particle sizes as ordinate and the particle sizes as abscissa (Figure 8.1).

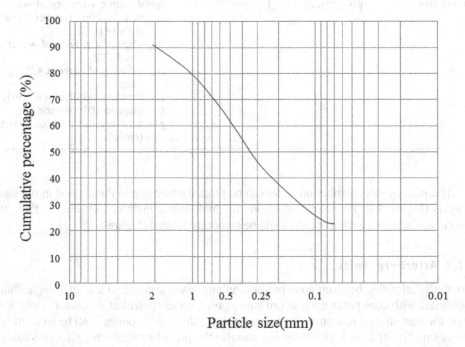

Figure 8.1 Particle size distribution (sieving method)

Table 8.5 Data analysis in test for particle size distribution (hydrometer method)

Calculation	Formula	Symbol
Mass percentage of particles smaller than the corresponding particle diameter (Hydrometer A)	$X = \dfrac{100}{m_d} C_G \left(R + m_T + n - C_D\right)$	m_d – mass of dry specimen (g); C_G – correction factor of G_s, refer to Table 8.6; m_T – correction of suspension temperature, refer to Table 8.7; $n°$ – correction of meniscus; C_D – correction of dispersant; R – reading of hydrometer A.
Mass percentage of particles smaller than the corresponding particle diameter (Hydrometer B)	$X = \dfrac{100 V_X}{m_d} C'_G \left[\left(R' - 1\right) + m'_T + n' - C'_D\right] \cdot \rho_{w20}$	C'_G – correction of G_s of Hm B, refer to the Table 8.6; m'_T – suspension temperature correction of Hm B, refer to Table 8.7; n' – correction of meniscus of Hm B; C'_D – correction of dispersant of Hm B; R' – reading of Hm B; V_X – volume of suspension ($=1000$ mL); ρ_{w20} – density of distilled water at 20 ° C ($=0.998232$ g/cm^3).
The particle diameter	$d = \sqrt{\dfrac{1800 \times 10^4 \cdot \eta}{\left(G_s - G_{wT}\right)\rho_{wT} g} \cdot \dfrac{L}{t}}$	$d°$ – particle diameter (mm); η – dynamic viscosity of water at T °C (test temperature) (kPa·s×10^{-6}); G_{wT} – Specific gravity of water at T °C; ρ_{wT} – density of water at 4 °C (g/cm^3); L – particle fall distance (cm); t – elapsed (fall) period (s); g – acceleration dues to gravity (cm/s^2).

The particle size distribution curve can be plotted according to the method in the sieve analysis (Figure 8.2). For joint-analysis via the hydrometer method and sieve method, the two curves based on each method should be smoothly jointed (Figure 8.3).

8.1.5 Atterberg limits

Plot the relationship between cone penetration and water content on a double-logarithmic coordinate with cone penetration as ordinate against water content as abscissa (Figure 8.4). Draw the best straight line fitting the plotted points. Those three points shall be on a straight line as line "A" (Figure 8.4). Otherwise, connect the point having the bigger water content with the remaining two points into two straight lines, respectively. Two water contents at

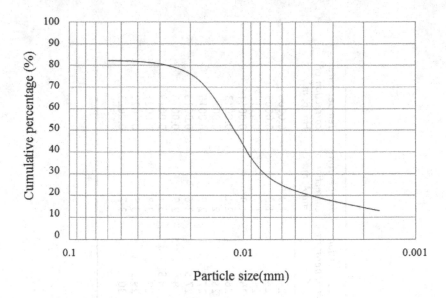

Figure 8.2 Particle size distribution (Hydrometer method)

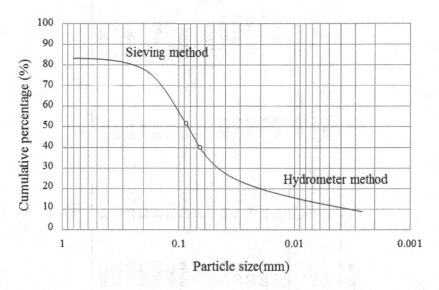

Figure 8.3 Particle size distribution (Joint analysis)

Table 8.6 Correction coefficient of specific gravity of soil

Specific gravity, G_s	Correction factor of G_s		Specific gravity, G_s	Correction factor of G_s	
	Hydrometer A, C_G	Hydrometer B, C'_G		Hydrometer A, C_G	Hydrometer B, C'_G
2.50	1.038	1.666	2.70	0.989	1.588
2.52	1.032	1.658	2.72	0.985	1.581
2.54	1.027	1.649	2.74	0.981	1.575
2.56	1.022	1.641	2.76	0.977	1.568
2.58	1.017	1.632	2.78	0.973	1.562
2.60	1.012	1.625	2.80	0.969	1.556
2.62	1.007	1.617	2.82	0.965	1.549
2.64	1.002	1.609	2.84	0.961	1.543
2.66	0.998	1.603	2.86	0.958	1.538
2.68	0.993	1.595	2.88	0.954	1.532

Table 8.7 Correction of suspension temperature

Suspension temperature (°C)	Temperature correction of Hm A, m_T	Temperature correction of Hm B, m'_T	Suspension temperature (°C)	Temperature correction of Hm A, m_T	Temperature correction of Hm B, m'_T	Suspension temperature (°C)	Temperature correction of Hm A, m_T	Temperature correction of Hm B, m'_T
10	−2	−0.0012	17	−0.8	−0.0005	24	1.3	0.0008
10.5	−1.9	−0.0012	17.5	−0.7	−0.0004	24.5	1.5	0.0009
11	−1.9	−0.0012	18	−0.5	−0.0003	25	1.7	0.001
11.5	−1.8	−0.0011	18.5	−0.4	−0.0003	25.5	1.9	0.0011
12	−1.8	−0.0011	19	−0.3	−0.0002	26	2.1	0.0013
12.5	−1.7	−0.001	19.5	−0.1	−0.0001	26.5	2.2	0.0014
13	−1.6	−0.001	20	0	0	27	2.5	0.0015
13.5	−1.5	−0.0009	20.5	0.1	0.0001	27.5	2.6	0.0016
14	−1.4	−0.0009	21	0.3	0.0002	28	2.9	0.0018
14.5	−1.3	−0.0008	21.5	0.5	0.0003	28.5	3.1	0.0019
15	−1.2	−0.0008	22	0.6	0.0004	29	3.3	0.0021
15.5	−1.1	−0.0007	22.5	0.8	0.0005	29.5	3.5	0.0022
16	−1	−0.0006	23	0.9	0.0006	30	3.7	0.0023
16.5	−0.9	−0.0006	23.5	1.1	0.0007			

2 mm penetration shall be obtained. If the difference between two smaller water contents is less than 2% then connect the midpoint of them and the point of high water content into a straight line as line B (Figure 8.4). If the difference between two smaller water contents is greater than 2%, repeat the test.

Report the water content corresponding to a cone penetration of 17 mm as the 17 mm liquid limit (Figure 8.5), a penetration of 10 mm as the 10 mm liquid limit and a penetration

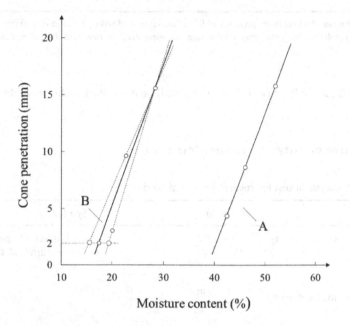

Figure 8.4 Relationship between cone penetration and water content

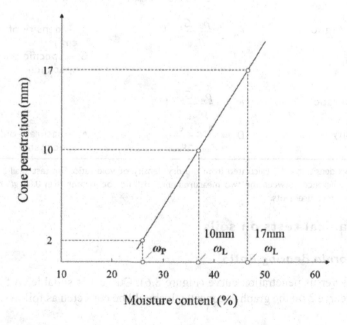

Figure 8.5 Relationship between cone penetration and water content

Table 8.8 Data analysis in test for Atterberg limits

Calculation	Formula	Symbol
Plasticity index	$I_p = \omega_L - \omega_p$	ω_L – liquid limit (%); ω_p – plastic limit (%).
Liquid index	$I_L = \dfrac{\omega_0 - \omega_p}{I_p}$	I_p – plasticity index.

Note: The liquid index shall reach its accuracy of 0.01. Parallel test of rolling method is necessary and the difference of parallel results should satisfy the requirements of water content test. Take the average of the results and record it.

of 2 mm as the plastic limit (w_p). The values shall be expressed as the percentage with accuracy of 0.1%.

8.1.6 Relative density of cohesionless soils

Table 8.9 Data analysis in test for relative density of sand

Calculation	Formula	Symbol
The minimum index density	$\rho_{dmin} = \dfrac{m_d}{V_d}$	m_d – mass of specimen (g); V_d – volume of specimen (g/cm^3).
The maximum index density	$\rho_{dmax} = \dfrac{m_d}{V_d}$	
Relative density	$D_r = \dfrac{\rho_{dmax}\left(\rho_d - \rho_{dmin}\right)}{\rho_d - \left(\rho_{dmax} - \rho_{dmin}\right)}$	ρ_d – natural dry density (g/cm^3).
Maximum void ratio	$e_{max} = \dfrac{\rho_w \cdot G_s}{\rho_{dmin}} - 1$	ρ_ω – density of water (g/cm^3); G_s – specific gravity of specimen.
Minimum void ratio	$e_{min} = \dfrac{\rho_w \cdot G_s}{\rho_{dmax}} - 1$	
Relative density	$D_r = \dfrac{e_{max} - e_0}{e_{max} - e_{min}}$	e_0 – natural void ratio of specimen.

Note: The relative density can be calculated from the dry density or void ratio. The test shall be carried out in parallel, and the difference between the two measurements shall not be greater than 0.03 g/cm^3 and take the average of the two measurements.

8.2 Mechanical tests on soils

8.2.1 California Bearing Ratio

Plot the stress versus penetration curve (Figure 8.6). Curve 1 is suitable on the graph. The beginning of curve 2 on the graph is concave. It should be corrected as follows: by changing

Table 8.10 Data analysis in CBR test

Calculation	Formula	Symbol
CBR, when the penetration value is 2.5 mm	$CBR_{2.5} = \dfrac{P}{7000} \times 100$	P – stress during penetration (kPa); 7000 – standard stress at 2.5 mm penetration (kPa).
CBR, when the penetration value is 5.0 mm	$CBR_{5.0} = \dfrac{P}{10500} \times 100$	10500 – standard stress at 5.0 mm (kPa).

Note: The bearing ratio reported for the soil is normally the one at 2.5 mm. When the ratio at 5 mm is greater, re-conduct the test. If the check test gives a similar result, use the bearing ratio at 5 mm penetration. Three parallel test groups shall be conducted and the average of the three results shall be taken. If the coefficient of variation of the three results is greater than 12%, remove the maximum deviation and the average of another two results shall be taken as the CBR.

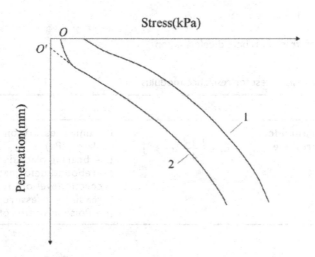

Figure 8.6 Curve of stress vs. penetration

the curvature point, all the lines intersect the ordinate at the point O', and the point O' is the corrected origin.

8.2.2 Modulus of resilience

Plot the relationship curve of the unit pressure against the rebound deformation (the unit pressure as abscissa and the rebound deformation as ordinate) in the lever pressure gauge method (Figure 8.7). The modulus of resilience of the specimen is calculated from the straight-line segment of the curve. For soft soil specimens, if the curve does not pass through the origin, the intersection of the initial straight-line segment and the ordinate is taken as the origin, and correct the rebound deformation under the pressure at each level.

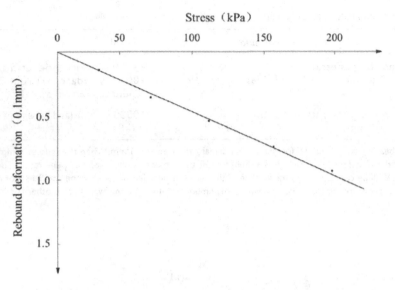

Figure 8.7 Curve of stress vs. rebound deformation

Table 8.11 Data analysis in test for resilience modulus

Calculation	Formula	Symbol
Resilience modulus under each level of pressure (kPa)	$E_e = \dfrac{p\pi D}{4l}\left(1-\mu^2\right)$	P – unit pressure on the bearing plate (kPa). D – bearing plate diameter (cm). l – rebound deformation under a certain level of pressure (pressure reading – pressure relief reading). μ – Poisson's ratio of soil, take 0.35.

8.2.3 Penetration test

Table 8.12 Data analysis in penetration test

Calculation	Formula	Symbol
Coefficient of permeability of constant head	$k_T = \dfrac{QL}{AHt}$	k_T – the coefficient of permeability of the specimen when the water temperature is T °C (cm/s); Q – the quality of water discharged (cm³); L – distance between manometers (cm); A – cross-sectional area of specimen (cm²); H – difference in head on manometers (cm); t – time (s).

Calculation	Formula	Symbol
Coefficient of permeability of variable head	$k_T = 2.3 \dfrac{aL}{A(t_2 - t_1)} \log \dfrac{H_1}{H_2}$	a – cross-sectional area of standpipe (cm²); 2.3 – conversion factor of ln and log; L – height of specimen (cm); t_1, t_2 – total time of the discharge (s); H_1, H_2 – Head before and after (cm).
Coefficient of permeability at standard temperature (20 °C)	$k_{20} = k_T \dfrac{\eta_T}{\eta_{20}}$	η_T – dynamic viscosity coefficient at T °C (kPa·s); η_{20} – dynamic viscosity coefficient at 20 °C (kPa·s).

Note: The permeability coefficient at standard temperature shall be calculated in accordance with related formulas. The permeability coefficient at the void ratio shall be obtained from the average of 3 or 4 results within the allowable deviation ($\leq 2 \times 10^{-n}$ cm/s).

8.2.4 Consolidation

Table 8.13 Data analysis in consolidation test

Calculation	Formula	Symbol
Initial void ratio of the specimen	$e_0 = \dfrac{(1+\omega_0)G_s \rho_w}{\rho_0} - 1$	e_0 – initial void ratio of the specimen.
Unit settlement of the specimen at the end of each load increment	$S_i = \dfrac{\sum \Delta h_i}{h_0} \times 10^3$	h_0 – initial height of specimen (mm); $\sum \Delta h_i$ – specimen deformation during each load increment (mm) (axial deformation minus apparatus deformation); 10^3 – unit conversion factor.
Void ratio of the specimen at the end of each load increment	$e_i = e_0 - \dfrac{1+e_0}{h_0} \Delta h_i$	Δh_i – deformation at a time interval.
Coefficient of compressibility for each load increment	$a_v = \dfrac{e_i - e_{i+1}}{p_{i+1} - p_i}$	a_v – coefficient of compressibility (MPa⁻¹); p_i – stress of each load increment (MPa).
Compressibility modulus for each load increment (MPa)	$E_s = \dfrac{1+e_0}{a_v}$	
Coefficient of volume compressibility for each load increment (MPa⁻¹)	$m_v = \dfrac{a_v}{1+e_0}$	

(Continued)

Table 8.13 (Continued)

Calculation		Formula	Symbol
Compression index or Expansion index		$C_c \text{ or } C_s = \dfrac{e_i - e_{i+1}}{\log p_{i+1} - \log p_i}$	C_c – compression index; C_s – expansion index, that is, swell-back index.
Coefficient of consolidation	Square Root of Time Method	$C_v = \dfrac{0.848\bar{h}^{-2}}{t_{90}}$	C_v – coefficient of consolidation (cm²/s); $\bar{h}$ – length of the drainage path (cm), is half the specimen height at the appropriate increment.
	Log Time Method	$C_v = \dfrac{0.197\bar{h}^2}{t_{50}}$	

Taking the void ratio as the ordinate and the pressure as the abscissa, plot the curve of void ratio versus pressure (Figure 8.8). Plot the logarithmic relationship curve between void ratio and pressure (Figure 8.9). Find the point O of the minimum radius of curvature R$_{min}$ on the e-log p curve, make the horizontal line OA over the O point, the bisector OD of the tangent line OB and the ∠AOB, and the extension line of the straight-line segment of the lower curve intersect at point E. The pressure value of the point corresponding to E is the predetermined pre-consolidation pressure p_c of the undisturbed soil specimen.

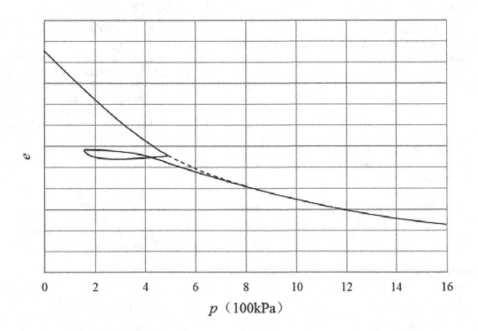

Figure 8.8 e – p curve

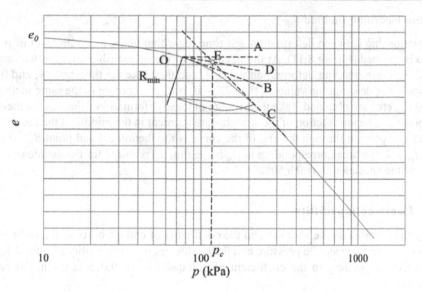

Figure 8.9 e – log p curve

(1) Square root of time method

Plot the relationship curve of deformation against time square root under a certain level of pressure (Figure 8.10). Extend the straight line at the beginning of the curve, the ordinate and d_s are the theoretical zero point, and make another line through d_s, so that the abscissa is 1.15 times of the horizontal line of the front line, the corresponding time square root of the intersection of line and the d- $\sqrt{t}$ curve is the time t_{90} required for the degree of consolidation of the specimen to reach 90%.

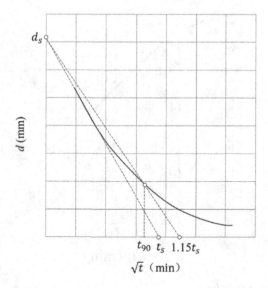

Figure 8.10 Time square root scale for t_{90}

(2) Time logarithm method

The relationship between deformation and time logarithm under a certain level of pressure should be plotted (Figure 8.11). At the beginning of the relationship curve, select any time t_1, find the corresponding deformation value d_1, and then take the time $t_2= t_1/4$, and find the corresponding deformation value d_2, then $2d_2 - d_1$ is d_{01}; According to the same method, d_{02}, d_{03} and d_{04}, etc., are obtained. Take the average of these deformation values as the theoretical zero-point d_s. The intersection of the straight-line segment in the middle of the curve and the tangent line passing through the end of the curve is the theoretical end point d_{100}, then $d_{50}=$ ($ds + d_{100}$) $/2$, the time corresponding to d_{50} is the time t_{50} required for the degree of consolidation of the specimen to reach 50%.

8.2.5 Loess collapsibility

Taking the pressure as the abscissa and the coefficient of collapsibility as the ordinate, plot the relationship between the pressure and the coefficient of collapsibility (Figure 8.12). The pressure corresponding to the coefficient of collapsibility of 0.015 is the initial collapse pressure.

8.2.6 Triaxial compression

1   UU: plot the curve of the deviator stress against axial strain. Failure point is at peak of the deviator stress on the curve (Figure 8.13). If there is no peak, failure is defined as deviator stress at 15% axial strain (Figure 8.14).

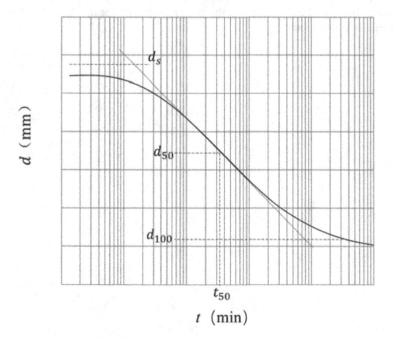

Figure 8.11 Time logarithm scale for t_{50}

Table 8.14 Data analysis in test for loess collapsibility

Parameter	Formula	Symbol
The collapsibility coefficient	$\delta_s = \dfrac{h_1 - h_2}{h_0}$	δ_s – coefficient of collapsibility; h_1 – height of specimen after deformation is stable under a certain pressure (mm); h_2 – height of specimen after the soaking collapsible deformation is stable under a certain pressure (mm).
The collapsibility coefficient under self-weight	$\delta_{zs} = \dfrac{h_z - h_z'}{h_0}$	δz_s – self-weight coefficient of collapsibility; h_z – height of specimen after deformation is stable under the pressure of saturated self-weight (mm); h_z' – height of specimen after the soaking collapsible deformation is stable under saturated self-weight pressure (mm).
The coefficient of leaching deformation	$\delta_{wt} = \dfrac{h_z - h_s}{h_0}$	δ_{wt} – leaching deformation coefficient; h_s – height of specimen caused by long-term filtration is stable after filtration under a certain pressure (mm).
The collapsibility coefficient at each load increment n	$\delta_{sp} = \dfrac{h_{pn} - h_{pw}}{h_0}$	δ_{sp} – coefficient of collapsibility at each load increment; h_{pw} – height of specimen after the soaking collapsible deformation is stable under all levels of pressure (mm); h_{pn} – height of specimen after deformation is stable under all levels of pressure (mm).

Note: At least two parallel tests are necessary. The difference of two parallel results shall not be greater than 0.03 g/cm³ and take the average of the two results.

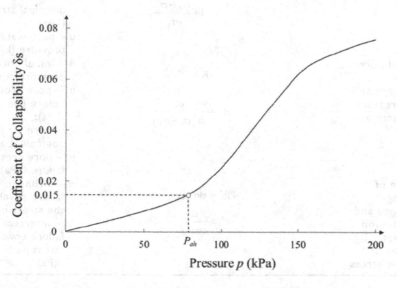

Figure 8.12 Relationship between pressure and coefficient of collapsibility

Table 8.15 Data analysis in triaxial compression test

	UU	CU	CD	Symbol
Specimen height after consolidation		$h_c = h_0\left(1-\dfrac{\Delta V}{V_0}\right)^{2/3}$		h_{0° – initial height of specimen (mm); ΔV – volume difference of specimen during a consolidation (cm³); V_0 – initial volume of specimen (mm³).
The cross-sectional area of specimen after consolidation		$A_c = A_0\left(1-\dfrac{\Delta V}{V_0}\right)^{2/3}$		A_0 – initial cross-sectional area of specimen (mm²).
Axial strain	$\varepsilon_1 = \dfrac{\Delta h}{h_0}\times 100$			ε_1 – axial strain (%); Δh – change in height of specimen during loading (mm); h_{0-} initial height of specimen (mm).
Corrected sectional area of the specimen	$A_a = \dfrac{A_0}{1-\varepsilon_1}$	$\varepsilon_1 = \dfrac{\Delta h}{h_0}$ $A_a = \dfrac{A_0}{1-\varepsilon_1}$	$A_a = \dfrac{V_c - \Delta V_i}{h_c - \Delta h_i}$	A_a – corresponding cross-sectional area (cm²); A_0 – initial area of specimen (cm²).
The deviator stress (principal stress difference)	$\sigma_1 - \sigma_3 = \dfrac{P}{A_a}\times 10$			σ_{1-} major principle stress (kPa); σ_{3-} minor principle stress (kPa); P – given applied axial load (N); 10 – unit conversion factor.
Effective principal stress ratio		$\sigma_1' = \sigma_1 - u$ $\sigma_3' = \sigma_3 - u$ $\dfrac{\sigma_1'}{\sigma_3'} = 1 + \dfrac{\sigma_1' - \sigma_3'}{\sigma_3'}$		σ_1' – effective major principal stress (kPa); σ_3' – effective minor principal stress (kPa); u – pore-water pressure (kPa).
The initial pore pressure coefficient and pore pressure coefficient at failure		$B = \dfrac{u_0}{\sigma_3'}$ $A_f = \dfrac{u_f}{B\left(\sigma_1 - \sigma_3\right)}$		B – initial pore water pressure coefficient; u_0 – pore pressure before compression (kPa); A_f – pore pressure coefficient at failure; u_f – pore pressure at failure (kPa).
The angle of shearing resistance and the cohesion intercept in terms of effective stress		$\varphi' = \sin^{-1} tg\alpha$ $c' = \dfrac{d}{\cos\varphi'}$		α – inclination of failure envelope on the stress path (°); $^\circ d$ – intercept of failure envelope on stress path curve (kPa).

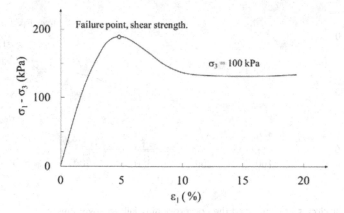

Figure 8.13 Deviator stress – axial strain curve

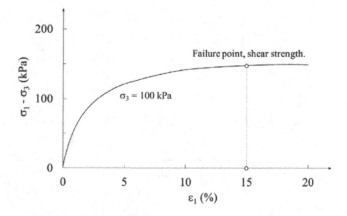

Figure 8.14 Deviator stress – axial strain curve

Take the shear stress as ordinate against axial stress as abscissa. Plot Mohr circles in terms of total stress with its center of $p\left(\dfrac{\sigma_{1f}+\sigma_{3f}}{2}\right)$ and its radius of $q\left(\dfrac{\sigma_{1f}-\sigma_{3f}}{2}\right)$ under different chamber pressure (Figure 8.15), and plot the corresponding failure envelopes. Derive the strength parameters under UU conditions.

2   CU: plot a graph of the deviator against axial strain (refer to UU conditions); plot a graph of the effective principal stress ratio against axial strain (Figure 8.16); plot a graph of the pore pressure against axial strain (Figure 8.17).

Plot the effective stress path curve (Figure 8.18) and calculate the angle (α) of shearing resistance and the cohesion intercept (d) in terms of effective stress according to the inclination and intercept of the failure envelope.

Failure point is defined by the peak of deviator stress or effective principal stress ratio. If there is no peak, failure point is defined by the dense parts of effective stress path or the value

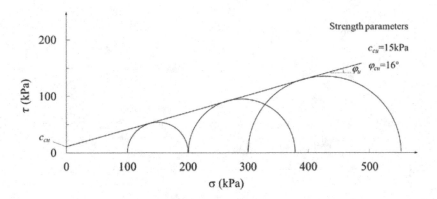

Figure 8.15 Mohr circles at failure and the corresponding failure envelopes

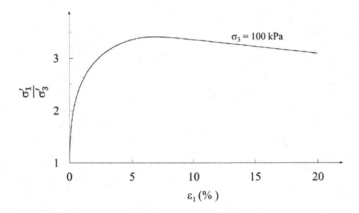

Figure 8.16 Effective stress ratio against axial strain.

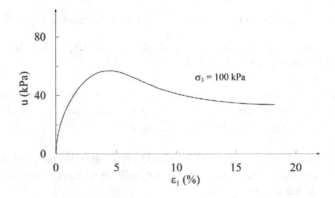

Figure 8.17 Pore pressure against axial strain

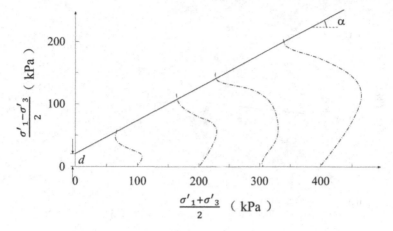

Figure 8.18 Effective stress path

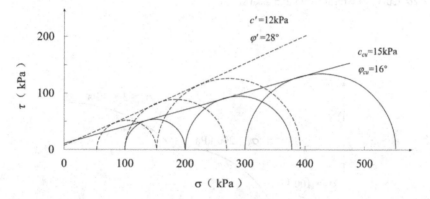

Figure 8.19 Envelope of consolidated-undrained shear strength

of deviator stress at an axial strain of 15%. Take the shear stress as ordinate against axial stress as abscissa. Plot Mohr circles in terms of total stress with its center of $p\left(\dfrac{\sigma_{1f}+\sigma_{3f}}{2}\right)$ and its radius of $q\left(\dfrac{\sigma_{1f}-\sigma_{3f}}{2}\right)$ under different chamber pressure (Figure 8.19), and plot the corresponding failure envelopes. Derive the total stress strength parameters. The angle of shear resistance and the effective cohesion shall be determined by Mohr circles in terms of effective stress with its center of p prime $\left(\dfrac{\sigma_1'+\sigma_3'}{2}\right)$ and its radius of q prime $\left(\dfrac{\sigma_1'-\sigma_3'}{2}\right)$.

3 CD: plot a graph of the deviator against axial strain (refer to UU conditions). Plot a graph of the effective principal stress ratio against axial strain (refer to CU conditions). The curve of body strain and axial strain shall be plotted (Figure 8.20). Plot the Mohr circle, and calculate the angle of shearing resistance and the cohesion intercept in terms of effective stress (refer to CU conditions).

4 Multistage triaxial compression test of soils under UU and CU conditions:

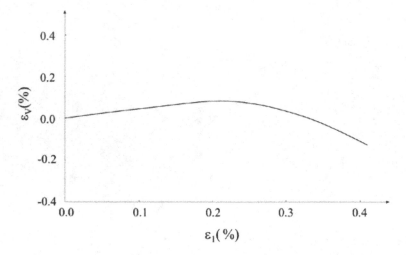

Figure 8.20 Curve of lateral strain and axial strain

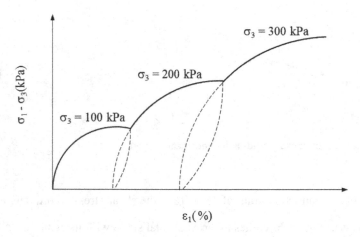

Figure 8.21 Relationship between the principal stress difference and the axial strain of unconsolidated-undrained shear

Multistage triaxial compression test of soils under UU: calculation and plotting shall be carried out in accordance with UU conditions. The axial strain of the specimen is calculated as cumulative deformation (Figure 8.21).

Multistage triaxial compression test of soils under CU: calculation and drawing shall be carried out in accordance with the CU conditions. In the calculation of the axial deformation, the stable height at the stage of chamber pressure after previous unload shall be taken as the initial height of specimen at the next stage. Calculate the axial strain under the chamber pressure at every stage.

8.2.7 Unconfined compression

Table 8.16 Data analysis in unconfined compression test

Calculation	Formula	Symbol
Axial strain	$\varepsilon_1 = \dfrac{\Delta h}{h_0}$	h_0 – initial height of specimen; Δh – change of specimen height.
The corrected specimen area	$A_a = \dfrac{A_0}{1 - \varepsilon_1}$	A_0 – initial cross-sectional area of specimen (cm²).
Axial stress of the specimen	$\sigma = \dfrac{C \cdot R}{A_a} \times 10$	C – coefficient of load ring (N/0.01 mm); R – reading of load dial (0.01 mm); σ – axial stress (kPa); 10 – unit conversion factor.
Sensitivity	$S_t = \dfrac{q_u}{q_u'}$	q_u – unconfined compressive strength of intact specimen (kPa); q_u' – unconfined compressive strength of remolded specimen (kPa).

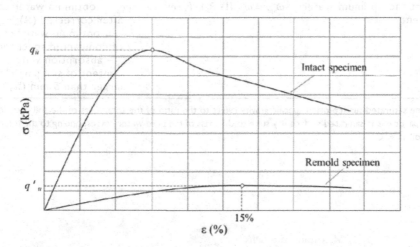

Figure 8.22 Typical relationship between axial stress and strain

Draw the axial stress-strain curve (Figure 8.22). Take the maximum axial stress on the curve as the unconfined compressive strength (q_u). If the peak value on the curve is not obvious, take the axial stress corresponding to 15% axial strain as the unconfined compressive strength $\left(q_u'\right)$.

8.2.8 Compaction

Plot the relationship curve of dry density and water content (Figure 8.23), take the corresponding vertical coordinate of the peak point on the curve as the maximum dry density and the corresponding abscissa as the optimum water content of the compacted specimen. If no peak on the curve, other test shall be rerun. Each soil specimen shall not be reused.

Plot the 100% saturation curve. Calculate values of water content for the condition of 100% saturation.

Table 8.17 Data analysis in compaction test

Calculation	Formula	Symbol
Dry density	$$\rho_d = \frac{\rho_0}{1+0.01\omega_i}$$	ω_i – water content of a specimen at a certain point (%); ρ_0 – density of wet soil (g/cm³).
Values of water content for the condition of 100% saturation	$$\omega_{sat} = \left(\frac{\rho_w}{\rho_d} - \frac{1}{G_s}\right) \times 100$$	ω_{sat} – specimen saturated water content (%); ρ_w – water density at 4 °C (g/cm³); ρ_d – dry density of specimen (g/cm³); Gs – soil particle specific gravity.
Correct the maximum dry density of specimen	$$\rho'_{d\,max} = \frac{1}{\dfrac{1-P_5}{\rho_{d\,max}} + \dfrac{P_5}{\rho_w \cdot G_{s2}}}$$	$\rho'_{d\,max}$ – maximum dry density of the specimen after correction (g/cm³); P_5 – mass percentage of soil larger than 5 mm (%); G_{s2} – saturated surface-dried specific gravity of soil particles larger than 5 mm.
Correct the optimum water content	$$\omega'_{opt} = \omega_{opt}(1-P_5) + P_5 \cdot \omega_{ab}$$	ω'_{opt} – optimum water content after corrected (%); ω_{opt} – optimum water content of compacted specimens (%); ω_{ab} – absorption water content of soil particles larger than 5 mm (%).

Note: The saturated surface-dried specific gravity refers to the ratio of the total mass of the soil particles when the soil particle is in saturated surface-dry state to the mass of the pure water corresponding to the total volume of the soil at 4 °C.

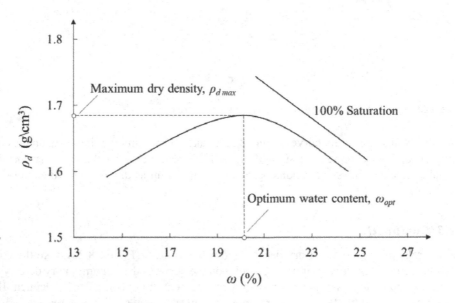

Figure 8.23 Relationship of dry density and water content

8.2.9 Consolidated-drained direct shear test

Plot the relation curve between shear stress and shear displacement (Figure 8.24). Select the peak point of the shear stress on the curve as shear strength. If there is no peak on the curve, take the shear stress corresponding to the shear displacement of 4 mm as the shear strength.

Plot the relationship curve between shear strength and normal stress (Figure 8.25) as a line of best fit through the plotted points. The slope gives the angle (φ) of shearing resistance, and the intercept gives the apparent cohesion (c).

8.2.10 Reversal direct shear test

Plot the relationship curve between shear stress and shear displacement (Figure 8.26). The first shear stress peak on the graph is the slow shear strength, and the final shear stress is the residual strength.

The curve of peak strength, residual strength and normal stress (Figure 8.27) and the residual internal friction angle (ϕ_r) and residual cohesion (C_r) should be determined by referring to the direct shear test.

Table 8.18 Data analysis in direct shear test

Calculation	Formula	Symbol
The failure time	$t_f = 50t_{50}$	t_f – total estimated elapsed time to failure (min); t_{50} – time required for the specimen to achieve 50% consolidation under the maximum normal stress increment (min).
Shear stress	$\tau = \dfrac{C \cdot R}{A_0} \times 10$	τ – shear stress (kPa); R – reading of a load ring (0.01 mm).

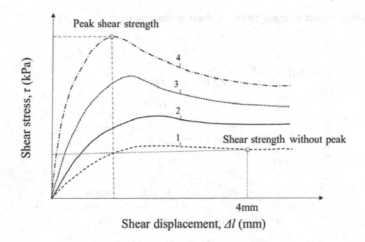

Figure 8.24 Typical curves of shear stress vs. shear displacement

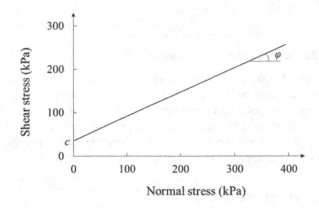

Figure 8.25 A Mohr-coulomb envelope

Table 8.19 Data analysis in reversal direct shear test

Calculation	Formula	Symbol
Shear stress	$\tau = \dfrac{C \cdot R}{A_0} \times 10$	τ – shear stress (kPa); R – reading of a load ring (0.01 mm).

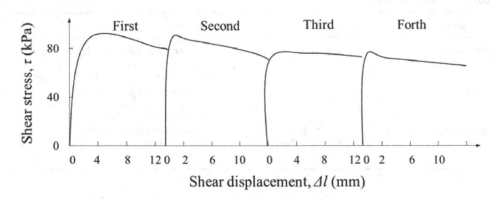

Figure 8.26 Typical curves of shear stress vs. shear displacement

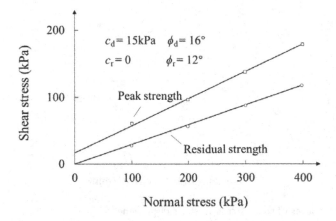

Figure 8.27 Relationship between shear strength and normal stress

8.2.11 Tensile strength

Table 8.20 Data analysis in test for tensile strength

Calculation	Formula	Symbol
Tensile strength	$\sigma_t = 4000\dfrac{P}{\pi d^2}$	σ_t – tensile strength (kPa); P – peak tensile load, failure load (N); d – diameter of failure surface (mm).

8.2.12 Free swelling test of clay

Table 8.21 Data analysis in free swelling test of clay

Calculation	Formula	Symbol
Free swell	$\delta_{ef} = \dfrac{V_{we} - V_0}{V_0} \times 100$	V_{we} – volume of the specimen after it has expanded in water (mL); V_0 – initial specimen volume, 10 mL.

8.2.13 Swelling ratio

Table 8.22 Data analysis in swelling ratio test

Calculation	Formula	Symbol
Swelling ratio under specific load	$\delta_{ep} = \dfrac{z_p + \lambda - z_0}{h_0} \times 100$	δ_{ep} – swelling ratio under a load (%); z_p – displacement meter reading after expansion and stability under a load (mm); z_0 – preloading displacement meter reading (mm); λ – compressive deformation under a load (mm); h_0 – initial height of the specimen (mm).
Swelling ratio at any time	$\delta_e = \dfrac{z_t - z_0}{h_0} \times 100$	δ_e – non-load expansion ratio at time t (%); Z_t – displacement meter reading at time t (mm).

8.2.14 Swelling pressure

Table 8.23 Data analysis in swelling pressure test

Calculation	Formula	symbol
Swell pressure	$P_e = \dfrac{W}{A} \times 10$	P_e – swell pressure (kPa); W – total equilibrium load applied to the specimen (N); A – specimen area (cm^2).

300 Test methods for soils and rocks

8.2.15 Shrinkage

Table 8.24 Data analysis in shrinkage test

Calculation	Formula	Symbol
Water content of the specimen at different times	$\omega_i = (\frac{m_i}{m_d} - 1) \times 100$	$\omega_{i\_}$ water content of the specimen at a certain moment (%); $m_{i\_}$ mass of the specimen at a certain moment (g); $m_{d\_}$ mass after drying (g).
Linear shrinkage	$\delta_{si} = \frac{z_t - z_0}{h_0} \times 100$	δ_{si} – linear shrinkage of the specimen at a certain moment (%); z_t – dial gauge reading at a certain moment (mm).
Volumetric shrinkage	$\delta_v = \frac{V_0 - V_d}{V_0} \times 100$	δ_v – volume's contraction e ratio (%); $V_{d\_}$ volume of specimen after drying (cm³).
Coefficient of shrinkage	$\lambda_n = \frac{\Delta\delta_{si}}{\Delta\omega}$	$\lambda_{n\_}$ vertical Coefficient of shrinkage; $\Delta\omega$ – the difference in water content between the two points in the first stage of the contraction curve (%); $\Delta\delta_{si\_}$ the difference between the two-point line shrinkage corresponding to $\Delta\omega$ (%).

Plot the relationship curve between the linear shrinkage and the water content rate (Figure 8.28), extend the straight-line segments of the first (I) and third stages (III) to the intersection, and the abscissa ω_s corresponding to the intersection point E is the shrinkage limit of the undisturbed soil.

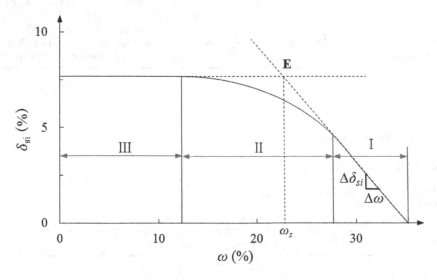

Figure 8.28 The curve between linear shrinkage and water content

8.3 Chemical tests on soils

8.3.1 Content of soluble salt

The test of the readily soluble salt mainly is to determine the total content of soluble salts (Table 8.25) as well as the contents of carbonate and bicarbonate ions (Table 8.26), chloride ion (Table 8.27), sulfate ion (Table 8.28), calcium and magnesium ions (Table 8.29, Table 8.30), sodium and potassium ions (Table 8.31).

Table 8.25 Data analysis in test for total soluble salt content

Calculation	Formula	Symbol
Total soluble salt	Without 2% sodium carbonate treatment: $$W = \dfrac{(m_2 - m_1)\dfrac{V_w}{V_s}(1+0.01\omega)}{m_s} \times 100$$ Treated with 2% sodium carbonate: $$W = \dfrac{(m - m_0)\dfrac{V_w}{V_s}(1+0.01\omega)}{m_s} \times 100$$ $m_0 = m_3 - m_1$ $m = m_4 - m_1$	W – total soluble salt (%); V_w – distilled water volume for leaching solution (mL); V_s – aspirated leaching solution volume (mL); m_s – air dried specimen mass (g); ω – air dried specimen water content (%); m_2 – evaporation dish plus drying residue mass (g); m_1 – evaporation dish mass (g). m_3 – mass of evaporation dish and sodium carbonate after evaporation (g); m_4 – mass of evaporation dish, sodium carbonate after evaporation and specimen (g); m_0 – sodium carbonate mass after evaporation (g); m – mass of sodium carbonate after evaporation and specimen (g).

Table 8.26 Data analysis in test for carbonate and bicarbonate content

Calculation	Formula	Symbol
Carbonate content	$$b\left(CO_3^{2-}\right) = \dfrac{2V_fc\left(H_2SO_4\right)\dfrac{V_w}{V_s}(1+0.01\omega)\times 1000}{m_s}$$ $CO_3^{2-} = b\left(CO_3^{2-}\right)\times 10^{-3}\times 0.060 \times 100\,(\%)$ $CO_3^{2-} = b\left(CO_3^{2-}\right)\times 60\,(mg/kg)$	$b\left(CO_3^{2-}\right)$ – mass molar concentration of carbonate (mmol/kg); CO_3^{2-} – carbonate content (% or mg/kg); V_1 – the amount of the standard solution of sulfuric acid titrated with phenolphthalein as an indicator (mL); V_s – aspirated volume of the specimen leaching solution (mL); 10^{-3}-Conversion factor; 0.060 – molar mass of bicarbonate (kg/mol); 60 – molar mass of bicarbonate (g/mol).

(Continued)

Table 8.26 (Continued)

Calculation	Formula	Symbol
Bicarbonate content	$$b\left(HCO_3^-\right)=\dfrac{2\left(V_2-V_1\right)c\left(H_2SO_4\right)\dfrac{V_w}{V_s}\left(1+0.01\omega\right)\times1000}{m_s}$$ $$HCO_3^-=b\left(HCO_3^{2-}\right)\times10^{-3}\times0.061\times100\,(\%)$$ $$HCO_3^-=b\left(HCO_3^-\right)\times61\,(mg/kg)$$	$b\left(HCO_3^-\right)$ – mass molar concentration of bicarbonate (mmol/kg); HCO_3^- – bicarbonate content (% or mg/kg soil); 10^{-3} – conversion factor; V_2 – the amount of the standard solution of sulfuric acid titrated with methyl orange as an indicator (mL); 0.061 – molar mass of bicarbonate (kg/mol); 61 – molar mass of bicarbonate (g/mol).

Note: The calculation is accurate to 0.01 mmol/kg and 0.001% or 1 mg/kg. Parallel titration error is no more than 0.1 mL and take the arithmetic mean.

Table 8.27 Data analysis in test for chloride content

Calculation	Formula	Symbol
Chlorine content	$$b\left(Cl^-\right)=\dfrac{\left(V_1-V_2\right)c\left(AgNO_3\right)\dfrac{V_w}{V_s}\left(1+0.01\omega\right)\times1000}{m_s}$$ $$Cl^-=b\left(Cl^-\right)\times10^{-3}\times0.0355\times100\,(\%)$$ $$Cl^-=b\left(Cl^-\right)\times35.5\,(mg/kg)$$	$b\,(Cl^-)$ – mass molar concentration of chloride (mmol/L); Cl^- – chlorine content (%, mg/kg); V_1 – leaching solution consumption volume of silver nitrate standard solution (mL); V_2 – pure water (blank) consumption of silver nitrate standard solution volume (mL); 0.0355 – molar mass of chloride (kg/mol).

Note: The calculation is accurate to 0.01 mmol/kg and 0.001% or 1 mg/kg. Parallel titration error is no more than 0.1 mL and take the arithmetic mean.

Table 8.28 Data analysis in test for sulfate content (EDTA Method)

Calculation	Formula	Symbol
Sulfate content	$$b\left(SO_4^{2-}\right)=\dfrac{\left(V_3+V_2-V_1\right)c\left(EDTA\right)\dfrac{V_w}{V_s}\left(1+0.01\omega\right)\times1000}{m_s}$$ $$SO_4^{2-}=b\left(SO_4^{2-}\right)\times10^{-3}\times0.096\times100\,(\%)$$ $$SO_4^{2-}=b\left(SO_4^{2-}\right)\times96\,(mg/kg)$$	$b\left(SO_4^{2-}\right)$ – mass molar concentration of sulfate (mmol/kg); SO_4^{2-} – sulfate content (% or mg/kg); V_1 – the amount of calcium magnesium and strontium magnesium mixture for EDTA standard solution in leaching solution (mL); V_2 – the amount of same volume of bismuth magnesium mixture (blank) for EDTA standard solution (mL); V_3 – the amount of calcium magnesium for EDTA standard solution in same volume of leaching solution (mL); 0.096 – molar mass of sulfate (kg/mol); c (EDTA) –concentration of EDTA standard solution (mol/L).

Note: The calculation is accurate to 0.01 mmol/kg and 0.001% or 1 mg/kg. Parallel titration error is no more than 0.1 mL and take the arithmetic mean.

Table 8.29 Data analysis in test for calcium ion content

Calculation	Formula	Symbol
Calcium ion content	$$b\left(Ca^{2+}\right)=\dfrac{V\left(EDTA\right)c\left(EDTA\right)\dfrac{V_w}{V_s}\left(1+0.01\omega\right)\times1000}{m_s}$$ $$Ca^{2+}=b\left(Ca^{2+}\right)\times10^{-3}\times0.040\times100\,(\%)$$ $$Ca^{2+}=b\left(Ca^{2+}\right)\times40\,(mg/kg)$$	$b\left(Ca^{2+}\right)$ – mass molar concentration of calcium ions (mmol/kg) or Ca^{2+} – calcium ion content (% or mg/kg); c (EDTA) – EDTA standard solution concentration (mol/L); V (EDTA) – EDTA standard solution dosage (mL); 0.04 – molar concentration of calcium ions (kg/mol).

Note: The calculation is accurate to 0.01 mmol/kg and 0.001% or 1 mg/kg. Parallel titration error is no more than 0.1 mL and take the arithmetic mean.

Table 8.30 Data analysis in test for magnesium ion content

Calculation	Formula	Symbol
Magnesium ion content	$$b\left(Mg^{2+}\right)=\dfrac{\left(V_2-V_1\right)c\left(EDTA\right)\dfrac{V_w}{V_s}\left(1+0.01\omega\right)\times1000}{m_s}$$ $$Mg^{2+}=b\left(Mg^{2+}\right)\times10^{-3}\times0.024\times100\,(\%)$$ $$Mg^{2+}=b\left(Mg^{2+}\right)\times24\,(mg/kg)$$	b (Mg^{2+}) – mass molar concentration of magnesium ions (mmol/kg) or Mg^{2+} – magnesium ion content (% or mg/kg); V_2 – the amount of calcium and magnesium ions to EDTA standard solution (mL); V_1 – the amount of calcium ions to EDTA standard solution (mL); c (EDTA) – EDTA standard solution concentration (mol/L); 0.024 – molar mass of magnesium ion (kg/mol).

Note: The calculation is accurate to 0.01 mmol/kg and 0.001% or 1 mg/kg. Parallel titration error is no more than 0.1 mL and take the arithmetic mean.

Table 8.31 Data analysis in test for sodium and potassium ion content

Calculation	Formula	Symbol
Sodium ion content	$$Na^+=\dfrac{\rho\left(Na^+\right)V_c\dfrac{V_w}{V_s}\left(1+0.01\omega\right)\times1000}{m_s\times10^3}\,(\%)$$ $$Na^+=b\left(Na^+\%\right)\times10^6\,(mg/kg)$$ $$b\left(Na^+\right)=b\left(Na^+\%\,/\,0.023\right)\times1000$$ $$K^+=\dfrac{\rho\left(K^+\right)V_c\dfrac{V_w}{V_s}\left(1+0.01\omega\right)\times1000}{m_s\times10^3}\,(\%)$$	Na^+, K^+ – contents of sodium and potassium in the specimen (% or mg/kg); b (Na^+), b (K^+) – molar concentrations of sodium and potassium in the specimen (mmol/kg); 0.023, 0.039 – molar masses of Na^+ and K^+ (kg/mol).
Potassium ion content	$$K^+=b\left(K^+\%\right)\times10^6\,(mg/kg)$$ $$b\left(K^+\right)=b\left(K^+\%\,/\,0.039\right)\times1000$$	

8.3.2 Content of organic matter

Table 8.32 Data analysis in test for organic matter content

Calculation	Formula	Symbol
Organic matter	$$O_m = \frac{c\left(Fe^{2+}\right)\left\{V'\left(Fe^{2+}\right)-V\left(Fe^{2+}\right)\right\}}{m_s}$$ $$\times \frac{0.003\times1.724\times\left(1+0.01\omega\right)\times100}{m_s}$$	O_m – organic matter (%); $c(Fe^{2+})$ – ferrous sulfate standard solution concentration (mol/L); $V'(Fe^{2+})$ – blank titration of ferrous sulfate (mL); $V(Fe^{2+})$ – the amount of ferrous sulfate for specimen measurement (mL); 0.003–1/4 Molar mass of ferrous sulfate standard solution concentration (kg/mol); 1.724 – the factor of organic carbon converted into organic matter.

Note: The calculation is accurate to 0.01%.

8.4 Physical tests on rocks

8.4.1 Water content

Table 8.33 Data analysis in test for water content of rock

Calculation	Formula	Symbol
Rock water content	$$\omega = \frac{m_1 - m_2}{m_2 - m_0}$$	ω – rock water content (%); m_0 – drying mass of the container (g); m_1 – total mass of specimen and container before drying (g); m_2 – total mass of specimen and container after drying (g).

Note: The calculation is accurate to 0.01%.

8.4.2 Grain density

Table 8.34 Data analysis in test for density of rock grains

Calculation	Formula	Symbol
Rock particle density	$$\rho_s = \frac{m_s}{m_1 + m_s - m_2} \cdot \rho_{WT}$$	m_s – dry rock mass (g); m_1 – the total mass of the bottle and solution (g); m_2 – the total mass of the bottle, solution and dry rock (g); ρ_{WT} – solution density at the same temperature as the test temperature (g/cm^3).

Note: The calculation is accurate to 0.01%.

8.4.3 Density

Table 8.35 Data analysis in density test (Direct measurement method)

Calculation	Formula	Symbol
Dry density of rock	$\rho_d = \dfrac{m_s}{AH}$	ρ_d – dry density of rock (g/cm³); m_s – dry specimen mass (g); A – cross-sectional area of specimen (cm²); H – the height of specimen (cm).

Note: The calculation is accurate to 0.01%.

Table 8.36 Data analysis in density test (Water displacement method)

Calculation	Formula	Symbol
Rock mass dry density	$\rho_d = \dfrac{m_s}{\dfrac{m_1 - m_2}{\rho_w} - \dfrac{m_1 - m_s}{\rho_p}}$	m – wet specimen mass (g); m_1 – wax-sealing specimen mass (g); m_2 – wax-sealing specimen mass in water (g);
Rock block wet density	$\rho = \dfrac{m}{\dfrac{m_1 - m_2}{\rho_w} - \dfrac{m_1 - m}{\rho_p}}$	ρ_w – water density (g/cm³); ρ_p – wax density (g/cm³); ω – rock water content (%).

Note: The calculation is accurate to 0.01%.

Table 8.37 Conversion between wet and dry densities of rock

Calculation	Formula	Symbol
Rock mass wet density converted to dry density	$\rho_d = \dfrac{\rho}{1 + 0.01\omega}$	ω – rock water content (%).

Note: The calculation is accurate to 0.01%.

8.4.4 Water absorption

Table 8.38 Data analysis in test for water absorption

Calculation	Formula	Symbol
Absorption rate	$\omega_a = \dfrac{m_0 - m_s}{m_s} \times 100$	m_0 – mass of the specimen after water immersion for 48 hours (g);
Saturated water absorption rate	$\omega_{sa} = \dfrac{m_p - m_s}{m_s} \times 100$	m_s – mass of dried specimen (g); m_p – mass of the saturated specimen in air (g);
Block dry density	$\rho_d = \dfrac{m_s}{m_p - m_w} \rho_w$	m_w – mass of the saturated specimen in water (g).
Particle density	$\rho_s = \dfrac{m_s}{m_s - m_w} \rho_w$	

Note: The calculation is accurate to 0.01%.

8.4.5 Slake durability

Table 8.39 Data analysis in test for slake-durability of rock

Calculation	Formula	Symbol
The slake durability index after second cycle	$I_{d2} = \dfrac{m_r}{m_s} \times 100$	I_{d2} – slake-durability index after 2nd cycle (%); m_s – dry mass of initial specimen (g); m_r – mass of specimen after 2nd cycle (g).

Note: The calculated value should take 3 significant digits

8.5 Mechanical tests on rocks

8.5.1 Swelling properties

Table 8.40 Data analysis in test for rock swelling property

Calculation	Formula	Symbol
Axial Free swell	$V_H = \dfrac{\Delta H}{H} \times 100$	V_H – axial Free swell (%); V_D – radial Free swell (%); ΔH – axial deformation value of specimen (mm);
Radial Free swell	$V_D = \dfrac{\Delta D}{D} \times 100$	H – specimen height (mm); ΔD – radial average deformation value of specimen (mm); D – specimen diameter or side length (mm).
Lateral constraint swelling ratio	$V_{HP} = \dfrac{\Delta H_1}{H} \times 100$	V_{HP} – lateral constraint swelling ratio (%); ΔH_1 – axial deformation value of laterally constrained specimen (mm); H – specimen height (mm).
Swelling pressure under conditions of zero volume change	$p_e = \dfrac{F}{A}$	F – axial load (N); A – sectional area of the specimen (mm²).

Note: The calculation is accurate to 0.01%.

8.5.2 Uniaxial compressive strength

Table 8.41 Data analysis in test for uniaxial compressive strength

Calculation	Formula	Symbol
Uniaxial compressive strength	$R = \dfrac{P}{A}$	R – uniaxial compressive strength of rock (MPa); P – failure load (N); A – cross-sectional area of specimen (mm²).
Softening coefficient	$\eta = \dfrac{R_w}{R_d}$	η – softening coefficient; R_w – average uniaxial compressive strength of saturated specimen (MPa); R_d – average uniaxial compressive strength of dried specimen (MPa).

Note: The calculated value of rock uniaxial compressive strength should take 3 significant digits, and the calculated value of rock softening coefficient should be accurate to 0.01.

8.5.3 Durability under freezing and thawing conditions

Table 8.42 Data analysis in test for durability under freezing and thawing conditions

Calculation	Formula	Symbol
Freeze-thaw mass loss rate	$M = \dfrac{m_p - m_{fm}}{m_s} \times 100$	m_p – saturated specimen mass before freezing and thawing (g); m_{fm} – saturated specimen mass after freezing and thawing (g); m_s – drying specimen mass before test (g);
Freeze-thaw coefficient	$K_{fm} = \dfrac{\overline{R}_{fm}}{\overline{R}_W}$	$\overline{R}_{fm}$ – average of uniaxial compressive strength of rock after freezing and thawing (MPa); $\overline{R}_W$ – average of rock saturated uniaxial compressive strength (MPa).

Note: The calculated value of rock freezing and thawing mass loss rate should retain 3 significant digits, and the calculated rock freezing and thawing coefficient should be accurate to 0.01.

8.5.4 Deformability in uniaxial compression

Table 8.43 Data analysis in uniaxial compression test for deformability

Calculation	Formula	Symbol
The average elastic modulus	$E_{av} = \dfrac{\sigma_b - \sigma_a}{\varepsilon_{1b} - \varepsilon_{1a}}$	σ_a – initial stress of linear portion of axial stress-strain curve (MPa); σ_b – finial stress of linear portion of axial stress-strain curve (MPa);
Average Poisson's ratio	$\mu_{av} = \dfrac{\varepsilon_{db} - \varepsilon_{da}}{\varepsilon_{1b} - \varepsilon_{1a}}$	ε_{la} – radical strain at σ_a; ε_{lb} – radical strain at σ_b; ε_{da} – radical strain at σ_a; ε_{db} – radical strain at σ_b.
The secant elastic modulus	$E_{50} = \dfrac{\sigma_{50}}{\varepsilon_{150}}$	σ_{50} – 50% of uniaxial compression strength (MPa);
The corresponding Poisson ratio	$\mu_{50} = \dfrac{\varepsilon_{d50}}{\varepsilon_{150}}$	ε_{d50} – axial strain when the axial stress is σ_{50}; ε_{l50} – radial strain when the axial stress is σ_{50}.

Note: The calculated value of the rock elastic modulus should take 3 significant digits, and the rock Poisson's ratio should be accurate to 0.01.

8.5.5 Strength in triaxial compression

Table 8.44 Data analysis in test for strength in triaxial compression

Calculation	Formula	Symbol
Maximum principal stress σ_1 under different confining pressures (σ_3)	$\sigma_1 = \dfrac{P}{A}$	P – axial failure load under different confining pressures (N); A – cross-sectional area of specimen (mm^2).
Shear strength parameter	$\sigma_1 = F_{3} + R$	F – slope of the relationship curve σ_1-σ_3; R – the intercept of the relation curve σ_1-σ_3 on the σ_1 axis is equivalent to the uniaxial compressive strength (MPa).

Calculation	Formula	Symbol
Mohr-Coulomb strength criterion parameter	$f = \dfrac{F-1}{2\sqrt{F}}$ $c = \dfrac{R}{2\sqrt{F}}$	f – coefficient of friction; c – cohesion (MPa).

According to the Mohr-Coulomb criterion, determine the shear strength parameters of rock, including the friction coefficient, f, and the cohesion, c, under triaxial stress (Figure 8.29).

The shear strength parameters can also be determined by other methods. Plot individual data points on the coordinate with major principal stress σ_1 and the confining pressure σ_3 (Figure 8.30), establish linear equation and calculate the shear stress parameters by using parameter F and R.

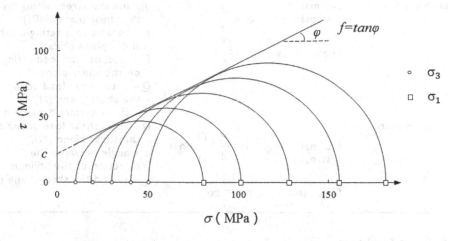

Figure 8.29 Mohr-Coulomb criterion

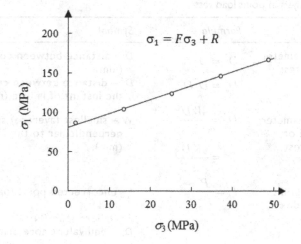

Figure 8.30 $\sigma_1 - \sigma_3$ curve

8.5.6 Indirect tensile strength by Brazilian test

Table 8.45 Data analysis in Brazilian test for indirect tensile strength

Calculation	Formula	Symbol
Rock splitting tensile strength	$\sigma_t = \dfrac{2P}{\pi Dh}$	σ_t – splitting tensile strength (MPa); P – failure load (N); D – specimen diameter (mm); h – specimen thickness (mm).

Note: The calculated value should take 3 significant digits.

8.5.7 Shear strength of rock joints

Table 8.46 Data analysis in test for shear strength of rock joints

Experiment method	Calculation	Formula	Symbol
Flat push method	Normal stress	$\sigma = \dfrac{P}{A}$	σ – normal stress acting on the shear plane (MPa); τ – shear stress acting on the shear plane (MPa); P – total normal load acting on the shear plane (N); Q – total shear load acting on the shear plane (N); A – shear surface area (mm²).
	Shear stress	$\tau = \dfrac{Q}{A}$	
Skew push method	Normal stress	$\sigma = \dfrac{P}{A} + \dfrac{Q}{A}\sin\alpha$	Q – total shear load acting on the shear plane (N); α – angle between the direction of the oblique load and the shear plane (°).
	Shear stress	$\tau = \dfrac{Q}{A}\cos\alpha$	

8.5.8 Point load test

Table 8.47 Data analysis in point load test

Calculation	Formula	Symbol
Equivalent core diameter using diametral test	$D_e^2 = D^2$ $D_e^2 = D^2 D'$	D – distance between contact points (mm); D' – distance between contact points at the instant of failure (mm).
Equivalent core diameter for axial, square or irregular lump test	$D_e^2 = \dfrac{4WD'}{\pi}$ $D_e^2 = \dfrac{4WD}{\pi}$	W – smallest (average) specimen width perpendicular to the loading direction (mm).
The uncorrected point load strength index	$I_s = \dfrac{P}{D_e^2}$	I_s – uncorrected point load strength index (MPa); P – failure load (N); D_e – equivalent core diameter (mm).

Calculation	Formula	Symbol
Corrected point load strength index	$D_e \neq 50$ mm (Data sufficient): $$I_{s(50)} = \frac{P_{50}}{2500}$$ $D_e \neq 50$ mm (Data insufficient): $$F = \left(\frac{D_e}{50}\right)^m$$ $$I_{s(50)} = FI_s$$	$I_{s(50)}$ – corrected point load strength index (MPa); P_{50} – corrected failure load refers to Figure 8.31 (MPa). F – size correction factor; m – size correction index (0.4–0.45) according to experience of similar rocks.
Point load strength anisotropy index	$$I_{a(50)} = \frac{I'_{s(50)}}{I''_{s(50)}}$$	$I'_{s(50)}$ – perpendicular to plane of weakness (MPa); $I''_{s(50)}$ – parallel to plane of weakness (MPa).

Note: When De $\neq 50$ mm, the calculated value should be corrected.

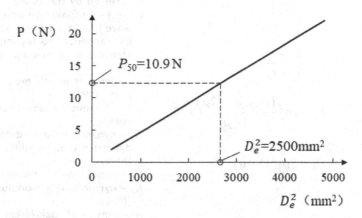

Figure 8.31 $P - D_e^2$ curve

8.5.9 Wave velocity by ultrasonic pulse transmission

Table 8.48 Data analysis in test for wave velocity

Calculation	Formula	Symbol
The longitudinal frequency of the transmitted transducer	$$f \geq \frac{2V_p}{D}$$	f – transducer transmitter frequency (Hz); V_p – rock longitudinal wave velocity (m/s); D – diameter of the specimen (mm).

(Continued)

Table 8.48 (Continued)

Calculation	Formula	Symbol
P and S-wave velocity of rock	$$v_p = \dfrac{L}{t_p - t_0}$$ $$v_s = \dfrac{L}{t_s - t_0}$$ $$v_p = \dfrac{L_2 - L_1}{t_{p2} - t_{p1}}$$ $$v_s = \dfrac{L_2 - L_1}{t_{s2} - t_{s1}}$$	v_s – shear wave velocity (m/s); L – distance between the centers of the transmitting and receiving transducers (m); t_p – propagation time of P-wave by direct transmission method (s); t_s – propagation time of S-wave by direct transmission method (s); t_0 – zero delay of instrument system (s); L_1 (L_2) – distance between transmitter launched by translucent method and two centers of the first (second) receiving transducer (mm); t_{p1} (t_{s1}) – propagation time of P (S) wave from transducer emitted by transducer to first transducer received by transducer (s); t_{p2} (t_{s2}) – propagation time of P (S) wave from transducer transmitted by transducer to second transducer received by transducer (s).
Various Kinetic and Elastic Parameters of Rock	$$E_d = \rho v_p^2 \dfrac{(1-\mu)(1-2\mu)}{1-\mu} \times 10^{-3}$$ $$E_d = 2\rho v_s^2 (1+\mu) \times 10^{-3}$$ $$\mu_d = \dfrac{\left(\dfrac{v_p}{v_s}\right)^2 - 2}{2\left[\left(\dfrac{v_p}{v_s}\right)^2 - 1\right]}$$ $$G_d = \rho v_s^2 \times 10^{-3}$$ $$\lambda_d = \rho\left(v_p^2 - 2v_s^2\right) \times 10^{-3}$$ $$K_d = \rho\dfrac{3v_p^2 - 4v_s^2}{3} \times 10^{-3}$$	E_d – dynamic elastic modulus of rock (MPa); μd – dynamic Poisson's ratio of rocks; G_d – dynamic rigid modulus or dynamics shear modulus of rocks (MPa); λ_d – rock Lame coefficient (MPa); K_d – dynamic bulk modulus of rock (MPa); ρ – density of rock (g/cm³).

Note: The calculated value should take 3 significant digits.

8.5.10 Rebound hardness

Table 8.49 Data analysis in test for rebound hardness

Calculation	Formula	Symbol
Rebound kinetic energy detection	$E = \frac{1}{2}CL^2$	E – standard energy of the Schmidt hammer (J); C – stiffness coefficient of the spring (N/m); L – stretch length of the spring (m).
Average rebound value	$\overline{N}_{test} = \sum_{i=1}^{10} N_{i\,test}/10$	$\overline{N}_{test}$ – average rebound value; $N_{i\,test}$ – rebound value of point i.

Correction for test results: when the angle between the axis of the rebounder and the horizontal direction is α (Figure 8.32), first calculate the average rebound value $\overline{N}_{test}$, and then find the correction value ΔN according to Table 8.50. Then the rebound hardness N is: $N = \overline{N}_{test} + \Delta N$.

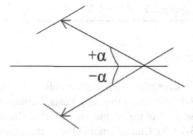

Figure 8.32 The direction of α

Table 8.50 Correction of rebound value when the angle between the axis of the rebounder and the horizontal direction is α

$\overline{N}_{test}$	α (°)							
	+90°	+60°	+45°	+30°	−30°	−45°	−60°	−90°
20	−6.0	−5.0	−4.0	−3.0	+2.5	+3	+3.5	+4
21	−5.9	−4.9	−4.0	−3.0	+2.5	+3	+3.5	+4
22	−5.8	−4.8	−3.9	−2.9	+2.4	+2.9	+3.4	+3.9
23	−5.7	−4.7	−3.9	−2.9	+2.4	+2.9	+3.4	+3.9
24	−5.6	−4.6	−3.8	−2.8	+2.3	+2.8	+3.3	+3.8
25	−5.5	−4.5	−3.8	−2.8	+2.3	+2.8	+3.3	+3.8
26	−5.4	−4.4	−3.7	−2.7	+2.2	+2.7	+3.2	+3.7
27	−5.3	−4.3	−3.7	−2.7	+2.2	+2.7	+3.2	+3.7
28	−5.2	−4.2	−3.6	−2.6	+2.1	+2.6	+3.1	+3.6
29	−5.1	−4.1	−3.6	−2.6	+2.1	+2.6	+3.1	+3.6
30	−5.0	−4.0	−3.5	−2.5	+2	+2.5	+3	+3.5
31	−4.9	−4.0	−3.5	−2.5	+2	+2.5	+3	+3.5
32	−4.8	−3.9	−3.4	−2.4	+1.9	+2.4	+2.9	+3.4
33	−4.7	−3.9	−3.4	−2.4	+1.9	+2.4	+2.9	+3.4

(Continued)

Table 8.50 (Continued)

$\overline{N}_{test}$	α (°)							
	+90°	+60°	+45°	+30°	−30°	−45°	−60°	−90°
34	−4.6	−3.8	−3.3	−2.3	+1.8	+2.3	+2.8	+3.3
35	−4.5	−3.8	−3.3	−2.3	+1.8	+2.3	+2.8	+3.3
36	−4.4	−3.7	−3.2	−2.2	+1.7	+2.2	+2.7	+3.2
37	−4.3	−3.7	−3.2	−2.2	+1.7	+2.2	+2.7	+3.2
38	−4.2	−3.6	−3.1	−2.1	+1.6	+2.1	+2.6	+3.1
39	−4.1	−3.6	−3.1	−2.1	+1.6	+2.1	+2.6	+3.1
40	−4.0	−3.5	−3.0	−2.0	+1.5	+2	+2.5	+3
41	−4.0	−3.5	−3.0	−2.0	+1.5	+2	+2.5	+3
42	−3.9	−3.4	−2.9	−1.9	+1.4	+1.9	+2.4	+2.9
43	−3.9	−3.4	−2.9	−1.9	+1.4	+1.9	+2.4	+2.9
44	−3.8	−3.3	−2.8	−1.8	+1.3	+1.8	+2.3	+2.8
45	−3.8	−3.3	−2.8	−1.8	+1.3	+1.8	+2.3	+2.8
46	−3.7	−3.2	−2.7	−1.7	+1.2	+1.7	+2.2	+2.7
47	−3.7	−3.2	−2.7	−1.7	+1.2	+1.7	+2.2	+2.7
48	−3.6	−3.1	−2.6	−1.6	+1.1	+1.6	+2.1	+2.6
49	−3.6	−3.1	−2.6	−1.6	+1.1	+1.6	+2.1	+2.6
50	−3.5	−3.0	−2.5	−1.5	+1	+1.5	+2	+2.5

References

[1]   The Ministry of Water Resources of the People's Republic of China. (1999) *Standard Test for Soil Test Method* (GB/T 50123-1999). China Planning Press, Beijing, China.

[2]   The Ministry of Water Resources of the People's Republic of China. (1999) *Standard Test for Soil Test Method* (GB/T 50123-1999). China Planning Press, Beijing, China.

[3]   Research Institute of Highway Ministry of Transport. (2007) *Test Methods of Soils for Highway Engineering* (JTG E40-2007). China Communications Press, Beijing, China.

[4]   CCCC Second Highway Consultants Co., Ltd. (2005) *Test Methods of Rock for Highway Engineering* (JTG E41-2005). China Communications Press, Beijing, China.

[5]   Ministry of Construction of the People's Republic of China. (2011) *Technical Specification for Inspecting of Concrete Compressive Strength by Rebound Method* (JGJ/ T 23-2011). China Architecture & Building Press, Beijing, China.

[6]   Editorial Committee of Engineering Geology Handbook. (2018) *Geological Engineering Handbook*. 5th ed. China Architecture & Building Press, Beijing, China.

[7]   Aydin, A. (2008) ISRM suggested method for determination of the Schmidt Hammer rebound hardness: Revised version. In: *The ISRM Suggested Methods for Rock Characterization, Testing and Monitoring: 2007–2014*. Springer, Cham, Germany.

[8]   Bai, X.C. (2008) *Tutorial of Geotechnical Testing*. Henan University Press, Zhengzhou, China.

[9]   Guo, Y.Y. & Mao, H.M. (2009) *Manual of Tests in Foundation Engineering*. Southwest Jiaotong University Press, Chengdu, China.

[10] Fu, X.M. & Deng, R.G. (2012) *Tests in Rock Mechanics*. Southwest Jiaotong University Press, Chengdu, China.

[11] Zhang, D.X. (2017) Correlation between indoor rock test results and rock mass strength. *World Nonferrous Metal*, (3), 188–188.

[12] Ren, J.X. (2015) *Geotechnical Testing*. Wuhan University of Technology Press, Wuhan, China.

[13] Yao, Z.S. & Cai, H.B. (2014) *Geotechnical Testing Techniques*. Wuhan University Press, Wuhan, China.

[14] Yuan, J.Y. (2011) *Tests on Rocks and Soils*. China Water & Power Press, Beijing, China.

[15] Gao, H.D. (2010) *Tutorial of Geotechnical Tests in Laboratory*. Beijing University of Technology Press, Beijing, China.

[16] Nie, L.Z. & Xiang, W. (2009) *Guidance of Soil Experiments*. China University of Geosciences Press, Wuhan, China.

[17] Shen, Y. & Zhang, W.H. (2013) *Geotechnical Testing Techniques*. Metallurgical Industry Press, Beijing, China.

[18] Xie, K.J., Wang, Y. & Jia, D.X. (2010) *Testing Manual of Concrete and Geomaterial*. China Electric Power Press, Beijing, China.

Part 3

Comparison of test standards

The geotechnical test is the basic content of rock and soil mechanics, which determines parameters for engineering design and verifies the correctness and practicality of theories. The geotechnical tests shall be carried out properly according to the related standards. These standards include both national and international standards, such as Chinese Standards (GB), American Standards (ASTM) and British Standards (BS), ISO standards (International Organization for Standardization) and ISRM suggested methods (International Society of Rock Mechanics). The standards issued by different organizations differ from each other and may lead to different test results. In this regard, a detailed comparison of standards from different nations/regions would delineate main differences between them, and this comparison would give guidance to practicing engineers when they are delivering a specified project in a certain region.

GB, ASTM and BS standards are compared for soil tests in Chapter 9, while the methods formulated in GB, ASTM and ISRM for rock tests are compared in Chapter 10. Methods to be compared for soil tests include determination of pH, soluble salts, organic matter content, water content, density, specific gravity, particle size distribution, permeability, compaction characteristics, relative density test of cohesionless soil, expansion rate and force of cohesive soils, California Bearing Ratio, consolidation, simple and reversal direct shear, unconfined compression and triaxial compression. Methods to be compared for rock tests include water content, water adsorption, bulk density, particle density, freezing-thawing, expansion, slake durability, acoustic, Brazilian splitting, uniaxial compression, triaxial compression, direct shear of rock joints, point load and Schmidt hammer rebound.

Each method is compared in the five aspects of scope of the method, apparatus to be used, requirements of sample, operation/procedures, and data processing. The similarities and differences between different standards on the same test are shown in Tables.

Chapter 9

Comparison of test methods for soil

The purpose of the soil test is to obtain the chemical, physical and mechanical properties of the soil sample. The experimentally obtained indexes can be used for theoretical analysis of soil mechanics and practical application of engineering. Chemical measurements of soil include determination of pH, soluble salts and organic content. Since ASTM does not include soil chemistry testing, this chapter only compares and analyzes the differences between GB and BS. In general, BS uses a more convenient method than GB in most chemical tests. Physical measurements of soil include determination of water content, density, specific gravity, particle size distribution, relative density, and California Bearing Ratio. The procedures for these tests are easier to follow in ASTM and BS than in GB. Mechanical measurements of soil include permeability, compaction characteristics, determination of consolidation properties, swelling ratio and pressure, simple and reversal direct shear, and unconfined and triaxial compression tests. For these tests, requirements from GB are strict while those in ASTM and BS are relatively flexible.

9.1 Chemical tests

9.1.1 pH value

Soil pH is an index of soil acidity and alkalinity. The pH of the acidic soil is less than 7, and that of the alkaline soil is greater than 7. When measuring the pH value, the soil is usually mixed with reagents to extract H^+ from the soil and determines the pH of the suspension. There are two ways to measure pH value of soil, water immersion and salt immersion. The former was obtained by mixing distilled water with the soil, whereas the latter was obtained by mixing a certain salt solution with the soil. The common reagent used for determining the salt immersion pH is potassium chloride solution or calcium chloride solution. When these solutions are mixed with the soil, K^+ or Ca^{2+} exchanges with Al^{3+} or H^+. The salt immersion pH value of the same soil is usually lower than the water immersion one. Water immersion is employed in GB and BS, while both water and salt immersion are suggested in ASTM.

The strategies for determining the pH are colorimetric and potentiometric methods. The colorimetric method determines the pH of the test solution by examining the color change of the indicator in the soil suspension. The directly measurement is to use a pH meter. The colorimetric method with a simple device is commonly used for the fast measurements in the field. However, the colorimetric method exhibits a relatively low accuracy as compared to the potential method. Both GB and BS include only the potential method, while ASTM introduces both colorimetric and potential methods.

Factors influencing the measured pH value of soil include the soil-water ratio, testing temperature of the solution and the reagents. The smaller the soil to water ratio is, the lower the concentration of the test solution is. With the increase in temperature, the ionic product constant of water increases and the ionization reaction of the substance in the water is enhanced, which affects the ion concentration of hydrogen in the solution. Different standards specify different soil-water ratios: GB requires a soil-water ratio of 1: 5, ASTM 1: 1, and BS 1: 2.5. ASTM explicitly requires the testing temperature of the solution ranging from 15 to 25 °C. Table 9.1 gives a detailed comparison of the related standards.

9.1.2 Content of soluble salt

The soluble salt test is to determine the content of total soluble salt, chloride ion, and sulfate. The soluble salt is determined by means of leaching the soil with water or acid solution and measuring the content of the soluble salts in the leaching solution.

The dissolution and crystallization of salts in soil affect the engineering properties of soil. As temperature decreases, crystallization of the sulfate leads soil to expand. As temperature rises, dehydration of sulfate would loosen the soil. This is why the total amount of easily soluble salt in railway embankments is required not to exceed 5%, and the sulfate not to exceed 2%. This section (Table 9.2–9.4) gives a detailed comparison of the test methods in GB and BS, in terms of apparatus, reagent, test preparation and test procedures.

9.1.3 Organic matter

Organic matter in soil refers to organic compounds which are mainly composed of carbon, nitrogen, hydrogen and oxygen, supplemented by sulfur, phosphorus and metal elements. In general, the presence of organic matter in soil leads to dispersity of soil. On the other hand, the water content has significant effect on the engineering soils. As the water content rises (up to 50% to 200%), the dry density (<1 g/cm^3), the strength and bearing capacity of soils decreases, and the compressibility and the shrinkage increases (>50%).

Common methods for determining organic matter in soil specimens include potassium dichromate volumetric method, hydration heat method and burning method. In the potassium dichromate volumetric method, acidic potassium dichromate standard solution was heated to react with organic compound in oil bath at 170–180 °C, and then the excess potassium dichromate is titrated with a ferrous sulphate solution. The content of organic carbon is calculated based on the consumption amount of potassium dichromate. The basic principle and main steps of water and thermal method are similar to those of potassium dichromate volumetric method. The rapid mixing of concentrated sulfuric acid and potassium dichromate solution accelerates the oxidation of organic matter. In the burning method, the dry soil is burned at a temperature of 350–1000 °C and the ratio of the lost mass to the dry soil mass is the content of organic matter. The comparison between GB and BS for the determination of organic matters is given in Table 9.5.

9.2 Physical tests

9.2.1 Water content

Water content is the ratio of the mass of water to the mass of the solid particles in a given mass of soil.

Table 9.1 Comparison of determination of the pH value

Standard	GB	ASTM	BS
Designation	GB 50123-1999	ASTM D4972-13	BS 1377-3: 1990
Scope	All kinds of soil		
Methods	Potentiometric	Potentiometric Colorimetric	Potentiometric
Apparatus	Acidometer, electric oscillator, electric magnetic blender, etc.	Acidometer pH test paper	Acidometer
Reagent	Mixed Reagent: Distilled water; No clear requirements of water quality.	1. Water: ASTM TYPE III or higher quality water; 2. Calcium chloride solution (c=0.01 mol/L): prepare calcium chloride stock solution (c=1 mol/L); dilute 20 mL of calcium chloride stock solution to gain 2 L with a pH of 5–7.	Same as GB
Buffer solution (for calibration of acidity meter)	Three solutions with different pHs: 1. 1000 mL solution (pH of 4.01) made from potassium hydrogen phthalate (10.21 g). 2. 1000 mL solution (pH of 6.87) made of disodium hydrogen phosphate (3.53 g) and potassium dihydrogen phosphate (3.39 g). 3. 1000 mL solution (pH of 9.18) made of borax (3.80 g). The solution shall be stored in a dry, airtight plastic bottle for a maximum duration of two months.	Three solutions with different pH values: 1. 1 L buffer solution (pH of 4.0 at 5 °C to 37 °C) made of potassium hydrogen phthalate (10.21 g). 2. 1 L solution (pH of 6.9 at 20 °C) made of potassium dihydrogen phosphate (3.40 g) and dipotassium hydrogen phosphate (3.55 g). 3. 1 L solution (pH of 10.1 at 20 °C) made of sodium carbonate (2.65 g) and sodium bicarbonate (2.10 g).	Two solutions with different pH value: 1. 500 mL of a pH value 4.0 solution made of potassium hydrogen phthalate (5.106 g), or a special buffer solution at pH 4.0. 2. 500 mL of solution made of borax (9.54 g) with a pH of 9.2, or a special buffer solution with a pH of 9.2.
Other solution	Saturated potassium chloride solution. Its use is not stated in the standard.	—	Saturated potassium chloride solution is required, and the solution is used for the protection of calomel electrodes.

(Continued)

Table 9.1 (Continued)

Standard		GB	ASTM	BS
Test procedures	Preparation of soil suspension	Preparation of test solution Weigh 10 g of air-dried specimen passing through a 2 mm sieve; put it in a jar and add 50 mL of distilled water. Soil-water ratio: 1 : 5 The way to mix: to slosh 3 min Standing time: 30 min	Preparation of test solution Aqueous solution with soil: Weigh 10 g of air-dried specimen passing through a 2 mm sieve, and add 10 mL of distilled water. Calcium chloride solution with soil: Weigh 10 g air-dried specimen passing through a 2 mm sieve and add 10 mL of calcium chloride solution. Soil-water ratio: 1 : 1 The way to mix: mix sufficiently Standing time: 1 h	Preparation of test solution Weigh 30 g of the air-dried specimen passing through a 2 mm sieve; put it in a beaker and add 75 mL distilled water. Soil-water ratio: 1 : 2.5 The way to mix: Stir for a few minutes Standing time: 8 h
	Calibration with pH meter	The pH meter shall be calibrated, prior to the test, by the standard buffer solution according to the instructions of the acidity meter.	—	It is only required that the electrode is completely cleaned with distilled water before it is placed in suspension. No further steps were specified.
	Procedures	1. Pour the specimen suspension into the beaker to 2/3 of its volume; place a stirring rod in the beaker; and place the beaker on an electric magnetic blender. 2. Place the test electrode in the suspension and ensure that the electrode is completely immersed in the suspension and does not touch the bottom of the beaker.	—	
	Stirring	With a magnetic blender for about 1 min.		No clear instruction
	Other requirements	—	When measuring pH, the suspension temperature shall be at room temperature (15–25 °C).	A second calibration is needed for the pH meter.
	Times of reading	Only one		Three consecutive readings in one minute. A reading difference less than 0.05 indicates a stable data.
	Accuracy of reading	0.01	0.1	0.1.
	Electrode treatment after test	Wash the electrode with distilled water and dry it with filter paper, or soak it in distilled water.	—	Same as GB

Table 9.2 Comparison of determination of total soluble salt

Standard		GB	BS
Designation		GB 50123-1999	BS 1377-3: 1990
Scope		All kinds of soil	Can also be used to test the content of soluble solids in water.
Apparatus		Analytical balance, water bath cauldron, electronic multi-purpose furnace, evaporating dish, oven, desiccator	Büchner funnel, vacuum filter bottle, vacuum source and vacuum tube, balance, oven, evaporating dish, desiccator, electric heating plate
Reagent		15% hydrogen peroxide solution, 2% sodium carbonate solution.	—
Solution		Soil specimen leachate	Groundwater
Operations	Dosage of solution	50–100 mL	Sufficient to produce 2.5 to 1000 mg of solid
	Determine the mass of evaporating dish	—	Bake at 180 ±10 °C for 30 min and weigh after cooling.
	Heating the test solution	Pour all specimens into an evaporating dish; cover the dish with a watch glass; and dry them on a water bath. When the residue appears yellowish brown, add hydrogen peroxide 1–2 mL and continue to dry until the color disappears.	Pour some solution into an evaporating dish and dry in a boiling water bath. It shall be done in a clean environment. As it is evaporated, another part of solution is poured into the dish until it is completely dried.
Drying the residue	Temperature	105–110 °C	180±10 °C
	Initial drying time	4–8 h	1 h
	Time interval for weighing	2–4 h	30 min
	Stop	Until the difference between two consecutive readings is not more than 0.1 mg	Until the difference between two consecutive readings is not more than 1 mg
	Other requirements	When the evaporated residue contains a large amount of crystal water, two evaporating dishes may be used. Add 50 mL of the leaching solution and 50 mL of pure water into each dish for being dried at 180 °C.	—
	Accuracy of weighing	0.0001 g	0.0005 g

(Continued)

Table 9.2 (Continued)

Standard		GB	BS
Data processing	Calculation formula	Not treated with 2% sodium carbonate: $$W = \frac{(m_2 - m_1)\dfrac{V_w}{V_s}(1+0.01\omega)}{m_s} \times 100$$ Treated with 2% sodium carbonate: $$W = \frac{(m - m_0)\dfrac{V_w}{V_s}(1+0.01\omega)}{m_s} \times 100$$ W – total soluble salt (%), V_w – volume of pure water for leachate (mL), V_s – volume of extracted leachate (mL), m_s – weight of air-dried specimen (g), ω – water content of air-dried specimens (%), m_0 – mass of sodium carbonate post evaporation (g), m_1 – mass of the evaporating dish (g), m_2 – mass of evaporation dish plus drying residue (g).	$$TDS = \frac{m_2 - m_1}{V} \times 10^6$$ TDS – total dissolved solids (mg/L), m_1 – quality of dry evaporation dishes (g), m_2 – total mass of dissolved solids added to the evaporating dish in the second or subsequent stages after drying at 180 °C (g), V – volume of water used for the specimen (mL).
	Calculation accuracy	—	Calculate to two significant digits. When testing multiple specimens, the average shall be calculated; If the difference exceeds 10% of the average, two more specimens shall be tested again.

Table 9.3 Determination of chloride ions

Standard		GB	BS-1	BS-2
Designation		GB 50123-1999	BS 1377-3: 1990	BS 1377-3: 1990
Scope		All kinds of soil	Soil will obtain chlorine ions when it is contacted with salt water or immersed in salt water.	Desert soil, or soil with chloride from unknown source
Apparatus		Analytical sieve, balance, electric oscillator, filtration equipment, centrifuge, analytical balance, acid burette, pipette, oven, conical flask, and volumetric flask	Balance, volumetric flask, glass graduated cylinder, pipette, beaker, burette, conical flask, funnel, screen, wash bottle, amber glass reagent bottle, plastic or metal bottle with wide-mouth screw cap, mechanical shaking device, oven, desiccator, filter funnel, test sieve, riffle box, pestle, and mortar	Balance, volumetric flask, glass graduated measuring cylinder, pipette, beaker, burette, closed conical flask, filter funnel, filter paper, sieve, pestle and mortar, and riffle box.
Reagent		1. Potassium chromate indicator: the mass ratio is 5%; 2. Silver nitrate standard solution: $c=0.02$ mol/L; 3. Sodium bicarbonate solution: $c=0.02$ mol/L; 4. Methyl orange indicator: mass ratio is 0.1%.	1. Silver nitrate solution: $c=0.1$ mol/L; 2. Thiocyanate solution: $c=0.1$ mol/L; 3. Nitric acid solution: $c=6$ mol/L; 4. 3,5,5-trimethyl-1-ethanol; 5. Halotrichite indicator.	Same as BS-1
Test operation of Leachate	Preparation	Air-dried specimen passed through a 2 mm sieve.	Dry specimen is passed through the 2 mm sieve and divided through the riffle box; 500 g for quantitative test and 50 g for qualitative examination of chloride.	10 g of dry specimen is passed through the 0.15 mm sieve and divided through the riffle box.
		50–100 g of soil (depending on the content of salt in the soil and the items of analysis), the ratio of soil to water is 1: 5.	500 g of soil; the ratio of soil to water is 1: 2 or 1: 1.	5 g of soil; add 30 mL of distilled water and 15 mL of nitric acid
		Filter by extracting air after shaking for 3 min. If the filtrate is turbid, it needs to be separated by centrifuge.	Shake for at least 16 h or overnight on the vibrator.	Heat until near to boiling point and keep the temperature for 10–15 min.
		25 mL of clarifying solution through filtering	100 mL of clarifying solution through filtering	All for testing
Qualitative examination of chloride		—	Take 50 g of soil, add the same amount of distilled water, stir intermittently for 4 h, and take the supernatant after suspension being precipitated. Acidify the liquid with nitric acid, add silver nitrate solution, and let it stand for 10 min. If there is no obvious turbidity, then the quantitative analysis test is not required.	—

(Continued)

Table 9.3 (Continued)

Standard	GB	BS-1	BS-2
Calibration of reagent	—	Calibration of thiocyanate solution: Take 25 mL of silver nitrate solution; add 5 mL of nitric acid solution and 1 mL of halotrichite indicator; titrate with thiocyanate solution. Until the color is changed from color less to pink, record the dosage of thiocyanate solution and calculate the concentration of the thiocyanate solution.	Same as BS-1
Analysis of leachate	1. Add 1–2 drop (s) of methyl orange indicator to the leachate and then add sodium bicarbonate solution drop by drop until the solution is pure yellow. 2. Add 5–6 drops of potassium chromate indicator, titrate with silver nitrate standard solution to produce brick red precipitate, and record the dosage of silver nitrate standard solution. 3. Take 25 mL of distilled water, repeat the steps 1 and 2, and record the amount of silver nitrate standard solution.	1. Add 5 mL of nitric acid solution to the leachate. 2. Titrate with silver nitrate solution until no white precipitation is produced, then add a small amount of silver nitrate and record the amount of silver nitrate solution. 3. Add 2 mL of 3,5,5-trimethyl-1-hexanol and shake vigorously to coagulate the precipitate. 4. Rinse with distilled water and collect the solution and the lotion together. Add 5 mL of the halotrichite indicator and titrate it with a calibrated thiocyanate solution until the solution turns from color less to brick red and the color is the same as that of the reagent when it's calibrated. Record the amount of thiocyanate solution.	Same as BS-1. But the test solution is an acidic solution, which is already under acidic conditions, so the first step can be removed and subsequent steps can be performed.
Data process	**Calculation formula** $$b(\text{Cl}^-) = \frac{(V_1 - V_2)c(\text{AgNO}_3)}{m_S}$$ $$\times \frac{V_W}{V_S}(1+0.01\omega)\times 1000$$ $$\text{Cl}^- = b(\text{Cl}^-)\times 10^{-3}\times 0.0355\times 100$$ $b(\text{Cl}^-)$ – the molarity of chlorine (mmol/kg), cl^- – the content of chlorine (%), V_1 – The volume of silver nitrate standard solution consumed by leachate (mL); V_2 – the volume of silver nitrate standard solution consumed by distilled water (blank) (mL); 0.0355–the molar mass of chloride (kg/mol).	1. The concentration of thiocyanate solution $$C : C = \frac{2.5}{V_1}$$ V_1 – amount of thiocyanate solution added in the calibration of reagents (mL) 2. Content of chloride ion = $K\,(V_{2-}10CV_3)$ V_2 – volume of silver nitrate solution added in analysis of leachate (mL), V_3 – volume of standard thiocyanate solution added in analysis of leachate (mL), and When ratio-water is 1 : 1, K=0.003546; When 1 : 2, K=0.007092.	1. Calculation of the concentration of thiocyanate solution is the same as that of BS-1; 2. Content of chloride ion = $0.07092\,(V_{2-}10CV_3)$
Calculation accuracy	Accurate to 0.01 mmol/kg and 0.001%. The difference in parallel titration shall not be greater than 0.1 mL, and the result is an arithmetic average.	The calculation of chloride ion content is accurate to 0.01%. If multiple specimens are tested and the difference between the results shall not exceed 0.1%, the average shall be the final result. If the difference is more than 0.1%, it needs to be tested again.	

Table 9.4 Determination of sulfate ions

Standard	GB	GB	BS	BS
Designation	GB 50123-1999	GB 50123-1999	BS 1377-3: 1990	BS 1377-3: 1990
Methods	EDTA complex capacity method	Turbidimetric method	Gravimetric method	Ion exchange method
Scope	Soil with a sulfate content of not less than 0.025%	Soil with sulfate content less than 0.025%	All kinds of soil	Not suitable for soil containing anions of other strong acids
Apparatus — Preparation of leachate	Analytical sieve, electric oscillator, filtration equipment, centrifuge, jar, volumetric flask, and glass rod.		Soil preparation: oven, balance, dryer, test sieve, riffle box, mortar and pestle, glass weighing bottle, red litmus paper. Acid soluble sulfate extraction: conical beaker, electric heating plate, Buchner funnel, vacuum filter, vacuum source, rubber vacuum tube, glass rod, washing bottles, glass filter funnel, burette, amber glass containers. Water-soluble sulphate extraction: In addition to the devices listed in the acid-soluble sulphate extraction, mechanical vibrating screen or blender, extraction bottle, pipette, and watch glass are required.	
Analysis of leachate	Balance, acid burette, pipette, conical flask, volumetric flask, measuring cup, horn spoon, oven, graduated cylinder, mortar and pestle	Photoelectric colorimeter or spectrophotometer, electric magnetic stirrer, measuring spoon, pipette, volumetric flask, sieve, oven, analytical balance	Gravimetric method: in addition to the instruments, sintered silica filter crucible or ignition crucible, electric muffle furnace and blue litmus paper are also required.	Ion exchange method: in addition to the instruments, glass ion exchange column and water receiver are required.
Reagent	1:4 hydrochloric acid solution, mixture of barium and magnesium, chrome-black T indicator, zinc-based solution (concentration of zinc ion: 0.01 mol/L), EDTA standard solution, 95% ethanol, 1:1 hydrochloric acid solution and 5% barium chloride solution	Suspension stabilizer, crystalline barium chloride, silver sulfate standard solution (concentration of sulfate: 0.1 mg/L)	Acid-soluble sulfate extraction: Hydrochloric acid (volume ratio: 10%), diluted ammonia solution, silver nitrate solution (mass ratio: 0.5%), concentrated nitric acid solution (concentration: 1.42 g/mL), and distilled water. Analysis: Barium chloride solution (mass ratio 5%)	Hydrochloric acid solution ($c=4$ mol/L), sodium hydroxide solution ($c=0.1$ mol/L), potassium hydrogen phthalate solution ($c=0.1$ mol/L), methyl orange indicator, silver nitrate solution (mass ratio is 0.5%), nitric acid solution (volume ratio is 5%), distilled water.

(Continued)

Table 9.4 (Continued)

Standard		GB	BS
Test operation	Preparation of leachate	Pass the air-dried soil through a 2 mm sieve; take 50–100 g of soil (depending on the amount of salt in the soil and the analysis items); the ratio of soil to water is 1:5. After 3 min oscillation, filtrate it through air extraction. If the filtrate is turbid, it must be separated by a centrifuge. Collect 25 mL of filtered clarifying solution.	After passing through a 2 mm sieve and a 15 mm riffle box, about 60 g of dried specimen is produced. Then pass it through 0.425 mm sieve and dry it. Take 50 g of soil specimen, add 100 mL of distilled water, put it on the oscillator, stir for 16 h, and filter.
			Take 50 mL of its clarification solution and add water to 300 mL.
			Take 50 mL of its clarification solution and add water to 100 mL.
		—	Extraction of Acidic solution After passing through a 2 mm sieve and a 15 mm riffle box, about 100 g of dried specimen is produced. Then by passing through 0.425 mm sieve and 7 mm riffle box, each dried specimen weighing about 10 g is produced. Take about 2 g of specimen, add 100 mL of hydrochloric acid, cover it with glass, boil it, keep warm for 15 min, and filter the test solution. Clean the beaker and residue thoroughly with distilled water until the cleaning solution contains no chlorides and pour it into the filtrate.
	Estimation of sulfate content	Take 5 mL of leachate, add 2 drops of 1:1 hydrochloric acid to the test tube, then add 5 drops of 5% barium chloride solution, shake well, and estimate the sulfate content according to Table 9.5. When the sulfate content is less than 50 mg/L, turbidimetric method is used, whereas EDTA complex capacity method is used instead.	—

| Analysis of leachate | 1. Put a certain amount of specimen leachate into a conical flask; dilute it with proper amount of distilled water; and then put a piece of Congo red test paper into it; drop (1 : 4) hydrochloric acid solution until the test paper become blue; drop 2–3 drops into it; heat it until it is boiled. When it is hot, add the excess barium magnesium mixture by the burette and shake it simultaneously until the expected dosage is added (note that the dripping amount shall be at least 50% excessive); continue to heat it for 5 min; remove and cool for 2 h. Then add 10 mL of ammonia buffer solution, a little chrome black T and 5 mL of 95% ethanol; shake well; then titrate with EDTA standard solution until the test solution is changed from red to azure; write down the dosage—V_1 (mL).

2. Add proper amount of distilled water to another conical flask; drop a piece of Congo red test paper; add (1 : 4) hydrochloric acid solution until the test paper is blue; then drop 2–3 drops. Add the same amount of barium magnesium mixture by burette accurately as the step 1; then add 10 mL of ammonia buffer solution, a little chrome-black T indicator; 5 mL of ethanol; shake well. Titrate with EDTA standard solution to change from red to azure as the end point; write down the dosage—V_2 (mL). | 1. Draw the standard curve: take 5 pieces of appropriate amount of sulfate standard solution and dilute them to 100 mL, so that their concentrations are 0.5, 1.0, 2.0, 3.0, 4.0 mg/100 mL respectively. Add 5 mL of suspension stabilizer and a full spoon of barium chloride crystals into them; stir them on a magnetic agitator for 1 min. Take distilled water as the blank control; measure the turbidity by purple filter on the photoelectric colorimeter; measure the light absorption value of suspension every 30 s in 3 min; obtain the value after being stable. Then take the sulfate content as the vertical coordinate and the corresponding absorbance value as the horizontal coordinate; draw the relation curve; and get the standard curve.

2. Determination of Sulfate Content: pipette 100 mL of the specimen leachate to place it in a beaker (When the content of sulfate is more than 4 mg/mL, take a small amount of leachate and dilute to 100 mL with distilled water). Then, operate the test according to the Estimation and Analysis steps, such as adding suspension stabilizer to the standard solutions; take the same specimen leachate as a reference; determine the absorbance value of the suspension; obtain the stable reading; check the corresponding sulfate content from the standard curve (mg/100 mL). | Test the solution with litmus paper, dilute it to 300 mL, boil the solution, stir it continuously and add 10 mL of barium chloride solution drop by drop until the sediment is formed. Keep the temperature of the solution below the boiling point for at least 30 min. Cool, filter and wash so that the washing solution does not contain chlorides. If using sintered silica filter crucible, transfer the test solution and sediment into it after it is ignited and weighed; then take them out; use electric muffle furnace to heat it to 800 °C until the mass is no longer reduced. If using an ignition crucible, transfer the filter paper and sediment to the crucible and heat it to coke the filter paper. Weigh the crucible; calculate the weight of sediment to be accurate to 0.001 g. | The prepared leachate is passed through an ion exchange column, and the column is flushed twice with 150 mL of distilled water, 75 mL per time. Add the indicator; titrate the test solution with standard sodium hydroxide solution; record the amount of sodium hydroxide solution to be accurate to 0.05 mL. |

(Continued)

Table 9.4 (Continued)

Standard	GB	BS
	3. Add the same volume of specimen leachate into another conical flask as step 1; add 5 mL of ammonia buffer solution; shake well; then add a little chrome black T indicator and 95% ethanol; shake well. Titrate the test solution with EDTA standard solution. When the test solution is changed from red to bright blue, it is to end. Record the dosage of EDTA standard solution and estimate it to be 0.05 mL. In this way, the dosage of calcium and magnesium to EDTA standard solution—V_3 (mL) in the same volume leachate is determined.	
Data processing	Calculation formula	

Calculation formula (GB):

$$b(SO_4^{2-}) = \frac{(V_3 + V_2 - V_1)c(EDTA)}{m_s}$$

$$\frac{V_w}{V_s}(1 + 0.01\omega) \times 1000$$

$$SO_4^{2-} = b(SO_4^{2-}) \times 10^{-3} \times 0.096 \times 100\,(\%)$$

$$or\ SO_4^{2-} = b(SO_4^{2-}) \times 96\text{mg/kg}$$

$b(SO_4^{2-})$ – molality of sulfate (mmol/kg),

SO_4^{2-} – content of sulfate (% or mg/kg),

V_1 – dosage of mixture of calcium magnesium and barium magnesium to EDTA standard solution in leaching solution (mL),

Calculation formula (BS):

$$SO_4^{2-} = \frac{m(SO_4^{2-})\frac{V_w}{V_s}(1+0.01\omega) \times 100}{m_s 10^3}\,(\%)$$

$$or\ SO_4^{2-} = (SO_4^{2-}\%) \times 10^6\ \text{mg/kg}$$

$$b(SO_4^{2-}) = (SO_4^{2-}\% / 0.096) \times 1000$$

SO_4^{2-} – content of sulfate (% or mg/kg),

$b(SO_4^{2-})$ – molality of sulfate (mmol/ kg), and

$m(SO_4^{2-})$ – content of SO_4^{2-} obtained from the standard curve (mg).

BS column:

1. Percentage of original soil specimens sieved by 2 mm sieve

$$= \frac{m_2}{m_1} \times 100$$

m_1 – initial mass of the specimen (g);

m_2 – mass of specimen after being sieved by 2 mm sieve (g).

2. Content of acid-soluble sulfate:

$$SO_3 = \frac{34.3m_4}{m_3}\,\%$$

m_3 – mass of each specimen used (g), and

m_4 – mass of the precipitate after drying (g).

Last BS column:

1. Content of water-soluble sulfate:

$$SO_3 = 0.8BVg / L$$

$$or\ SO_3 = 0.16BV\,(\%)$$

2. Content of sulfate in groundwater specimens: $SO_3 = 0.4B\ V$

B – concentration of sodium hydroxide solution (mol/L), and

V – volume of sodium hydroxide solution (L).

V_2 — the dosage of mixture of barium and magnesium in the same volume of leachate to standard solution (mL),

V_3 — the dosage of mixture of calcium and magnesium in the same volume of leachate to standard solution (mL),

0.096 — molar mass of sulfate (kg/mol), and

c(EDTA) — concentration of EDTA standard solution (mol/L).

3. Content of water-soluble sulfate: $SO_3 = 6.86m_4$ (g/L) or $SO_3 = 1.372m_4$ (%)

4. Traditionally, the content of sulfate is expressed as SO_3. If SO_4 is hoped to express the content, multiply the result of the formula by 1.2.

| Calculation accuracy | 1. The molality of sulfate is accurate to 0.01 mmol/kg. The content of sulfate is accurate to 0.001% or 1 mg/kg.
2. Parallel titration is required. The deviation of titration shall not be greater than 0.1 mL. Take the average. | Same as EDTA complex capacity method (1). | 1. Percentage of original soil specimens sieved by a 2 mm sieve is accurate to 1%.
2. The content of sulfate is accurate to 0.01% or 0.01 g/L, and parallel specimens shall be operated. If the difference between the two test results is more than 0.2% or 0.2 g/L, and the test shall be done again. If not, the average is the final result. | Same as gravimetric method (2). |

Table 9.5 Comparison of determinations of the organic matter of soils

Standard		GB	BS
Designation		GB 50123-1999	BS 1377-3: 1990
Scope		Soil with an organic matter content of no more than 15%	All kinds of soil
Apparatus		1. Analytical balance: weigh 200 g, readable to 0.0001 g. 2. Oil bath pot with wire cage, vegetable oil. 3. Heating equipment: drying oven, electric furnace. 4. Other equipment: thermometer (0–200 °C, scale: 0.5 °C), standard sieve of 0.15 mm, etc.	1. Drying oven: temperature can be maintained at 50 ±2.5 °C. 2. Balance: 2 balances with accuracy of 1 g and 0.001 g respectively; no specification of measuring range. 3. Standard sieve: two sieves with aperture size of 2 mm and 2 μm respectively; with a receiver. 4. Riffle boxes with the opening width of 7 mm and 15 mm respectively. 5. Other equipment: glass boiling tube, blue litmus paper, etc.
Requirements of Reagent		1. Potassium dichromate standard solution: accurately weigh 44.1231 g of potassium dichromate which is dried at 105–110 °C and porphyrized in advance; dissolve it in 800 mL of pure water (heating if necessary); slowly add 1000 mL of concentrated sulfuric acid while constantly stirring; cool and transfer it into a 2 L volumetric flask; dilute with distilled water. (c=0.075 mol/L) 2. Ferrous sulphate standard solution: weigh 56 g of ferrous sulphate; dissolve it in distilled water; add 30 mL of 3 mol/L sulfuric acid, and then dilute it to 1 L. 3. Phenanthroline indicator: 1.845 g of phenanthroline and 0.695 g of ferrous sulphate are dissolved in 100 mL of distilled water and stored in a brown bottle.	1. 1 L of potassium dichromate solution (c=0.167 mol/L); 2. Ferrous sulphate solution: about 140 g of ferrous sulphate is dissolved in sulfuric acid solution (c=0.25 mol/L) in order to obtain 1 L solution; 3. Concentrated sulfuric acid: its density is 1.84 g/mL; 4. Phosphoric acid solution: its volume concentration is 85% and mass concentration is 1.70–1.75 g/mL; 5. The indicator is made by dissolving 0.25 g of sodium diphenylamine sulfonate in 100 mL of distilled water; 6. Dilute hydrochloric acid: dilute 250 mL of concentrated hydrochloric acid (ρ=1.18 g/mL) to 1L with distilled water; 7. Lead acetate paper: filter paper soaked in 10% lead acetate solution; 8. 1 L of sulfuric acid: c≈1 mol/L.
Operation of test	Calibration of ferrous sulphate solution	<u>Oxidant</u> Take 3 specimens of 10 mL of potassium dichromate standard solution and titrate them respectively. <u>Indicator</u> 3–5 drops of phenanthroline indicator <u>Characteristic after complete reaction</u> The color of solution changes from yellow to green and then suddenly to orange-red.	At first, take 10 mL of potassium dichromate standard solution; add 20 mL of concentrated sulfuric acid solution to it; dilute it with 200 mL of distilled water, and then titrate it. After the reaction is completely finished, drop 0.5 mL of potassium dichromate solution into it and continue to titrate. 10 mL of sulfuric acid and 1 mL of indicator The color of solution changes from blue to green

Result of test	The error of the three parallel tests shall not exceed 0.05 mL. The concentration of ferrous sulphate standard solution can be calculated by the average. Record the amount of ferrous sulphate solution which shall be accurate to 0.05 mL.
Preparation of specimen	Take dry soil specimens. When the content of organic carbon in the specimen is less than 8 mg, remove the plant roots and sieve it through 0.15 mm sieve. Take the initial specimen and dry it at 50 ± 2.5 °C. Weigh the specimen accurately to 0.1%. The large particles, except for the stones, are crushed and passed through 2 mm sieve. Remove the stones and weigh the mass of the specimen accurately to 0.1%. Obtain 100 g of specimen by using a 15 mm riffle box and crush it; pass it by 0.425 mm sieve; obtain the test specimen by using a 7 mm riffle box.
Detection and treatment of sulfide	— 1. Detection of sulfide: take 5 g of specimen, put it into the test tube; add 20 mL of dilute hydrochloric acid in it; then heat and make the steam pass through the test paper of lead acetate to detect whether hydrogen sulfide is formed. If the test paper turns black, the specimen contains sulfide. 2. Treatment of sulfide: take 50 g of specimen, put it into conical flask, then add 1 mol/L sulfuric acid solution until no hydrogen sulfide is generated. Filter the test solution and wash the filter residue with hot distilled water until the washing solution is not acidic (the color of blue litmus paper does not change color). Dry and weigh the specimen, and the weighing result is accurate to 0.01 g.
Detection and treatment of chloride	— 1. Detection of chloride: take 50 g of specimen, place it in a conical flask, add equal mass of distilled water; intermittently stir it for 4 h. Filter the specimen suspension; take 25 mL of the supernatant; add nitric acid solution and silver nitrate solution. Leave it to stand for 10 min, after this if the suspensions become turbid, the specimen contains chloride which needs to be treated. 2. Treatment of chloride: take 50 g of specimen; wash it with distilled water until there is no turbidity to generate, when the silver nitrate solution is dropped into the wash solution. Dry and weigh the specimen, accurate to 0.01 g.

(Continued)

Table 9.5 (Continued)

Standard	GB	BS
Determination of organic carbon	**Specimens** Take soil specimen (0.1–0.5 g) and a specimen of pure sand; oxidize them respectively. **Dosage of potassium dichromate:** 10 mL **Oxidation conditions of specimens** For 5 min under oil bath of 170–180 °C **Titration of potassium dichromate** Dilute the solution to 60 mL; add phenanthroline indicator. Titrate it with ferrous sulphate solution. Until the color of solution changes from yellow to green and then suddenly to orange-red, the titration is terminated. Record the dosage of ferrous sulphate solution and the reading is estimated to 0.05 mL.	Take an appropriate amount of specimen which has been dealt with. (0.2–5.0 g) Add 20 mL of concentrated sulfuric acid (when concentrated sulfuric acid is dissolved in water, it will release heat) and place it on an adiabatic surface for 30 min. Add 200 mL of distilled water, 10 mL of phosphoric acid, and 1 mL of indicator; shake them well; titrate them with ferrous sulphate solution; then rotate the beaker until the color of solution changes from blue to green. Add 0.5 mL of potassium dichromate solution, the color of the solution is changed back to blue, and continue titrating it with ferrous sulphate solution until the color of solution changes from blue to green. Record the amount of ferrous sulphate solution which is accurate to 0.05 mL.
Data processing Calculation	$$C(FeSO_4) = \frac{C(K_2Cr_2O_7)V(K_2Cr_2O_7)}{V(FeSO_4)}$$ $$O_m = \frac{C(Fe^{2+})\left[V'(Fe^{2+}) - V(Fe^{2+})\right]}{m_s}$$ $$\times 0.003 \times 1.724 \times (1 + 0.01\omega) \times 100$$ O_m – content of organic matter (%), $C(Fe^{2+})$ – concentration of ferrous sulphate standard solution (mol/L), $V'(Fe^{2+})$ – content of blank titration of ferrous sulphate (mL), and $V(Fe^{2+})$ – content of specimen determination of ferrous sulphate (mL).	1. Volume of potassium dichromate solution used for oxidizing organic matters (V) : $V = 10.5\left(1 - \dfrac{y}{x}\right)$ y– dosage of ferrous sulphate solution in titration of specimen suspension (mL), and x–dosage of ferrous sulphate solution in titration of potassium dichromate solution (mL). 2. Percentage of original soil specimen passing through 2 mm test sieve $= \dfrac{m_2}{m_1} \times 100\,(\%)$ m_1 – mass of initial specimen (g), and m_2 – mass of the left specimen after passing through 2 mm sieve (g). 3. Content of organic matter $= \dfrac{0.67V}{m_3}$ m_3 – mass of the specimen used for the determination of organic matter (g).
Accuracy	to 0.01%	to 0.1%.
Parallel test	—	Parallel test on two specimens is necessary. If the difference between the two test results exceeds 2%, it shall be done again; if not, the average is used as the final result.

This test method is applicable for coarse-grained, fine-grained, organic and frozen soils. Parameters to be measured in this test include the mass of wet soil specimen and the mass of dry soil specimen.

There are many methods for determining the water content of soil, such as drying method and direct heating method. The drying method is to dry the soil sample to a constant weight in an oven at a certain temperature. The method can provide accurate results, but the test time is long. The heating method is to heat the sample in the container until the sample is dried with a constant weight. This method can be rapidly carried out even in field, but it is not as accurate as the drying method. The differences in drying method are shown in Table 9.6.

9.2.2 Density

Density of soil is closely correlated with other engineering properties. The methods for measuring density of soil include direct method and a wax-sealing method. The main difference between these two methods is the way to measure the specimen volume. The direct method is suitable for specimens in regular geometrical shape. Wax-sealing method is used for irregular samples. Here the volume of specimen is generally measured by immersing the specimen coated with wax into water. GB, ASTM and BS standards all present the details for direct method.

For undisturbed samples, the direct measurement method in GB uses a ring cutter with known size and mass, and the volume of the specimen is calculated by the size of the ring cutter. This method is simple to operate, but it is difficult to cut the soil sample with low water content. ASTM and BS use a caliper to measure the volume and calculate the specimen volume, but the measurement positions are different in these two methods. For remolded samples, the method is similar to that of undisturbed specimen in GB, ASTM and BS. The only difference is that GB uses ring cutters, whereas ASTM and BS use tubes or molds. See Table 9.8 for a detailed comparison of standards.

GB, ASTM and BS follow the same test steps in wax-sealing method. The main differences between them are in sample size, wax-sealing operation, and precautions after immersion. In terms of sample size BS needs the most and GB asks for the least. When making wax-sealing, GB requires a layer of wax, whereas ASTM requires two layers. Water-insoluble materials are used to fill the pores of the sample before waxing, and the concave corners are needed to be waxed with a brush in BS. Compared to ASTM and GB, the test operations in BS are relatively advanced. Before immersing samples in water, both GB and ASTM require that the water temperature is determined to properly calculate the volume of the wax-sealed sample; the bubbles at the bottom of the sample shall be eliminated in BS. After immersion in water, GB requires that the mass of the wax-sealed sample in the air is measured again to avoid effect of the water immersion in the seal sample. See Table 9.9 for a detailed comparison of the wax-sealing in these standards.

9.2.3 Specific gravity of soil solids

The specific gravity of soil particles, also known as the relative density of soil particles, refers to the mass ratio of soil particles to the same volume of pure water at 4 °C. The specific gravity can be used for the calculation of other physical properties of soil. Its value depends

Table 9.6 Drying Method for determining water content of soil

Standard		GB	ASTM	BS
Designation		GB 50123-1999	ASTM D2216-10	BS 1377-2: 1990
Scope		Coarse-grained soil, fine-grained soil, organic soil and frozen soil	Soil other than those with a high content of salt	All kinds of soil
Apparatus		1. Electric oven: the temperature can be controlled at 105–110 °C. 2. Balance: Weigh 200 g and it is readable to 0.01 g; weigh 1000 g and it's readable to 0.1 g.	1. Electric oven: It is best to use a forced ventilation oven the temperature can be controlled at 105–115 °C. 2. Balance: same as GB.	1. Electric oven: same as GB 2. Balance: when fine-grained soil is weighed it is readable to 0.01 g; when medium-grained soil is weighed it is readable to 0.1 g; when coarse-grained soil is weighed it is readable to 1 g.
Mass of specimen		1. Generally, take 15–30 g of representative specimen. 2. 50 g of organic soil, sand soil and monolithic frozen soil. 3. The layered and reticular frozen soil shall be cut by the method of quartering and then take 200–500 g of specimen (The mass of specimen is determined according to the degree of structural uniformity. If its structure is uniform, take less specimen; if not, take more.).	1. According to different maximum particle size of the specimen and test method, the mass of specimens is different. See Table 9.7 for details. 2. If the specimen is less than 200 g, large particles shall be removed.	1. Fine-grained soil: take at least 30 g. 2. Medium grain soil: take at least 300 g. 3. Coarse grain soil: take at least 3 kg.
Operation of test	Drying temperature	105–110 °C	110±5 °C	Same as GB
		If organic matter content is more than 5%, 65–70 °C.	For soils with high organic matter content, 60 °C.	No more than 80 °C
	Drying time	1. Clay, silt: not less than 8 h; 2. Sand: not less than 6 h.	60 °C 1. Generally, 12–16 h is needed. If the specimen is not dried, continue to bake for at least 2 h until the mass difference of specimens in two consecutive tests meet the requirement for accuracy; 2. If forced ventilation is used to dry the sand soil, only 4 h is needed.	The mass of the specimen is weighed once every 4 h until the mass difference of the specimens in two consecutive tests is less than 0.1% of the original specimen.

Other requirements	The layered and reticular frozen soil is made into a smooth paste after it has melted. Then take a piece of specimen from the mushy soil and determine its water content.	—

Process of data

Common soil

$$\omega = \left(\frac{m_0}{m_d} - 1\right) \times 100$$

ω – water content (%),
m_d – dry soil mass (g), and
m_0 – wet soil mass (g).

Layered or frozen soil

$$\omega = \left[\frac{m_2}{m_1}(1 + 0.01\omega_h) - 1\right] \times 100$$

ω – water content (%),
m_1 – mass of frozen soil specimen (g),
m_2 – mass of paste specimen (g), and
ω_h – water content of paste specimen (%).

For soil with a high salt content

—

$$\omega = \left(\frac{m_2 - m_3}{m_3 - m_1}\right) \times 100\%$$

ω – water content (%),
m_1 – mass of container (g),
m_2 – total mass of wet specimen and container (g), and
m_3 – total mass of dry specimen and container (g).

—

When there is a large amount of soluble salt in the specimen, the soluble salt will be precipitated under high temperature. The drying mass of the specimen is the sum of the mass of solid particle and the mass of soluble salt. The water content (ω_f) of the fluid in the specimen shall be calculated according to the following formula:

1. If the salt content p (the mass of soluble salt in a unit mass of fluid) in the fluid (mixed solution of water and salt in the pores) is known, then:

$$\omega_f = \frac{1000\omega}{1000 - p\left(1 + \dfrac{\omega}{100}\right)}$$

2. If the concentration of soluble salt q (the mass of soluble salt in a unit volume of fluid) and the density of fluid are known, then:

$$\omega_f = \frac{1000\omega}{1000 - \dfrac{p}{\rho_f}\left(1 + \dfrac{\omega}{100}\right)}$$

Requirement of accuracy	0.1%	1% (method A) 0.1% (method B)	Readable to two significant figures

Table 9.7 Specimen size and balance in ASTM

Maximum particle size (100% passing)	Method A Water content recorded to ±1%		Method B Water content recorded to ±0.1%	
Sieve size (mm)	Specimen mass	Balance readability (g)	Specimen mass	Balance readability (g)
75.0	5 kg	10	50 kg	10
37.5	1 kg	10	10 kg	10
19.0	250 g	1	2.5 kg	1
9.5	50 g	0.1	500 g	0.1
4.75	20 g	0.1	100 g	0.1
2.00	20 g	0.1	20 g	0.01

on the mineral compositions of the soil. The general inorganic soil particles normally have a specific gravity ranging from 2.6 to 2.8, the organic soil particles have the range of 2.4 to 2.5, and peat from 1.5 to 1.8.

The commonly used methods for determining specific gravity of soil particles include pycnometer, float, siphon cylinder, and gas jar methods. The scope of each method is listed in Table 9.10. See Table 9.11 for specific comparison. Because of the highest accuracy of the GB test results, it is recommended to test according to GB.

9.2.4 Particle size distribution

Soils is composed of solid particles with different sizes. Coarse-grained soil has large water permeability and is non-viscous or less viscous; the fine-grained soil has little water permeability and a large viscosity. In engineering, soil fraction is used to delineate ranges of soil particles. The sizes which divide soil fractions are called limiting particle sizes. As shown in Table 9.13, GB, ASTM and BS use different limiting particle sizes. The composition of particles in a soil can be presented with a curve of particle size distribution, where x axis presents the limiting sizes and y axis presents the cumulative mass percentage.

The methods for determining particle size distribution include sieve analysis, densitometer and pipette methods. Content of coarse soil fraction is determined by sieving method. Densitometer and pipette methods are used for fine fractions. The sieving method employs a set of sieves. The densitometer method allows the sample to settle freely. According to the Stokes formula, the diameter of soil particle is determined via measuring the sinking speed of the particles. The pipette method obeys the same rule as the densitometer method. Sieving and densitometer methods are included GB, ASTM and BS, and the pipette method is also included in GB.

Regarding sieving method, the main differences among these standards are the hole sizes of sieves as different standards use different limiting particle size. The strategy of choosing sieves is shown in Figure 9.1. The specific procedures for conducting sieving analysis in GB and ASTM are shown in Figure 9.2 and Figure 9.3, respectively. Standards of BS consist of wet and dry sieving methods. The test procedures in BS is simpler than that in ASTM, but more complicated than that in GB. Refer to Table 9.14 for a detailed comparison

Table 9.8 Comparison of direct measurement methods of density of soil

Standard		GB	ASTM	BS
Designation		GB 50123-1999	ASTM D7263-09	BS 1377-2: 1990
Scope		Fine-grained soil	Soil that can retain its shape	Clay soil with regular shape
Apparatus		Cutting ring	Caliper	
Shape of specimen		—	vertical cylinder, cuboid	
Requirements	Undisturbed specimen	Diameter No need to measure; the same as the inner diameter of the cutting ring---61.8 or 79.8 mm. Height No need to measure; the same as the height of the cutting ring---20 mm. —	Make at least 3 measurements at quarter points of height. Measure at least 3 times but there is no specification about measuring position. Side length of cuboid Length, width and height are measured at least 3 times respectively, and the measuring position is not specified.	Measure the diameters perpendicular to each other at both ends and the middle, for 6 times Measure at least 3 times and the measuring position is along the side of the cylinder with an interval of 360°/n, where "n" is the times of measurement. The times of measurement are not specified. The measuring position is at the edge of each surface and the center point.
	Remolded specimen	Same as that of undisturbed specimen	After the soil is extruded from the tube, it is measured in the same way as the method of cylinder, or the inner diameter and length of the tube are measured to replace those of the cylinder.	1. Measure the inner diameter of the mold or tube to replace the diameter of soil. 2. The height of the specimen is calculated by measuring the distance of the soil surface from the end of the tube.
Data processing	Accuracy	—	Readable to 4 significant digits	Accuracy to 0.1 mm
	Calculation formula for density	$\rho_0 = \dfrac{m_0}{V}$ ρ_0 – density (g/cm³), m_0 – mass of specimen (g), and V – volume of specimen (cm³).	Cuboid specimen: $\rho = \dfrac{1000m}{LBH}$ Cylindrical specimen: $\rho = \dfrac{4000m}{\pi D^2 L}$ ρ – density of specimen (g/cm³), m – mass of specimen (g), L, B, H – the length, width and height of the cuboid specimen (cm), and D, L – diameter and height of cylindrical specimen (cm).	

(Continued)

Table 9.8 (Continued)

Standard	GB	ASTM	BS
Calculation formula of dry density	$\rho_d = \dfrac{100\rho}{100+\omega}$ ρ_d – dry density of specimen (g/cm^3), ρ – density of specimen (g/cm^3), and ω – water content of specimen (%).		
Requirement of accuracy	<u>Density</u>: Accurate to 0.01 g/cm^3 <u>Water content</u>: —	Readable to 4 significant digits Readable to 4 significant digits	Same as GB Readable to 2 significant digits
Parallel tests	Parallel test on two specimens is necessary and the difference between the two results shall not be more than 0.03 g/cm^3. The average of the two results is the final result.	—	—

Table 9.9 Comparison of wax-sealing methods for density of specimens

Standard	GB	ASTM	BS
Designation	GB 50123-1999	ASTM D7263-09	BS 1377-2: 1990
Scope	Hard soil that is easily broken and irregular in shape	Specimen that the wax liquid cannot penetrate its surface	Block specimen with suitable size
Apparatus	Heater to melt wax, balance	Balance, container for melting wax, wire cage	Balance, bracket for weighing in water; equipment for melting wax
Required materials	Hard paraffin		In addition to the materials required in GB, water-insoluble materials such as putty or plastic clay are still required.
Requirement of specimen	Size The volume is not less than 30 cm³.	1. The minimum size is 30 mm, and the size of the largest particle in the specimen shall be less than one tenth of the minimum size of the specimen. 2. When the size of specimen is more than 72 mm, the maximum particle size shall be less than one sixth of the minimum size of the specimen.	The sides are roughly equal and at least 100 mm.
	Shape Sharp edges and angles on the surface shall be remove.	Same as GB; concave angle shall be avoided.	
Preparation	—	The specimen can be trimmed in the environment with a high humidity in order to reduce the effect on water content.	Fill the voids of the specimen with a water-insoluble material (putty or plastic clay); do not fill the holes left by the stones.
Melt waxing	—	It is best to melt in water bath at a constant temperature.	Use a control device for constant temperature to melt the wax. If not consisting of this device, use a woodworking gelatin pan.
Waxing	1. The specimen is slowly immersed in the wax liquid whose temperature is just over the melting point, and then it shall be immediately taken out; 2. When there are bubbles in the wax film, they shall be pierced with a needle and then filled with wax liquid.	Same as GB, and it shall be coated with two layers of wax.	At first, use a brush to paint a layer of wax to the concave surface of the specimen and the hole left by the stones. After then, the operation is the same as GB.
Immersion	When the wax-sealing specimen is in water, the temperature of the distilled water shall be determined.		When the wax-sealed specimen is in water, there must be no bubbles at the bottom.

(Continued)

Table 9.9 (Continued)

Standard	GB	ASTM	BS
Parameter to weigh	1. Mass of wet specimen--m_0; 2. Mass of wax-sealing specimen before being placed in water--m_n; 3. Mass of wax-sealing specimen in distilled water – m_{nw}; 4. Mass of the wax-sealing specimen after being taken out of the water. If the mass of wax-sealing specimen is different before and after it is put into water, the test shall be done again with new specimen.	Same as GB (1, 2, 3)	1. Mass of wet specimen--m_s; 2. Mass of specimen after its voids being filled – m_f; 3. Mass of wax-sealing specimen – m_w; 4. Mass of wax-sealing specimen in water – m_g.
Accuracy of weighing	Accurate to 0.01 g	Readable to 4 significant digits	Accurate to 1 g
Wet density	$$\rho_0 = \dfrac{m_0}{\dfrac{m_n + m_{nW}}{\rho_{wT}} - \dfrac{m_n - m_0}{\rho_n}}$$ ρ_{wT} – density of pure water at T °C (g/cm³), and ρ_n – density of wax (g/cm³).		$$V_s = (m_w - m_g) - \left(\dfrac{m_w - m_t}{\rho_P}\right)$$ $$\rho = \dfrac{m_s}{V_s}$$ ρ_P – density of wax (g/cm³), and V_s – volume of specimen (cm³).
Dry density	$$\rho_d = \dfrac{\rho_0}{1 + \omega_0}$$ ρ_d – dry density (g/cm³), ρ_0 – wet density (g/cm³), and ω_0 – water content of specimen.		
Accuracy	Accurate to 0.01 g/cm³	Readable to 4 significant digits	Same as GB
Parallel tests	Parallel tests on two specimens is necessary, and the difference of two results shall not exceed 0.03 g/cm³.	—	—

Table 9.10 Scope of methods for determination of specific gravity of soil particles

Standard	GB	ASTM	BS
Small pycnometer	Soils consisting of particles smaller than 5 mm	Soils consisting of particles smaller than 4.75 mm	Soils consisting of particles smaller than 2 mm
Float method	Soils consisting of particles greater than 5 mm; the mass content of particles greater than 20 mm is less than 10%.	Soils consisting of particles greater than 4.75 mm.	—
Siphon cylinder method	Soils consisting of particles greater than 5 mm; the mass content of soil particles greater than 20 mm is greater than 10% of the total mass of soil.	—	—
Gas jar method	—	—	Soils consisting of particle size greater than 37.5 mm is less than 10% of the total mass of soil.
Large pycnometer	—	—	Non-cohesive soil consisting of particle size smaller than 20 mm.

of sieving method among the standards. See Table 9.18 for detailed differences of particle size distribution analysis in different standards.

9.2.5 Atterberg limits of soil

The Atterberg limits refer to a set of water content of soil as the soil changes from one consistency state to another. It includes liquid limit, plastic limit and shrinkage limit. Liquid limit is the water content at which the soil changes from flowing state to plastic state. Plastic limit refers to the water content where the soil transfers from plastic state to semi-solid state, and the shrinkage limit is when the soil from semi-solid to solid state. Among them, liquid and plastic limits can be used for classification of fine-grained soil. Commonly used methods for determining the Atterberg limits include cone penetrometer method, Casagrande method, rolling method and shrinking method (Table 9.19). GB uses a cone with the weight of 76 g while BS uses a cone of 80 g. Correspondingly, the penetration depth indicating the liquid limit in these two standards are 17 mm and 20 mm, respectively. See Table 9.20 for a detailed comparison. Casagrande method is to place the sample on the disc where the sample is notched into two halves. The disc falls at a certain rate to hit the bottom plate, so that the two halves of the specimens are folded to a length of 13 mm with 25 hits. The water content satisfying this situation is the liquid limit. Casagrande method is included in GB, ASTM and BS. The difference between these standards is hitting number required to determine the liquid limit. BS requires that the hitting numbers in two successive tests is the same. ASTM requires that the difference of the hitting number in two successive tests shall not exceed 2. GB has no requirement in this regard (Table 9.22).

Rolling method in the most used method for determining the plastic limit of soil. This method has been involved in GB, ASTM and BS standards. However, the requirements

Table 9.11 Comparison of pycnometer methods in different standards

Standard	GB	ASTM	BS
Designation	GB 50123-1999	ASTM D854-14	BS 1377-2: 1990
Scope	Soils consisting of particles smaller than 5 mm	Soils consisting of particles smaller than 4.75 mm	Soils consisting of particle size smaller than 2 mm
Apparatus	Long-necked or short-necked pycnometer, balance, thermostatic water-bath, sand bath	Pycnometer, balance, oven, temperature measuring device, degassing device, insulated container	Small pycnometer, water bath, vacuum pump, oven, balance
Specimen	15 g of drying specimen	Dry or wet specimen. The required amount depends on the estimated specific gravity and type of specimen, see Table 9.12.	Two specimens, 5–10 g for each
Calibration of pycnometer	<u>Temperature</u> No requirement, but the appropriate temperature difference shall be within 5 °C <u>Accuracy</u> 1. Mass is accurate to 0.001 g; 2. Temperature is accurate to 0.5 °C; 3. Mass of the bottle is measured once; 4. Parallel tests on two specimens are necessary for the total mass of the bottle and water at each temperature, and the difference shall not exceed 0.002 g. <u>Calibration of results</u> Draw the relationship curve between temperature and the total mass of bottle and water.	Room temperature ± 4 °C, 15–30 °C 1. Mass is accurate to 0.01 g; 2. Temperature is accurate to 0.1 °C; 3. Mass of the bottle shall be measured for 5 times, and the average shall be taken. The standard deviation shall not exceed 0.02 g; 4. Total mass of the bottle and water at each temperature needs to be measured once. Calculate the volume of pycnometer at each temperature by the following formula: $$V_p = \frac{M_{pw,c} - M_p}{\rho_{w,c}}$$ $M_{pw,c}$ – total mass of bottle and water at the test temperature (g), M_p – mass of bottle (g), and $\rho_{w,c}$ – density of water (g/mL) at the test temperature. Take the average as the final result, and the standard deviation shall not exceed 0.05 mL.	—

Making mud	No need to distinguish between wet soil and dry soil; Pour the soil into the pycnometer and then add water into it.	Distinguish between wet soil and dry soil: 1. Wet soil shall be added with water to make it into mud, and then the mud is poured into the pycnometer. 2. For common dry soil, the method is the same as one in GB; if the soil cannot be crushed after being dried, it is made according to the method for wet soil.	Same as GB
Water needed	Inject half a bottle of distilled water	Inject water to 1/3–1/2 of the bottle	Inject water to cover soil
Degassing	Use one of the following two methods: 1. Boil: The sand shall be boiled for not less than 30 min, and the silt and clay for not less than 1 h. 2. Vacuum pumping shall not be less than 1 h.	Use one of the following three methods: 1. The boiling method is used only, and the duration shall not be less than 2 h; 2. The vacuum pumping method is used only, and the duration shall not be less than 2 h; 3. Both heating and vacuum pumping method are used; and the duration shall not be less than 1 h.	Only one method is provided: Vacuum pumping for 1 h, then agitate the liquid with a Chattaway spatula or shake the pycnometer. Repeat the GB test steps until no bubbles are generated. The total pumping time shall be several hours, preferably one night.
Temperature balance	The water bath is carried out until the temperature is stable, and the upper suspension in the bottle is clarified.	Place the pycnometer, temperature measuring device and water-filled bottle in an insulated and closed container for one night to achieve temperature balance.	The water bath lasts for at least 1 h or until the temperature in the bottle is stable.
Accuracy	Mass to 0.001 g; temperature to 0.5 °C	Mass to 0.01 g; temperature to 0.1 °C To	Mass to 0.001 g; temperature not specified
Mass of bottle and water	Check the relationship curve between temperature and total mass of bottle and water to get the total mass of bottle and water at test temperature.	The total mass of bottle and water at test temperature is obtained from the following formula: $M_{pw,t} = M_p + V_p \cdot \rho_{w,t}$	—
Calculation	The specific gravity of the soil is obtained from the following formula: $$G_s = \frac{m_d}{m_{bw} + m_d - m_{bws}} \cdot G_{iT}$$ G_s – specific gravity of soil particle, m_d – mass of dry soil (g), m_{bw} – total mass of bottle and water (g), m_{bws} – total mass of bottle, soil and water (g), and G_{iT} – specific gravity of pure water or neutral liquid at T °C.		
Parallel tests	Parallel test on two specimens is necessary, and the difference between the two results shall not be more than 0.02. The average is the final result.	—	The difference between the results of the two specimens shall not exceed 0.03 mg/m³, and the average shall be the final result.

Table 9.12 Required amount of specimen in ASTM for pycnometer test

Type of soil	Dry mass of the specimen when using a 250 mL hydrometer (g)	Dry mass of the specimen when using a 500 mL hydrometer (g)
SP, SP-SM	60±10	100±10
SP-SC, SM, SC	45±10	75±10
Silt or clay	35±5	50±10

Table 9.13 Limiting particle sizes (mm) in different standards

Standard		GB	ASTM	BS
Clay		$d \leq 0.005$		$d \leq 0.002$
Silt	Fine	$0.005 < d \leq 0.075$		$0.002 < d \leq 0.0063$
	Medium			$0.0063 < d \leq 0.02$
	Coarse			$0.02 < d \leq 0.063$
Sand	Fine	$0.075 < d \leq 0.25$	$0.075 < d \leq 0.425$	$0.063 < d \leq 0.2$
	Medium	$0.25 < d \leq 0.5$	$0.425 < d \leq 2$	$0.2 < d \leq 0.63$
	Coarse	$0.5 < d \leq 2$	$2 < d \leq 4.75$	$0.63 < d \leq 2$
Gravel	Fine	$2 < d \leq 5$	$4.75 < d \leq 19$	$2 < d \leq 6.3$
	Medium	$5 < d \leq 20$	—	$6.3 < d \leq 20$
	Coarse	$20 < d \leq 60$	$19 < d \leq 75$	$20 < d \leq 63$
Cobbler		$60 < d \leq 200$	$75 < d \leq 300$	$63 < d \leq 200$

for conducting the test vary from standard to standard. Refer to Table 9.25 for a detailed comparison.

Both GB and BS take shrinking dish as the method to determine the shrinkage limit of soil. The main difference between these two standards is the method to determine the volume of the specimen after drying. GB uses the wax-sealing method to determine the volume of the specimen, while BS employs the mercury overflow method (Table 9.26).

9.2.6 Relative density of cohesionless soil

The relative density is an index of the compactness of the cohesionless soil. It refers to the ratio of the difference between the void ratio at the loosest state (e_{max}) and that at the natural state (e_0) to the difference between the void ratio of the loosest state (e_{max}) and that the densest state (e_{min}). The calculation is as follow:

$$D_r = \frac{e_{max} - e_0}{e_{max} - e_{min}}$$

As the void ratio cannot be directly determined, this formula is transformed into:

$$D_r = \frac{(\rho_d - \rho_{dmin})\rho_{dmax}}{(\rho_{dmax} - \rho_{dmin})\rho_d}$$

(a)

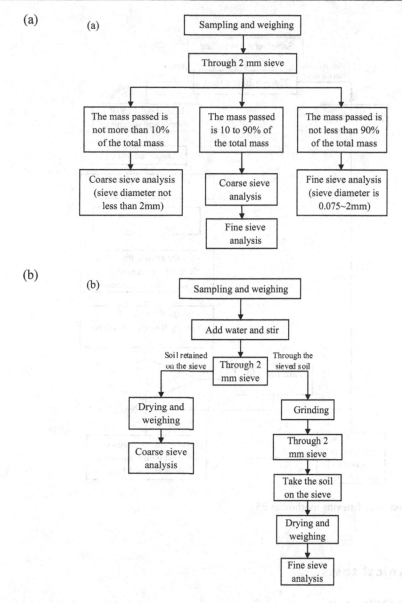

Figure 9.1 (a) Process of sieving method in GB, and (b) Sieving of fine-grained fractions in GB

Therefore, the test for relative density is actually to determine the maximum dry density (ρ_{dmin}), the minimum dry density (ρ_{dmax}) and the natural dry density (ρ_d).

For obtaining minimum dry density, the test methods are different in different standards are different. GB is applicable to soil consisting of particles smaller than 5 mm. In GB, the soil is evenly filled into the measuring cylinder by using a funnel. ASTM is suitable for soil consisting of particles smaller than 75 mm. BS is suitable for soil with particles smaller than 37.5 mm. A detailed comparison of these standards is given in Table 9.27 and Table 9.29.

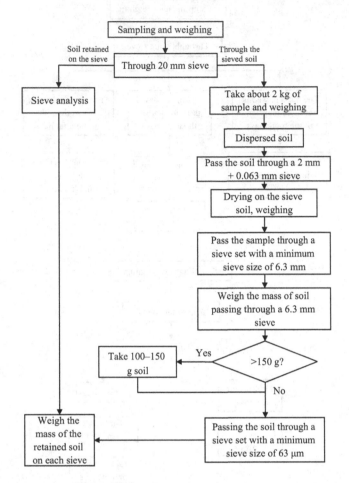

Figure 9.2 Procedures of sieving method in BS

9.3 Mechanical tests

9.3.1 Permeability test

The permeability tests refer to constant-head method and falling head method. GB standard involves both constant-head and variable-head methods, BS presents only the constant-head. The constant-head test was included in ASTM for determining the permeability coefficient of granular soil (ASTM D2434-68, 2006), but it was withdrawn in 2015. The falling-head test is not introduced in ASTM.

For the constant-head test, the main difference between GB and BS lies in the applicable scope. In GB, constant-head test is used to determine the permeability coefficient of coarse-grained soil. BS is suitable for soils with the permeability coefficient of 10^{-5}-10^{-2} m/s. See the detailed comparison between GB and BS in Table 9.31.

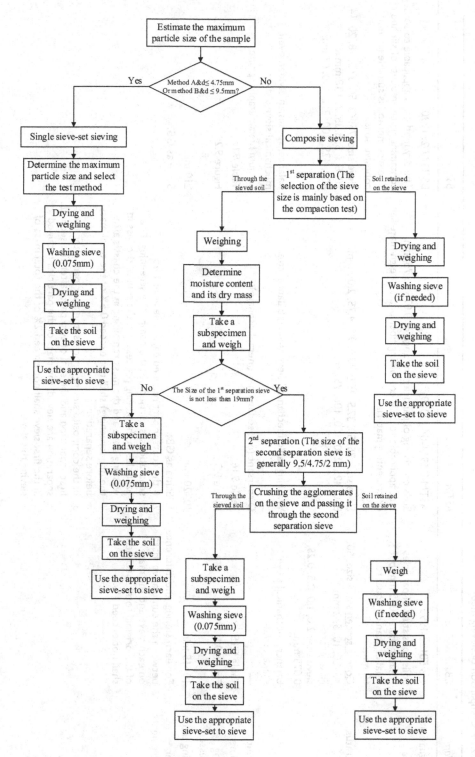

Figure 9.3 Procedures of sieving method in ASTM

Table 9.14 Comparison of sieving methods

Standard	GB	ASTM	BS
Designation	GB 50123-1999	ASTM D6913-04 (2009)	BS 1377-2: 1990
Scope	Soils consisting of particles greater than 0.075 mm but smaller than or equal to 60 mm.	Soils consisting of particles in range of 0.075 to 75 mm. Not applicable to soils containing fiber peat, foreign material or cementing material.	The dry method is applicable to soils containing trace amounts of silt and clay. The wet method is suitable for non-cohesive soils.
Apparatus	Coarse screen sieve size: 60, 40, 20, 10, 5, 2 mm Fine screen sieve size: 2.0, 1.0, 0.5, 0.25, 0.075 mm.	Sieve size: 75, 50, 37.5, 25.0, 19.0, 9.5, 4.75, 2.00 mm; 850, 425, 250, 150, 106, 75 μm.	Sieve size: 75, 63, 50, 37.5, 28, 20, 14, 10, 6.3, 5, 3.35, 2, 1.18 mm; 600, 425, 300, 212, 150, 63 μm.
	Balance, mechanical sieve shaker, oven	Washing tank, mechanical sieve shaker, balance, oven	Balance, specimen divider, oven, mechanical sieve shaker
Reagent	—	Dispersant (e.g. sodium hexametaphosphate)	Sodium hexametaphosphate
Specimen	Table 9.15	Table 9.16	Table 9.17
Sieving processes	Figure 9.1	Figure 9.3	Figure 9.2
Shaking time	10–15 min	10–20 min	≥10 min
Other requirements	For each sieving, the difference between the cumulative mass on each sieve and the mass of the original total specimen shall not exceed 1%.	1. Same as GB; 2. If the coarser part retained each time after two separations is carried out the composite screening, fine particles in the sieving process (i.e. the sum of the loss in washing sieve and the mass remaining in the chassis after the dry screening) shall not exceed 0.5% of the total mass before separation. 3. In the composite sieving process, the finer part after the first separation and the sum of two parts after the second separation are sieved, the mass of the particles retained in the first sieve shall not exceed 2% of the total mass of sieving this time.	Same as GB.

Accuracy of weighing	0.1 g	1. For method A, readable to 3 significant digits; 2. For method B, readable to 4 significant digits.	0.1%
Data processing	Calculate the percentage of the specimen less than a certain particle size to the total mass of the specimen; Calculate the gradation index (non-uniformity coefficient and curvature coefficient) if necessary.	Calculate the dry mass of the specimen retained on the Nth sieve; Calculate the screening percentage of the Nth sieve; Calculate the percentage of the specimen retained on the first sieve; Calculate the correction factor of the composite sieve.	Calculate the mass percentage of material retained on each test sieve; Obtain the percentage of specimen passing through the 63 µm test sieve by difference, and check by weighing the specimen in the receiver; Calculate the cumulative percentage of specimen passing through each sieve (in terms of mass).
Overall accuracy	—	Method A: 1% Method B: 0.1%	1%

Table 9.15 Mass requirement for specimen in GB sieve analysis

Particle size (mm)	Specimen mass (g)
<2	100–300
<10	300–1000
<20	1000–2000
<40	2000–4000
<60	>4000

Table 9.16 Mass requirement for specimens in ASTM sieve analysis

Maximum particle size of material (99% or more passes)		Minimum dry mass of specimen	
Alternative sieve designation	Maximum particle size (mm)	Method A results reported to nearest 1%	Method B results reported to nearest 0.1%
No. 40	0.425	50 g	75 g
No. 20	2.00	50 g	100 g
No. 4	4.75	75 g	200 g
3/8 in.	9.5	165 g	—
3/4 in.	19.0	1.3 kg	—
1 in.	25.4	3 kg	—
3/2 in.	38.1	10 kg	—
2 in.	50.8	25 kg	—
3in.	76.2	70 kg	—

Table 9.17 Mass requirement for specimens in BS sieve analysis

Maximum size of material present in substantial proportion (more than 10%), test sieve aperture (mm)	Minimum mass of sampling to be taken for sieving (kg)
63	50
50	35
37.5	15
28	6
20	2
14	1
10	0.5
6.3	0.2
5	0.2
3.35	0.15
2 or smaller	0.1

Table 9.18 Differences in particle size distribution analysis of different standards

Standard	GB	ASTM	BS
Designation	GB 50123-1999	ASTM D7928-16	BS 1377-2: 1990
Scope	Soils consisting of particles smaller than 0.075 mm.	Soils consisting of particles in the range of 0.0002–0.075 mm; Not applicable to soils containing fiber peat, foreign material or cementing material.	Soils consisting of particles from coarse sand to clay.
Conditions to use	The proportion of fine-grained soil is not less than 10%.	The proportion of fine-grained soil is not less than 5%.	Same as GB
Apparatus	Densitometer (Type A and B), measuring cylinder, washing sieve, washing sieve funnel, balance, boiling equipment, thermometer, stopwatch, blender, electric conductivity meter	Densitometer (151 H and 152 H), settling cylinder, grading sieve, thermometer, balance, oven, insulation device, dispersing device, soil suspension blender, timing device, soil suspension drying container, blender (manual), desiccator, mortar	Densitometer, glass graduated glass, thermometer, analytical sieve, balance, oven, evaporating dish, measuring cylinder, mechanical sieve shaker, stopwatch, desiccator, steel ruler, centrifuge, washing bottle, thermostat, pipette, electric heating plate, Buchner funnel, glass vacuum filter bottle, filter paper, vacuum source, vacuum tube
Reagent	4% sodium hexametaphosphate solution, 5% acidic silver nitrate solution, 5% acidic barium chloride solution	Sodium hexametaphosphate, isopropanol.	Hydrogen peroxide, sodium hexametaphosphate, methylated essential oil.
Specimen	1. The test specimen is air-dried specimen which has passed through the 2 mm sieve, and its dry mass is 30 g. 2. If the soluble salt content is greater than 0.5%, some salt shall be removed. 3. 200–300 g specimens are taken and screened by 2 mm sieve. Calculate the proportion of soil retained on the. The soil retained on the sieve is taken to determine the air-dried water content and then the mass of the air-dried specimen required for the test is calculated.	1. Wet specimen (recommended) or air-dried specimen. 2. As in GB (2), if the soluble salt content in the soil is high, some salt shall be removed, but no definite limits and methods are specified. 3. The test specimen is passed through a 2 mm sieve, and the mass of required specimen shall be calculated based on the capacity of the selected densitometer (45 g for 151 H, 55 g for 152 H), the estimated water content and the estimated pass rate of the 0.075 mm sieve.	1. Sampling: 100 g of sand; 50 g of silt; 50 g of clay. 2. If the specimen contains organic matter, hydrogen peroxide solution shall be used to process the specimen in advance. If the content of organic matter is more than 0.5%, parallel test is necessary through a blank control. 3.

(Continued)

Table 9.18 (Continued)

Standard	GB	ASTM	BS
Apparatus calibration	-	1. The densitometer shall be dry and clean. 2. Calibration of temperature and density: during the test, an accompanying measurement is carried out in the control tube filled with standard solution. 3. Calibration of Meniscus: Insert the densitometer into the solution containing the dispersant. The meniscus correction value is taken as the difference between the reading at the top of the meniscus and the reading of the intersection of solution and rod. 4. Effective depth is used to calculate the falling distance of particles at each reading.	1. Weigh the mass of densitometer to 0.1 g. 2. Calibration of dimension: calibrate by drawing the relationship curve of the effective depth and the corresponding reading of densitometer. 3. Calibration of Meniscus: same as ASTM
Dispersion	Add 200 mL water; soak overnight; boil for 40 min; pass through 0.075 mm sieve. Pour the sieved suspension into the measuring cylinder; add 10 mL sodium hexametaphosphate; add distilled water until the solution is 1 L.	Add 5.0±0.1 g of sodium hexametaphosphate; add 100 ml of pure water; and disperse the specimen with a stirring device or an air dispersing device. After the dispersion, add distilled water until the solution is 1 L.	Add 100 mL of dispersant, shake well, stir or vibrate, pass through a 63 μm sieve, place the suspension and the specimen which are under the sieve in a 1 L graduated cylinder, add distilled water until the solution is 1 L.
Temperature balance	—	Place it in a thermostatic water bath for overnight.	Place it in a thermostatic water bath or incubator for at least 1 h to reach a balance of temperature.
Mixed slurry	Use a blender. Stir for 1 min up and down along the depth of the suspension.	Shake well with a blender or through flipping: 1. Stirring time is the same as the time in GB, and the rate of flipping is 60 times per minute and the time of flipping is not less than 1 min.	Shake well by flipping, and the requirement for flipping is 60 times for 2 minutes.
Check before using the densitometer	—	1. Check whether there is flocculation or not. If there is, re-prepare the specimen. 2. When there is foam above the suspension, isopropanol can be added to dissipate the foam.	If there is foam above the suspension, two drops of methylated essential oil can be added to disperse the foam.
Time to read	0.5, 1, 2, 5, 15, 30, 60, 120, 1440 min	1, 2, 4, 5, 30, 60, 240, 1440 min	0.5, 1, 2, 4, 8, 30, 120, 480, 1440 min
Requirements of densitometer	1. Keep the floating bubbles of the densitometer at the center of the measuring cylinder.	1. Put the densitometer in the solution at 15 s and 20 s before reading is obtained.	1. Put the densimeter in solution at 15 s before reading is obtained.

2. The reading is based on the surface of the meniscus.
3. The reading of densitometer A is accurate to 0.5 and that of densitometer B is accurate to 0.0002.
4. After each reading, the densitometer shall be removed and placed in a graduated cylinder filled with distilled water.
5. Carefully put or take out the densitometer to avoid stirring the suspension.

2. The reading of HI51 densitometer is accurate to 0.00025 and the reading of HI52 densitometer is accurate to 0.25 g/L.
3. The densitometer shall be taken out slowly for a duration of 5 to 10 s.
4. The densitometer is put into the test water immediately and rotated to be clean after it is taken out.

2. Record the readings at 0.5, 1, 2, and 4 min at a time. Then take out the densitometer, place it in a dispersant solution and record the readings. Reinsert the densitometer into the specimen suspension and record the reading of densitometer at 8 min, 30 min, 2 h, 8 h and 24 h.
3. Same as GB (5).

Determining and Reading of temperature		

1. Same as GB.
2. The thermometer shall not cause interference in the suspension.

Accurate to 0.5 °C

1. In the first 15 min, observe and record the temperature of the suspension after each reading is obtained.
2. Same as GB.
3. If the change of temperature is more than ±1 °C, determine the reading again.

Calculation

1. Densimeter A:

$$X = \frac{100}{m_d} C_G \times (R + m_T + n - C_D)$$

2. Densimeter B:

$$X = \frac{1000V}{m_d} \times C_G \times \left[(R' - 1) + m'_T + n' - C'_D \right]$$

$$\cdot \rho_{w20}$$

3. Diameter of particle:

$$d = \sqrt{\frac{1800 \times 10^4 \cdot \eta}{(G_s - G_{wT}) \rho_{wTg}} \cdot \frac{L}{t}}$$

1. Densimeter HI51:

$$N_m = \left(\frac{G_s}{G_s - 1} \right) \left(\frac{V_{sp}}{M_d} \right) \rho_c \times (r_m - r_{m,d}) \times 100$$

2. Densimeter HI52:

$$N_m = 0.6226 \left(\frac{G_s}{G_s - 1} \right) \left(\frac{V_{sp}}{M_d} \right) \rho_c$$

$$\times (r_m - r_{m,d}) \times \left(\frac{100}{1000} \right)$$

3. Diameter of particle:

$$D_m = 10 \sqrt{\frac{18\mu}{\rho_w g (G_s - 1)} \cdot \frac{H_m}{t_m}}$$

1. True reading of the hydrometer:

$$R_h = R'_h + C_m$$

Diameter of particle:

$$D = 0.005531 \sqrt{\frac{\eta Hr}{(\rho_s - 1)t}}$$

Reading correction:

$$R_d = R'_h - R'_o$$

2. Proportion of soil consisting of particles smaller than D:

$$K = \frac{100 \rho_s}{m(\rho_s - 1)} R_d$$

Accuracy —

The proportion of particles is accurate to 1%

Table 9.19 Methods for determining Atterberg limits

Atterberg limits	GB	ASTM	BS
Liquid limit	Cone penetrometer, Casagrande method	Casagrande method	Cone penetrometer, Casagrande method
Plastic limit	Cone penetrometer, rolling method	Rolling method	Rolling method
Shrinkage limit	Shrinking dish	—	Shrinking dish

Table 9.20 Comparison of cone penetrometer

Standard	GB	BS
Designation	GB 50123-1999	BS 1377-2: 1990
Scope	Soils consisting of particles smaller than 0.5 mm and organic matter content less than 5%.	Soils consisting of particles smaller than 0.425 mm
Apparatus	Mass of cone–76 g; angle of cone–30°; specimen container has an inner diameter of 40 mm and a height of 30 mm	Penetrometer, stainless steel cone (mass of 80 g and angle of 30 °), metal container (inner diameter of 55 mm and height of 40 mm)
Specimen	250 g of undisturbed or 200 g of air-dried specimen passing through 0.5 mm sieve. Add water until soil becomes paste and soak it overnight.	1. Original soil: take 500 g of natural soil and remove the coarse particles if there are any. Add water until soil becomes paste and soak it overnight. 2. Sieved soil: enough specimens are sieved with the 0.425 mm sieve and ensure that there is at least 300 g of soil which passes through the sieve.
Treatment of specimen	The specimen is thoroughly stirred. No specified stirring time; sufficient rubbing is required for dry soil.	Four-point method: take 300 g of soil; One-point method: take 100 g of soil; The specimen is mixed with a palette knife for at least 10 min.
The usage of cone	1. The cone is painted with petroleum jelly. 2. The tip of the cone shall touch the surface of the specimen. 3. After releasing the cone for 5 seconds, record the penetrating depth of cone. 4. Remove the petroleum jelly from the tip of the cone; take the specimen near the cone to determine its water content. 5. At least three specimens with different water contents shall be determined. The depths of the three measurement shall be within 3–4 mm, 7–9 mm and 15–17 mm, respectively.	1. General requirements are the same as (2, 3) in GB; 2. Four-point method: 1) For the first determination, the cone penetrating depth shall be within 15 mm; 2) Add soil and make the soil level flat; read again; 3) If the difference between the penetration depths in two successive measurements is less than 0.5 mm, record the average. If less than 1 mm and more than 0.5 mm, add soil and make the soil level flat for the third determination. If the differences among all readings are less than 1 mm, the average of the three readings is recorded as the penetrating depth of the single specimen. Otherwise the specimen shall be changed and then test again; 4) At least four specimens are needed. Each reading shall be between 15–25 mm. 3. One-point method: 1) The penetrating depth shall be about 20 mm. 2) The soil around the tip of cone is taken to determine the water content.

Standard	GB	BS
Accuracy of reading	—	Weight: accurate to 0.01 g; Cone penetration: to 0.1 mm.
Data processing	The graph of water content vs. penetration depth shall be a straight line. The water content corresponding to 17 mm penetration is taken as the liquid limit; that corresponding to 2 mm is the plastic limit.	1. Four-point method: 1) Same as GB; 2) Liquid limit: cone penetration 20 mm. 2. One-point method: The determined value is multiplied by a correction coefficient (see Table 9.21) to get the liquid limit.
Calculation accuracy	The water content is accurate to 0.1%	

Table 9.21 Correction coefficient for calculating the liquid limit for BS cone penetrometer method

Cone penetration/mm	Correction coefficient		
	< 35%	35% to 50%	>50%
15	1.057	1.094	1.098
16	1.052	1.076	1.075
17	1.042	1.058	1.055
18	1.030	1.039	1.036
19	1.015	1.020	1.018
20	1.000	1.000	1.000
21	0.984	0.984	0.984
22	0.971	0.968	0.967
23	0.961	0.954	0.949
24	0.955	0.943	0.929
25	0.954	0.934	0.909
Plasticity	Low	Intermediate	High

Table 9.22 Comparison of Casagrande method in GB, ASTM and BS

Standard	GB	ASTM	BS
Designation	GB 50123-1999	ASTM D4318-10[e1]	BS 1377-2: 1990
Scope	Soils consisting of particles less than 0.5 mm	Soils consisting of particles less than 0.425 mm	
Apparatus	Casagrande disc, notcher	Casagrande disc, notcher, gauge	Casagrande disc, notcher, gauge
Specimen	250 g of natural soil specimen or 200 g of air-dried soil that has passed through 0.5 mm sieve. Add water until soil become paste and soak it overnight.	1. Wet method: Take 150–200 g of 425 µm sieved soil and adjust its water content to be suitable for the test. Soak it overnight. 2. Dry method: dry the soil at a temperature lower than 60 °C.	1. Natural soil: Take 500 g of natural soil and remove the coarse particles, if any. Add water until soil become paste and soak it overnight. 2. Sieved soil: take 300 g of soil which has passed through the 0.425 mm sieve.

(Continued)

Table 9.22 (Continued)

Standard	GB	ASTM	BS
Calibration of apparatus	The copper disc is lifted until it is 10 mm in height.	The copper disc is lifted until it is 10 ± 0.2 mm in height.	Same as GB.
Operation of test	1. The center of specimen is 10 mm in thickness. 2. Use a notcher to cut the specimen along the diameter of the copper disk and through the center of the snail wheel so that a V-shaped groove is formed. 3. Rotate the handle at a speed of two revolutions per second to make the copper disc rise and fall repeatedly. When it falls, it is required to hit the base until the specimen on both sides of the groove is combined together and the total length is 13 mm. The total times of hits is recorded, then the water content can be determined. 4. There shall be 4–5 specimens with different water content, and 15–35 hits are needed to combine specimens.	1. General requirements: Same as (1, 2) in GB; Same as (3) in GB; but rotate the handle at a speed of 1.9–2.1 revolutions per second. 2. Multi-point method: Adjust the water content at least three times. They are required to fall separately and hit the base. The times of hits shall be distributed in three ranges of 25–35, 20–30, and 15–25. 3. One-point method: The specimens with the same water content are required to fall and hit the base several times until the difference in times of hits of two tests is no more than two.	1. General requirements: Same as (2, 3) in GB; Add a small amount of specimen and test again until the times of hits is the same for two consecutive tests. In this way, a test is finished and the water content of the specimen is determined. 2. Four-point method: 1) There shall be 50 hits in the first test; 2) Test at least 4 specimens with different water content; 3) The times of hit shall be in the range of 10 to 50 and they shall be evenly distributed. 3. One-point method: It is necessary to adjust the water content of the specimen until the times of hit is 15–35, preferably close to 25. Then start to hit the copper disc.
Data processing	1. Draw the relationship curve between water content and times of hits; 2. When the times of hits is 25, the water content is taken as the liquid limit.	1. Multi-point method: same as GB. 2. One-point method: the measurement needs to be corrected by the coefficient in Table 9.23.	Same as ASTM; the correction coefficients are shown in Table 9.24 for one-point method.
Accuracy	Accurate to 1%.		

Table 9.23 Correct coefficient for liquid limit test in ASTM (Casagrande method)

Number of bumps	Factor	Number of bumps	Factor
20	0.973	26	1.005
21	0.979	27	1.009
22	0.985	28	1.014
23	0.990	29	1.018
24	0.995	30	1.022
25	1.000		

Table 9.24 Correct coefficient for liquid limit test in BS (Casagrande method)

Number of bumps	Factor	Number of bumps	Factor	Number of bumps	Factor	Number of bumps	Factor
15	0.95	21	0.98	26	1.00	31	1.02
16	0.96	22	0.99	27	1.01	32	1.02
17	0.96	23	0.99	28	1.01	33	1.02
18	0.97	24	0.99	29	1.01	34	1.02
19	0.97	25	1.00	30	1.02	35	1.03
20	0.98						

Table 9.25 Comparison of rolling method

Standard	GB	ASTM	BS
Designation	GB 50123-1999	ASTM D4318-10[gl]	BS 1377-2: 1990
Scope	Soils consisting of particles smaller than 0.5 mm	Soils consisting of particles smaller than 0.425 mm	Same as ASTM
Apparatus	frosted glass plate, caliper	frosted glass plate, gauge	glass plate, a thin rod
Specimen	Take 100 g of representative specimen which has passed through the 0.5 mm sieve. Add water, stir well and soak overnight.	Weigh 20 g of soil used in the liquid limit test, reduce the water content.	Same as ASTM
Method of reducing the water content	Not specified	Air-dry	
Main operation	1. 8–10 g of soil. 2. 3–5 g of the fractured soil strips to determine the water content. 3. Two parallel tests.	1. 1.5–2.0 g. 2. After the specimen is broken, 6 g of specimen is needed for water content test. 3. Same as (3) in GB.	1. 2.5 g for each, totally 8 specimens. 2. The water content of each specimen is determined after rolling.
Requirements of rolling	1. Starts to break at a diameter of 3 mm. 2. Rolling pressure shall be evenly applied. The soil strip shall not be hollow, and its length shall not be more than the width of palm.	1. Cracks at a diameter of 3.2 mm. 2. Rate of rolling is 80–90 rolls per minute. 3. Rolling shall finish within 2 min.	Repeat the specimen until it breaks at a diameter of 3 mm.
Data processing	1. Two parallel tests are necessary. 2. When the result is less than 40%, the difference between parallel tests shall be no more than 1%; when the result is greater than 40%, the difference shall be no more than 2%.	1. Same as (1) in GB. 2. The difference is no more than 0.5%.	Same as ASTM

Table 9.26 Comparison of the shrinking dish method

Standard	GB	BS
Designation	GB 50123-1999	BS 1377-2: 1990
Scope	Soils consisting of particles smaller than 0.5 mm	Soils consisting of particles smaller than 0.425 mm
Apparatus	Shrinking dish (diameter of 45–50 mm, height of 20–30 mm), caliper	Shrinking dish (diameter of 57 mm; height of 12 mm), glass plate or transparent acrylic, porcelain evaporation dish
Specimen	200 g	50 g
Specimen installation	1. Layer the specimen into the shrinking dish. 2. After each filling, slap the bottom of the shrinking dish on the test table until the air is exhausted. 3. Flatten the surface after the dish is filled.	The main requirements are the same as those in GB. Additional requirements: 1. Three layers and the amount of each layer occupies 1/3 the volume of the shrinking dish. 2. Same as in GB (2). Layers of blotting paper or similar materials are added to allow the soil to flow to the edge of the plate and release any bubbles that appear.
Measurement of volume	Wax-sealing method	Overflow mercury method
Calculation	$\omega_n = \omega - \dfrac{V_0 - V_d}{m_d} \cdot \rho_w \times 100$ ω_n – shrinkage limit of soil (%), ω – the water content during preparation (%), V_0 – the volume of wet specimen (cm^3), and V_d – the volume of dried specimen (cm^3). Accurate to 0.1%	Accurate to 1%

9.3.2 Compaction test

The compaction test is to determine the maximum dry density and the optimum water content of the disturbed soil under a certain amount of compaction work. The compaction test includes light and heavy compactions according to the compaction work. The size of sample chamber, the compactor, the number of sample layers and the number of shots per layer are all important factors affecting the test results. The specific requirements of the compaction test in GB, ASTM and BS are different (Table 9.32). Both GB and ASTM require compacting multiple unused samples with different water contents, and the compacted samples are not allowed to be reused in the subsequent tests. BS allows reuse of non-friable sample for later tests. A detailed comparison is given in Table 9.33.

9.3.3 Swelling ratio and swelling pressure of cohesive soil

Expansive soil, also called "swellable soil", expands in volume after being immersed in water and shrinks when losing water. The swelling ratio and swelling pressure are indicators for evaluating engineering properties of expansive soil. Swelling ratio refers to the ratio of

Table 9.27 Determination of minimum dry density

Standard	GB	ASTM	BS
Designation	GB 50123-1999	ASTM D4254-16	BS 1377-4: 1990
Scope	Soil with particles smaller than 5 mm, the content of fraction of 2–5 mm is not more than 15%.	Cohesionless soil with less than 15% of particles smaller than 75 μm. Method 1: content of 37.5–75 mm fraction < 30%. Method 2: soil consisting of particles less than 19 mm. Method 3: content of 2.0–9.5 mm fraction < 10%.	Sand: content of 0.063–2 mm fraction < 10%. Gravel: content of 0.063–37.5 mm fraction < 10%.
Apparatus	Measuring cylinder, long neck funnel, conical plug, blade	Drying oven, test sieve. Methods 1 and 2: Standard molds, balances, pouring devices, rigid thin-walled tubes, mixing pans, large metal spoon, hair-bristled dusting brush and metal straightedge. Method 3: Glass graduated cylinder, balance	1. Sand: measuring cylinder, rubber bung, balance, drying oven, test sieve. 2. Gravel: metal mold, metal tray, spoon, straightedge, balance, test sieve, drying oven.
Sampling	700 g of dried specimen	The mass of specimen is related to its maximum particle size. See Table 9.28 for details.	1. Sand: 1000 ± 1 g 2. Gravel: At least 50% larger than the internal volume of the mold.
Accuracy	The unit is g/cm^3	The result is readable to three or four significant digits. The unit is Mg/m^3 or g/cm^3.	0.01 Mg/m^3

Table 9.28 Required mass of specimen in ASTM

Maximum particle size (100% passing) (in., mm)	Mass of required specimen (kg)	Placing device for soil specimens	The size of the mold (ft^3, cm^3))
3, 75	34	shovel or oversized spoon	0.500, 14200
3/2, 38.1	34	spoon	0.500, 14200
3/4, 19.0	11	spoon	0.100, 2830
3/8, 9.5	11	Pouring device with a nozzle of 1 in. (25 mm) diameter	0.100, 2830
No. 4, 4.75 or smaller	11	Pouring device with a nozzle of 1/2 in. (13 mm) diameter.	0.100, 2830

the sample height after immersion to its original height, under confined conditions. Swelling pressure is the internal stress generated by the soil when it is immersed in water. Swelling pressure is numerically equal to the pressure required to maintain the sample height when the soil is immersed in water. Both GB and ASTM contain tests for swelling rate and swelling pressure, while BS contains only the test for swelling pressure. The differences between GB and ASTM for the swelling rate test and swelling pressure test are shown in Tables 9.36–9.37.

Table 9.29 Comparison of maximum dry density test

Standard	GB	ASTM	BS
Designation	GB 50123-1999	ASTM D4253-16	BS 1377-4: 1990
Scope	Soil consisting of particles smaller than 5 mm and the fraction of 2–5 mm is less than 15%.	Soil consisting of particles smaller than 75 mm; the fraction of smaller than 0.075 mm is less than 15%.	Sand: Soil consisting of particles smaller than 6.3 mm; the fraction of smaller than 0.063 mm is less than 10%. Gravel: Soil consisting of particles smaller than 37.5 mm; the fraction of smaller than 0.063 mm is less than 15%.
Apparatus	Metal cylinder, vibrating fork, hammer.	Mold, dial indicator, balance, hoist, drying oven, sieve, calibration bar, vibration table (electromagnetic vibration table, eccentric or cam driven vibration table), mixing pans, metal scoop, hair-bristled dusting brush, stopwatch, micrometer.	In addition to ASTM, a set of sieves are necessary.
Mass of specimen	2000 g	It is related to the maximum particle size of the specimen. See Table 9.30 for details.	Sand: two specimens, each about 3 kg. Gravel: two specimens, each about 8 kg.
Test operation	1. The specimen is poured into the metal cylinder three times, 1/3 of the volume of the cylinder each time. 2. The vibrating fork hits both sides of the cylinder at a speed of 150 to 200 times per minute. 3. While the vibrating fork is tapping, hammer the surface of the specimen with a hammer, 30 to 60 times per minute.	1. Specimen is placed into the mold at one time. 2. Requirements of vibration: vertical vibration with double amplitude. (1) 0.33 ± 0.05 mm; 60 Hz; 8 ± 0.25 min; (2) 0.48 ± 0.08 mm; 50 Hz; 10 ± 0.25 min.	1. Specimen mounted in three layers for compaction. 2. Compaction of each layer with a vibrating hammer for at least 3 min.
Data processing	The unit is g/cm^3	Readable to four significant digits. Mg/m^3 or g/cm^3.	0.01 Mg/m^3

Table 9.30 Required mass of specimen in ASTM for maximum dry density

Maximum particle size (100% passing), in. (mm)	Mass of specimen required (kg)	Size of mold to be used, ft^3 (cm^3)
3 (75)	34	0.500 (14200)
3/2 (38.1)	34	0.500 (14200)
3/4 (19.0)	11	0.100 (2830)

Table 9.31 Comparison of determination of permeability of soils

Name of standard	GB	BS
Designation	GB 50123-1999	BS 1377-5: 1990
Scope	Coarse-grained soil	Soils with permeability coefficient of about 10^{-2}-10^{-5} m/s
Apparatus	Constant-head test device: metal-sealed cylinder (inner diameter of 10 cm; height of 40 cm), metal perforated plate, strainer, manometer tube and water supply bottle, drain pipe (adjustable)	Constant-head test device: permeation cell (inner diameter of 75–100 mm), reservoir (with adjustable height), drain tank (immovable), metal perforated plate, strainer, manometer tube
Specimen	1. 3–4 kg air-dried soil. 2. Maximum particle size <1/10 the diameter of the sample chamber.	1. Air-dried sample with no more than 10% passing through the 63 μm sieve. 2. Maximum particle size < 1/12 the diameter of the sample chamber.
Preparation	Fill it with water until the water level of the manometer tube is the same as the overflow hole.	Fill it with water until the water level of the manometer tube is the same as that of the upper reservoir.
Readings	1. Record the level of water in the manometer tube after it is stable. 2. Record the amount of seepage at prespecified interval. 3. Determine the temperature of water at the inlet and outlet and the average is taken.	1. Record the levels of water in the manometer tubes. 2. Record the amount of seepage in a given time interval or record the time when the water in the cylinder reaches a certain position. 3. Record the temperature of the water in the discharge reservoir.
Calculation formula under the condition of single hydraulic gradient	1. Permeability coefficient of the specimen at the test temperature: $$k_T = \frac{QL}{AHt}$$ k_T – permeability coefficient (cm/s) of specimen when the temperature of water is T °C, Q – the amount of seepage in t seconds (cm³), L – distance between the centers of the manometer tubes at both ends (cm), A – area of cross section of the specimen (cm²), H – average of the difference of water level (cm), and t – time (s). 2. Permeability coefficient k_{20} of specimen at standard temperature: $$k_{20} = k_T \frac{\eta_T}{\eta_{20}}$$ η_T–coefficient of dynamic viscosity of water at T °C (kPa·s), and η_{20}–coefficient of dynamic viscosity of water at 20 °C (kPa·s).	1. Calculate the rate of flow, q_1, q_2 etc. at one hydraulic gradient. $$q_1 = \frac{Q_1}{t} \text{ etc.}$$ 2. Q_1, Q_2, etc. is the volume of water collected from the outlet reservoir during each time period t (in sec). Calculate the average rate of flow, q, for the set of readings at one hydraulic gradient. 3. hydraulic gradient i: $$i = h/y$$ h – the difference between the two manometer levels (in mm), and y – the difference between the corresponding gland points (in mm). 4. Permeability coefficient at standard temperature k: $$k = \left(\frac{q}{i}\right)\left(\frac{R_t}{A}\right)$$ Rt – Temperature correction factor, and A – Area of cross section of the specimen (mm²).
Calculation accuracy	—	to two significant figures.

9.3.4 California Bearing Ratio

California Bearing Ratio (*CBR*) was first proposed by the California Highway Bureau. It is defined as the ratio in percent and at a penetration of either 0.1 or 0.2 in. of: (1) stress required to penetrate a soil mass to (2) stress required to penetrate a standard material (crushed aggregate) using standard equipment and procedures. It is one of the main strength indexes of roadbed and pavement materials. *CBR* tests are included in GB, ASTM and BS.

CBR tests stipulated in GB and ASTM are similar, except some minor differences in sample preparation. A detailed comparison between these two standards is given in Table 9.38.

9.3.5 Consolidation

The compressibility of soil refers to the rearrangement of the soil particles with the reduced soil volume under external force. The compression test is mainly carried out by using an oedometer, which gives a compression curve (a curve representing the relationship between stress and void ratio of a soil as obtained from a consolidation test). Based on the compression curve, the coefficient of compressibility and compression index can be obtained. Compression index is the slope of the linear portion of the compression curve on a semi-log plot. Coefficient of compressibility is the secant slope, for a given pressure increment, of the compression curve.

Consolidation of soil is a special case of the saturated soil under pressure, which refers to the process in which the internal water is discharged and the pore water pressure is dissipated. The soil consolidation test uses oedometer to determine the height of the soil specimen at different times under a fixed pressure. The plots from consolidation test is typically a curve of vertical displacement vs. time. This curve gives consolidation coefficient of the specimen.

Consolidation tests can be performed via stress-controlled or strain-controlled apparatus. Both methods are included in GB, ASTM and BS. A detailed comparison of consolidation test in different standards is given in Table 9.39.

9.3.6 Direct shear test

The shear strength of soil refers to the ultimate strength of soil resistance to shear failure. The shear strength of soil is used for earth pressure calculation, slope stability verification, engineering design, etc.

This direct shear test for determining shear strength of soil is included in GB, ASTM and BS. GB stipulates three types of direct shear, namely quick shear, consolidated quick shear, and slow shear. The quick shear is to apply horizontal thrust to shear the specimen immediately after applying vertical pressure, and the specimen does not need to be consolidated. The consolidated quick shear first drains the specimen to a stable state under a certain load, and then horizontal thrust is applied at a large rate to shear the specimen. While the slow shear test is to apply horizontal thrust to shear at a low rate to allow drainage.

In BS, the specimen is divided into fine-grained soils and sands. The fine-grained soils require the specimen to be consolidated. The test procedure is similar to that in ASTM, but the formula for calculating the maximum shear rate of the specimen is different. There is no need to consolidate the specimen in sandy soil direct shear test. It requires a certain shear rate to cause shear failure of the specimen in 5 to 10 min. Note that the temperature of the test environment is specified in BS. See a detailed comparison of the standards in Table 9.40.

Table 9.32 Basic requirements of compaction in different standards

Standard	Hammer type	Applicable conditions	Mold Diameter (mm)	Mold High (mm)	Hammer mass (kg)	Fall high (mm)	Number of layers	Hits per layer	Hammer point distribution angle (°)
GB 50123-1999	Light compaction	d_{max} <5 mm, clay soil	102	116	2.5	305	3	25	53.5
	Heavy compaction	d_{max} <20 mm	152		4.5	457	5	56	45
		d_{max} <40 mm					3	94	
ASTM D698-12	Light compaction	X (d>4.75 mm)≤25%	101.6	116.4	2.495	305	3	25	
		X (d>9.5 mm)≤25%							
		X (d>19 mm)≤30%	152.4					56	
ASTM D1557-12	Heavy compaction	X (d>4.75 mm)≤25%	101.6		4.5364	457.2	5	25	
		X (d>9.5 mm)≤25%							
		X (d>19 mm)≤30%	152.4					56	
BS1377-4	Light compaction	d_{max}≤37.5 mm X (d≤20 mm) ≥95%	105	115.5	2.5	300	3	27	Evenly distributed
		d_{max}≤63 mm X (d≤37.5 mm) ≥90% 95%≥X (d≤20 mm)≥70%	152	127				62	
	Heavy compaction	d_{max}≤37.5 mm X (d≤20 mm)≥95%	105	115.5	4.5	450	5	27	
		d_{max}≤63 mm X (d≤37.5 mm)≥90% 95%≥X (d≤20 mm)≥70%	152	127				62	

Note: X ( ) refers to the mass ratio of soil in accordance with the conditions in brackets.

Table 9.33 Comparison of compaction tests

Standard	GB	ASTM	BS
Designation	GB 50123-1999	ASTM D698-12 ASTM D1557-12	BS 1377-4: 1990
Apparatus	Compactor, sample cylinder, balance, platform scale, standard sieve, and specimen ejector.	Compaction cylinder, rammer, specimen extruder, drying oven, balance, straightedge, standard sieve, mixing tools.	Standard sieves, balances, apparatus for water content determination, equipment for chopping hard cohesive soil, compaction cylinders, and rammer.
Sampling	20 kg for light compaction; 50 kg for heavy compaction	23 kg for cylinder of 101.6 mm diameter; 45 kg for 152.4 mm diameter	See Table 9.34.
Specimen preparation	Sample by quartering; air-dry and crush the sample; No specified treatment.	See Table 9.35	See Table 9.34. Same as ASTM
Required number of specimens	5	Same as GB	Non-fragile soil, 1; Fragile soil, 5
Moisture content of specimen	One specimen is close to the optimal water content (it is estimated as the plastic limit); two are higher than the optimum; two are lower than the optimum. The difference shall be around 2%.		1. Sandy or gravelly soil: the difference of water content shall be 1% or 2%. 2. Cohesive soils: the difference shall be 2 to 4%.
Compaction cylinder	A layer of lubricant shall be applied to the inner wall of the cylinder.	—	
Hammering point	1. Hammering points shall be evenly distributed at a certain angle. 2. For heavy compaction, an extra hit is required at the point in each lap.	Same as (1) in GB	Hammering points shall be evenly distributed on the surface of the specimen without specific requirements for the pattern.
Compaction of each layer	Each layer of specimen shall be equal in height.	-	Same as (1) in GB
Calculation of dry density	$\rho_d = \dfrac{\rho_0}{1+0.01\omega_i}$ ρ_d – dry density of specimen (g/cm^3), ρ_0 – density of specimen (g/cm^3), and ω_i – water content of a specimen at a certain point (%).		
Plotting	According to the test data, plot dry density vs. water content, as well as the saturation line. The formula of saturation line: $\omega_{sat} = (\dfrac{\rho_w}{\rho_d} - \dfrac{1}{G_s}) \times 100$ ω_{sat} – the saturated water content of the specimen (%), ρ_w – density of water at 4 °C (g/cm^3), ρ_d – dry density of specimen (g/cm^3), and G_s – specific gravity of soil particles.		Formula: $\rho_d = \dfrac{1 - \dfrac{V_a}{100}}{\dfrac{1}{\rho_s} + \dfrac{W}{100\rho_w}}$

Table 9.34 Requirement of samples in BS

Size of specimen	Process method	Mass of single specimen (kg)	
		Non-fragile specimen	Fragile specimen
d≤20 mm	All for testing	6	2.5
d≤37.5 mm X (d≤20 mm) ≥95%	Soil passing through 20 mm sieve is used.	6	2.5
d≤37.5 mm 95%≥X (d≤20 mm) ≥70%	All for testing	15	6
d≤63 mm X (d≤37.5 mm) ≥95% X (d≤20 mm) ≥70%	Soil passing through a 37.5 mm sieve is used.	25	6
d≤63 mm 95%≥X (d≤37.5 mm) ≥90% X (d≤20 mm) ≥70%	Remove the specimen retained on the 37.5 mm sieve and replace it with a specimen of equal mass retained on the 20 mm sieve.	15	6

Note: X () Refers to the proportion of the mass of the soil that meets the conditions in brackets.

Table 9.35 Requirement of static time of specimen in ASTM

Classification	Minimum static time (h)
GW, GP, SW, SP	No requirement
GM, SM	3
Other soil	16

Table 9.36 Comparison of swelling ratio in different standards

Item	GB		ASTM
	With load	Without load	
Designation	GB 50123-1999		ASTM D4546-14
Apparatus	Oedometer, cutting ring (diameter 61.8 mm or 79.8 mm, height 20 mm), dial gauge (range 10 mm, graduation of 0.01 mm; or displacement sensor with 0.2% accuracy)		In addition to the requirements in GB, the device shall be capable of maintaining the specified loads for long periods of time.
Specimen	With no specific requirement		1. min. diameter 50 mm; 2. min. height 20 mm; 3. height of the specimen shall be at least 6 times greater than the largest particle size of the specimen. 4. Variations in length or diameter shall not exceed 5%.

(Continued)

Table 9.36 (Continued)

Item	GB		ASTM
	With load	Without load	
Sample installation	A thin filter paper is needed between the sample and the permeable plate.		It is not advised to use filter paper.
Water injection requirements	Inject water from the top, and the water head is 5 mm higher than the sample.		Submerge the sample.
Reading intervals	At an interval of 2 h		0.5, 1, 2, 4, 8, 15, 30 min, 1, 2, 4, 8, 24 h.
Measurement accuracy	Swelling displacement shall be accurate to 0.01 mm.		Height of specimen and diameter of mold shall be measured to the nearest 0.025 mm or better.
Temperature requirements	—		Temperature: 22 ± 5 °C.
Calculation	$\delta_{ep} = \dfrac{z_p + \lambda - z_0}{h_0} \times 100$ δ_{ep} – the swelling ratio under certain load (%), z_p – the reading of the displacement meter when soil stops swell under certain load (mm), z_0 – the reading of displacement meter before loading (mm), λ – compression deformation under a certain load (mm), and h_0 – the initial height of samples (mm).	$\delta_e = \dfrac{z_t - z_0}{h_0} \times 100$ δ_e – the swelling ratio before loading at time t (%), and z_t – the reading of the displacement meter at time t (mm).	$h_1 = h - \Delta h_1$ $\varepsilon_s = \dfrac{100 \Delta h_2}{h_1}$ ε_s – swelling strain (%), Δh_2 – change in specimen height: swell or collapse caused by wetting, mm), h_1 – specimen height immediately prior to wetting, mm, h – initial height of specimen, mm, and Δh_1 – specimen compression after stress application and immediately prior to wetting, mm.
Accuracy	—		0.1%

Table 9.37 Comparison of swelling pressure tests in different standards

Item	GB	BS
Designation	GB 50123-1999	BS 1377-5: 1990
Water Injection requirements	Inject water from the top, and the water surface is 5 mm higher than the sample.	Submerge the sample
Parameters record	Record the maximum equilibrium load applied	Record loads and time.
Maximum expansion	0.01 mm	
Cease the test	The specimen stops swell under certain load for 2 h.	Examine the curve of cumulative load vs. time.

Item	GB	BS
Temperature requirements	—	The ambience temperature shall be within 22 ± 5 °C. The variation of temperature of the oedometer, specimen, and immersion reservoir must not exceed ±2 °C throughout.
Calculation	$$P_e = \frac{W}{A}$$ P_e – swelling pressure (kPa), W – the total equilibrium load imposed on the sample (N), and A – sample area (cm³).	
Accuracy	—	Keep two significant figures

Table 9.38 Comparison of CBR between GB, ASTM and BS

Items	GB	ASTM	BS
Designation	GB 50123-1999	ASTM D1833-16	BS 1377-4: 1990
Scope	1. Disturbed soil consisting of particles ≤ 20 mm; 2. When compacting sample in three layers, the grain sizes shall be ≤ 40 mm.	If the specimen contains large particles, the grain size of the soil shall be corrected.	Particles>20 mm shall be limited within 25%.
Apparatus	Test cylinder, hammer, guide cylinder, standard sieve, expansion measuring device, perforated top plate with adjusting rod, penetration instrument (penetration rod with diameter of 50 mm, length of 100 mm), water tank, platform scale, stripper, etc.	Loading machine, permeation measuring device, molds, spacer disk, rammer, expansion-measuring apparatus, additional loads, penetration pistons (with diameter of 49.63 ± 0.13 mm, and with length≥101.6 mm)	Standard sieve, CBR mold, instrument for compaction, perforated plate, perforated expansion plate, tripod, dial gauge, annular surcharge discs, a cylindrical metal plunger (with nominal cross-sectional area of 1935 mm², diameter of 49.65±0.10 mm, length of 250 mm),
Sample Preparation	1. Sieve the soil on a 20 mm or 40 mm sieve for sample preparation. 2. Prepare samples according to the heavy compaction test, and determine the maximum dry density and the optimum water content, and prepare three specimens with optimal water content. 3. To prepare samples of three different dry density, prepare nine samples; the dry density of the sample shall be 95–100% of the maximum dry density.	Same as GB	1. Sieve the soil on a 20 mm test sieve for sample preparation. 2. Samples can be prepared in six ways, including compression with tamping, compression in layers, rammer compaction to a specified density, vibrating compaction to a specified density, rammer compaction with specified effort and vibrating compaction with specified effort.

(Continued)

Table 9.38 (Continued)

Items	GB	ASTM	BS
Soak Requirements	1. The water surface shall be 25 mm above the specimen; 2. Immerse the specimen allowing the access of water to its top and bottom. 3. Soak in water for 4 full days. 4. After soaking, measure the expansion; take out the sample, absorb the surface moisture of the sample, and allow it to drain downward for 15 min before removing the device. 5. seating weigh of 5 kg.	1. Same as GB (2, 3, 4) 2. Seating weight shall not be less than 4.54 kg.	1. Soak the specimen bottom up. 2. Be soaked for 4 days. 3. Same as GB (4). 4. No specified seating weight.
The seating load of the penetration piston.	45 N	Less than 44 N	Estimate the *CBR* value. For *CBR* value < 5%, 10 N. For *CBR* value between 5% and 30%, 50 N. For *CBR* value > 30%, 250 N.
Rate of Penetration	1–1.25 mm/min	1.27 mm/min	1 ± 0.2 mm/min
Depth of Penetration	10–12.5 mm	13 mm	<7.5 mm
Reading	At time intervals of 20, 40, 60 sec etc.	At intervals of penetration, 0.64, 1.3, 1.9, 2.5, 3.18, 3.8, 4.45, 5.1, 7.6, 10, 13 mm.	At an interval of penetration 0.25 mm.
Standard Stress	For penetration of 5 mm, the standard stress is 7000 kPa; For 5 mm, the standard stress is 10500 kPa.	For penetration of 2.54 mm, the standard stress is 6.9 MPa; For 5.08 mm, 10 MPa.	For 2.5 mm, 13.2 kN; For 5 mm, 20 kN.
CBR Value	If the ratio at 5 mm penetration is greater than that at 2.5 mm, retest. If the check test gives a similar result, use the bearing ratio at 5 mm penetration.		Compare the *CBR*s at 2.5 mm and 5 mm. Take the higher one as the *CBR* value.

9.3.7 *Reversal direct shear test*

The reversal direct shear test is often used to determine the residual strength of soil. After the shear stress reaches its peak value, the continuous increase of displacement will lead to the gradually reduced strength. Eventually, the shear stress reaches a certain stable value, which is the residual strength. It is can be used for stability analysis of old or ancient landslides.

Table 9.39 Consolidation test contrast

Items	GB	ASTM	BS
Designation	GB 50123-1999	ASTM D2435/D2435M-11	BS 1377-5: 1990
Scope	Applicable to saturated clay.	It is generally applied to saturated, undisturbed and fine-grained soil naturally deposited in water, but it is also suitable for compacted soil and undisturbed soil formed by other causes such as weathering or chemical alteration.	Applicable to saturated or near-saturated soil and undisturbed natural sedimentary soils collected from the ground in the form of cores or blocks.
Apparatus	Oedometer (cutting ring, guard ring, water permeable plate, water tank, pressurized upper cover), load device, deformation-measuring equipment	Load device, oedometer, porous disc, filter screen, specimen trimming device, diameter, and balances.	Oedometer (consolidation ring, porous disc, oedometer, diameter, loading device).
Calibration of instruments	Oedometer and load device shall be calibrated regularly.	The measured axial deformations shall be corrected whenever the deformation of equipment itself exceeds 0.1% of the initial specimen height.	1. Deformation of the apparatus may be significant when testing stiff soils but can often be ignored for soft soils. 2. The test apparatus needs to be calibrated by replacing the soil sample with a metal plate.
Loading sequence	The loading sequence shall be 12.5, 25, 50, 100, 200, 400, 800, 1600, 3200 kPa.	The standard loading schedule shall consist of a load increment ratio which is obtained by approximately doubling the total axial stress on the soil to obtain values of about 12, 25, 50, 100, 200 kPa, etc.	A loading sequence is specified as follows: 6, 12, 25, 50, 100, 200, 400, 800, 1 600, 3 200 kPa.
The initial pressure	The initial pressure depends on the type of the soil, and it is appropriate to be 12.5 kPa, 25 kPa or 50 kPa.	—	The initial pressure depends on the type of soil, and the first loading increment shall be greater than the swelling pressure.
The maximum pressure	1. Shall be greater than the effective pressure which occurs in situ because of the overburden and proposed construction.	—	Same as GB (1).

(Continued)

Table 9.39 (Continued)

Items	GB	ASTM	BS
Seating load	1 kPa	1. Generally 5 kPa; 2. For expansive soil, the seating load needs to be increased; 3. If a total axial stress of 5 kPa will causes significant consolidation, reduce the seating load to 3 kPa or less.	-
Loading Requirements	1. When it is necessary to measure the sedimentation rate and the consolidation coefficient, it is advised to measure the height change of the sample in the following time intervals after applying each stage of pressure: 6 s, 15 s, 1 min, 2 min 15 s, 4 min, 6 min 15 s, 9 min, 12 min 15 s, 16 min, 20 min 15 s, 25 min, 30 min 15 s, 36 min, 42 min 15 s, 49 min, 64 min, 100 min, 200 min, 400 min, 23 h, 24 h, until the specimen is stable. 2. When it is not necessary to measure the sedimentation rate, the height change of the sample is measured as the stability standard after each increment of load for 24 h. For the specimen with the compression coefficient is measured only, after each load increment, when the deformation reading is 0.01 mm/h, record the height change as the stable standard.	Choose one of the following two methods: 1. The standard load increment duration shall be approximately 24 h. For at least two load increments including at least one load increment after the pre-consolidation stress has been exceeded; record the axial deformation, at time intervals of approximately 0.1, 0.25, 0.5, 1, 2, 4, 8, 15 and 30 min, and 1, 2, 4, 8 and 24 h. 2. For each increment, record the axial deformation, at time intervals of approximately 0.1, 0.25, 0.5, 1, 2, 4, 8, 15, 30 min, and 1, 2, 4, 8 and 24 h.	Only one method is introduced: 1. A typical test comprises four to six increments of loading; 2. Each held constant for 24 h; 3. Take readings of the compression gauge at suitable time intervals of 0, 10, 20, 30, 40, 50 s 1, 2, 4, 8, 15, 30 min 1, 2, 4, 8, 24 h. Plot the compression gauge readings against logarithm of time, and maintain the pressure until the plotted readings indicate that primary consolidation has been completed; then increase the pressure to the next value.

Items	GB	ASTM	BS
Unloading	1. If the rebound test is necessary, unload the specimen when the consolidation is stable under a certain pressure. 2. Read the rebound after 24 h.	1. Unloading to the seating load 5 kPa. 2. Unloading process is in reverse to the loading process; each successive stress level is as large as one fourth of the proceeding stress level. 3. Once the change in axial deformation has reduced to less than 0.2% in an hour, record the end-of-test axial deformation and remove the specimen.	1. Normally the number of unloading stages shall be at least half the number of loading stages. 4. Read the compression gauge at convenient time intervals. Plot the readings so that the completion of swelling can be identified.
Temperature	—	1. The standard test temperature shall be 22 ± 5 °C. 2. The temperature of the oedometer, test specimen, and immersion reservoir shall not vary more than ±2 °C throughout the test.	Record daily highest and lowest air temperatures near the apparatus (accurate to 1 °C)
Post-test treatment	Determine the water content of the tested specimen.	1. Measure the height of the specimen to the nearest 0.01 mm by taking the average of at least four evenly spaced measurements over the top and bottom surfaces of the specimen. 2. Determine the final water content.	Weigh the specimen and determine the water content.
Data processing	Calculating: initial void ratio, settlement and void ratio after each pressure increment, coefficient of compression and the compressive modulus of a certain pressure range, coefficient of volume compressibility, the compression index, and the rebound index within a certain pressure range. Curve: void ratio in linear scale against the applied pressure in a logarithmic scale. Determine the pre-consolidation pressure.	Calculation: dry mass, dry density, volume, height, initial and final void ratio, initial and final saturation degrees, change in specimen height, rate of consolidation, coefficient of consolidation, and pre-consolidation pressure.	Calculation: initial water content, initial bulk density, initial dry density, the initial void ratio, and compressibility. Curve: compression in linear scale against the applied pressure in logarithmic scale.

Table 9.40 Comparison of direct shear test in different standards

Items	GB	ASTM	BS
Designation	GB 50123-1999	ASTM D3080/ D3080M-11	BS 1377-7: 1990
Scope	1. Slow shear is for fine-grained soil; 2. Quick shear is for fine-grained soils with permeability coefficient less than 10^{-6} cm/s	All kinds of soil	
Apparatus	Shear device, cutting ring, displacement device, etc.	Shear device, shear box, loading device, normal force measurement device, displacement meter, cutting ring, etc.	Shear box, loading device, displacement meter, Vernier caliper, stop watch, balance, etc.
Specimen	Cylinder specimen	Circular or square specimen	Cuboid specimen
	At least four specimens.	At least three	Three
	Diameter of 61.8 mm and thickness of 20 mm.	1. The minimum specimen diameter for circular specimens, or width for square specimens shall be 50 mm or not less than ten times the maximum particle size. 2. The minimum initial specimen thickness shall be 13 mm, but not less than six times the maximum particle size. 3. The ratio of diameter (width) to thickness shall be greater than 2:1.	The thickness of specimen shall not be less than ten times the maximum particle size.
	1200 g	—	The samples need to be deaired.
Environment	—	—	Air temperature shall be maintained at ± 4 °C.
Seating load	—	5 kPa	—
Cease of consolidation	The consolidation is ceased when the deformation is not more than 0.00 5 mm/h.	Same as GB	Plot the vertical deformation against elapsed time to define the primary consolidation and cease the consolidation.

Items	GB	ASTM	BS
Before shearing	Remove the alignment screws or pins from the shear box.	Use the gap screws to separate the shear box halves to approximately the diameter of the maximum particle size or 0.64 mm as a minimum default value for fine grained materials.	For fine-grained soils, lift the upper shear box by a half-turn of the screws. For sandy soils the lift might need to be slightly greater but it shall not exceed than 1 mm.
Shearing rate	1. In slow shear test, the shearing rate shall be less than 0.02 mm/min; 2. In consolidate quick shear test, quick test and sand-soil direct shear test, the shearing rate shall be 0.8 mm/min to cause failure within 3–5 min.	According to the consolidation curve under the maximum normal stress, total estimated elapsed time to failure t_f can be determined. d_f is the estimated relative lateral displacement at failure. Determine the appropriate maximum displacement rate R_d from the equation: $R_d = d_f / t_f$, the shearing rate shell be less than R_d.	1. For the consolidated quick shear test, the shearing rate is fixed to cause failure within 5–10 min. 2. For the slow shear test, the shearing rate is the same as that in ASTM.
Estimated elapsed time to failure	For slow shear: $$t_f = 50t_{50}$$ t_{50}: time in minute required for the specimen to achieve 50% consolidation.	Time logarithm: $$t_f = 50t_{50};$$ Time square root method: $$t_{90}t_f = 11.6t_{90}$$ t_{50} – time required for the specimen to achieve 50% consolidation under the maximum normal stress increment, and t_{90} – time required for the specimen to achieve 90% consolidation under the maximum normal stress.	Root time curve: $$t_f = 50t_{100}$$ t_{100} – time required for the specimen to achieve 100% consolidation.
Data reading	At an interval of lateral displacement of 0.2–0.4 mm.	at relative lateral displacements of about 0.1, 0.2, 0.3, 0.4, 0.5, 1, 1.5, 2, 2.5, 3, and then every 2% relative lateral displacement.	At an interval of horizontal displacement of 0.1 mm.
Cease of the test	At a further shear displacement of 4 mm after peak.	The specimen shall be sheared to a strain of at least 10%.	Same as GB.

Table 9.41 Reversal direct shears test comparison

Item	GB	BS
Designation	GB 50123-1999	BS 1377-7: 1990
Scope	Cohesive soil	Any type
Apparatus	Strain-controlled shear device, cutting ring, displacement meter	Shear box, loading device, displacement meter, Vernier caliper, stop watch, balances
Specimen	1. Top and bottom surfaces of the specimen shall be parallel to the weak layer in the middle height. 2. for reconstructed specimen, the allowable difference in density of the specimens in the same group shall be less than 0.03 g/cm^3.	Same as the specimen used in direct shear test.
Test Operating	The vertical deformation is measured at an interval of 1h. The consolidation is completed when the deformation in one hour is not more than 0.005 mm.	Plot the vertical deformation against the elapsed time. The consolidation ceased when the primary consolidation is reached.
	Remove the alignment screws or pins from the shear box.	For fine-grained soils, lift the upper shear box by a half-turn of the screws. For sandy soils the lift might be slightly greater but it shall not exceed 1 mm.
	0.02 mm/min for clay, and 0.06 mm/min for silt.	Determine the appropriate maximum displacement rate R_d from the following equation: $$R_d = \frac{d_f}{t_f}$$ where, R_d is the maximum shearing rate. t_f, the estimated time to failure, and d_f, the estimated lateral displacement to failure.
	Before peak stress, the shear force and shear displacement shall be recorded at an interval of shear displacement of 0.2–0.4 mm. After peak stress, record the data every 0.5 mm.	At an interval of shear displacement of 0.1 mm. At least 20 readings shall be taken before the peak.
	Stop forwarding the shear box at a shear displacement of 8–10 mm and return the shear box to its original position.	Stop forwarding until the full travel of the apparatus has been reached.
	1. Shearing rate shall be less than 0.6 mm/min; 2. Wait for half an hour prior to the next shearing.	Reverse the shear box by a motor driver until the two halves of the shear box return to their original alignment. Adjust the rate of reversing so that the reversing operation takes place over a period of time is equal to the time from the beginning of shearing to the peak stress. Reverse within a few minutes, by hand winding. Allow to stand for at least 12 h to enable pore pressure equilibrium to be re-established.

Item	GB	BS
Data Processing	1. Calculate the shear stress; plot the curve of shear stress against shear displacement. The first shear stress peak on the graph is taken as the peak shear strength, and the final shear stress is the residual strength. 2. Plot the curve of residual strength against vertical pressure to determine the friction angle and cohesion.	1. Calculate the initial water content, the initial bulk density, the initial voids ratio, the initial degree of saturation, the voids ratios at the end of the consolidation stage and at the end of shearing, and the shear stress and the normal stress. 2. Plot the curve of shear strength against cumulative shear displacement. Determination of peak shear strength and residual strength are the same as GB. 3. Plot the curve of vertical deformation against shear displacement. 4. Plot peak strength and residual strength against normal stress to determine the friction angles and cohesions.

In reversal direct shear test, the shear box was pushed back and forth so that the specimen is sheared repeatedly for several times until a stable shear stress value is reached. The test is included in both GB and BS. In ASTM, the ring shear test (D6467-13) is suggested. A detailed comparison of reversal direct shear in different standards is summarized in Table 9.41.

9.3.8 Unconfined compressive

Unconfined compressive strength is the maximum axial stress that a saturated soil can resist without lateral confining. The ratio of unconfined compressive strength of undisturbed soil to that of remodeled soil is called sensitivity. The higher the sensitivity of a soil is, the stronger structure the soil has. This test is included in GB, ASTM and BS as shown in Table 9.42.

9.3.9 Undrained shear strength in triaxial compression

In triaxial compression test, the axial stress σ_1 is applied under the condition that the cylindrical specimen is subjected to a confining pressure σ_3. According to the drainage condition, the triaxial compression test can be divided into unconsolidated undrained test (UU), consolidated undrained test (CU) and consolidated drainage test (CD). UU test doesn't allow the sample water to be extruded during the application of ambient and axial pressures. CU test allows the specimen to drain for being consolidated under a certain surrounding pressure, but does not allow drainage during axial compression. It can give the effective shear strength parameters and pore pressure coefficient. CD test allows drainage in both consolidation and axial compression stages. It gives the effective shear strength. All these three types of tests are included in GB, ASTM and BS. Table 9.43 summarizes the differences in UU test in different standards, while Table 9.44 compares the CU tests.

Table 9.42 Comparison of unconfined compressive strength test in different standards

Item	GB	ASTM	BS
Designation	GB 50123-1999	ASTM D2166/D2166M-16	BS 1377-7: 1990
Scope	Saturated cohesive soil.	Cohesive materials	Saturated, non-fissured cohesive soil.
Apparatus	Strain-controlled unconfined compression device (composed of dynamometer, compression frame, lifting equipment), axial displacement meter, balance.	Compression device, deformation indicator, stop watch, balance	Compression device, deformation indicator, force-measuring device, stop watch
Specimen	35–50 mm	1. Specimens shall have a minimum diameter of 30 mm and the largest particle contained shall be smaller than one tenth of the specimen diameter. 2. For specimens consisting of particles with diameter of 72 mm or larger, the largest particle size shall be smaller than one sixth of the specimen diameter.	38–100 mm
	height-to-diameter ratio: between 2 and 2.5	—	2.0
Preparation of Test Specimens	A thin layer of petroleum jelly shall be applied to the surface of specimen.	—	—
Axial strain rate	1 to 3% per min	0.5 to 2% per min	Same as ASTM
Reading	Prior to 3% axial strain, record it at a strain interval of 0.5% (or 0.4 mm). Beyond 3%, record it at a strain interval of 1% (or 0.8 mm).	usually 10 to 15 points for composing a stress-strain curve.	at least 12 sets of readings in order to plot the stress-strain curve.
Time to failure	8–10 min	≤15 min	—
Failure criteria	1. after the axial loading reaches a maximum value, continue the axial strain of 3–5%. 2. in the case where no peak loading is observed, an axial strain of 20% is needed before stop the test.	In the case where no peak loading is observed, an axial strain of 15% is needed before stop the test.	Same as ASTM

Data Processing

Axial strain

Axial strain: $\varepsilon_l = \dfrac{\Delta h}{h_0}$

The average cross-sectional area: $A_a = \dfrac{A_0}{1 - \varepsilon_l}$

Compressive stress: $\sigma = \dfrac{P}{Aa}$

Accuracy

1. Calculate the axial strain to the nearest 0.1%.

2. Calculate the compressive stress to three significant figures or nearest 1kPa.

Calculate the axial strain and the compressive stress to two significant figures.

Sensitivity

$S_t = \dfrac{q_u}{q_u}$

S_t – sensitivity,

q_u – unconfined compressive strength of intact specimen (kPa), and

q_u – unconfined compressive strength of remolded specimen (kPa).

Table 9.43 Comparison of unconsolidated-undrained test

Item	GB	ASTM	BS
Designation	GB 50123-1999	ASTM D2850-15	BS 1377-7: 1990
Scope	Fine-grained soil and coarse-grained soils with particle diameter≤20 mm.	All kinds of soil	All kinds of soil
Specimen	Shape: Cylinder Diameter: 1. The diameter of the specimen shall be 35–101 mm. For fissured specimen, the diameter shall exceed 60 mm. 2. When the diameter is less than 100 mm, the maximum particle diameter shall be less than one-tenth of the specimen diameter. 3. When the diameter of the specimen exceeds 100 mm, the maximum particle diameter is not more than 1/5 of the diameter of the specimen. Height-to-diameter ratio: 2–2.5	Specimens shall have a minimum diameter of 33 mm. The largest particle size shall be smaller than one sixth the specimen diameter.	The diameter of the specimen shall be 38–110 mm. The size of the largest soil particle shall not be greater than one-fifth of the specimen diameter.
Setting up	1. Place the impervious plate, the sample and the impervious sample cap on the base of the compression chamber. 2. Place the specimen and fill the compression chamber with distilled water. 3. Contact the sample with the force-measuring device, install the deformation indicator, and zero setting. 4. Close the drain valve, open the surrounding pressure valve and apply ambient pressure.	1. Encase the specimen in the rubber membrane. 2. Bring the axial load piston into contact with the specimen. During this procedure, take care not to apply an axial stress to the specimen exceeding approximately 0.5% of the estimated compressive strength. men cap. 3. Attach the pressure-maintaining and measurement device and fill the chamber with the chamber fluid. Wait approximately ten minutes after the application of chamber pressure to allow the specimen to stabilize under the chamber pressure prior to application of the axial load.	2 1. Place the specimen centrally on the base pedestal of the triaxial cell, ensuring that it is in correct vertical alignment. 2. Fill the triaxial cell with water, ensuring that all the air is displaced through the air vent. 3. Pressurize the triaxial cell. 4. Adjust the loading machine to bring the loading piston within a few millimeters of its seating on the specimen top cap. 5. Adjust the machine further to bring the loading piston just into contact with the seating on the top cap.

Strain rate	0.5–1.0%/min	The rate of strain shall be chosen so that the time to failure does not exceed from 15 min, normally 0.3 to 1% per min.	Select a rate of axial deformation such that failure is produced within a period of 5 min to 15 min.
Stop the test	The axial strain has reached at least 15–20%.	1. An axial strain of 15% is normally necessary; 2. If there is a peak stress observed, an extra axial strain of 5% is necessary; 3. When the deviatoric stress has peaked then dropped by more than 20%.	1. Continue the test until the maximum value of the axial stress has been passed and the peak is clearly defined; 2. Or until an axial strain of 20% has been reached.

Calculation

Axial strain:

$$\varepsilon_i = \frac{\Delta h_i}{h_0} \times 100$$

ε_i – axial strain (%),
Δh_i – change in height of specimen during loading (mm), and
h_0 – initial height of test specimen (mm).

Average cross-sectional area:

$$A_a = \frac{A_0}{1 - \varepsilon_i}$$

A_a – average cross-sectional area (cm²), and
A_0 – initial average cross-sectional area of the specimen (cm²).

$$\sigma_1 - \sigma_3 = \frac{CR}{A_a} \times 10 \qquad \sigma_1 - \sigma_3 = \frac{P}{A_a}$$

σ_1 – major principal total stresses (kPa),
σ_3 – minor principal total stresses (kPa),
C – dynamometer rate coefficient (N/0.01 mm or N/mV),
R – the reading of force-measuring device (0.01 mm), and
10 – unit conversion factor.

$\sigma_1 - \sigma_3$ – measured principal stress difference (deviatoric stress) (kPa), and
P – measured applied axial load (kN).

(Continued)

Table 9.43 (Continued)

Item	GB	ASTM	BS
	—	**Correction for Rubber Membrane:** The following equation shall be used to correct the deviatoric stress for the effect of the rubber membrane if the error in deviatoric stress because of the strength of the membrane exceeds 5%: $\Delta(\sigma_1 - \sigma_3)_m = \dfrac{4E_m t_m \varepsilon_1}{D}$ $\Delta(\sigma_1-\sigma_3)_m$ – membrane correction to be subtracted from the measured principal stress difference, D – diameter of specimen (m), E_m – Young's modulus for the membrane material (kPa), t_m – thickness of the membrane (m), and ε_1 – axial strain.	The correction is obtained directly from Figure 9.4, at the strain corresponding to failure, for specimens 38 mm diameter fitted within a membrane of latex rubber 0.2 mm thick when a predominantly barreling type of deformation occurs. For specimens of any other diameter, D (in mm), and latex rubber membranes of any other thickness, t (in mm) (which may be made up of more than one membrane), multiply the correction derived from Figure 9.4 by a factor equal to $\dfrac{38}{D} \times \dfrac{t}{0.2}$
The Young's modulus of the rubber membrane	—	The Young's modulus of the rubber membrane (E_m) may be determined by hanging a 15.0 mm (0.6 in.) circumferential strip of membrane over a thin rod, placing another rod through the bottom of the hanging membrane, and measuring the force per unit strain obtained by stretching the membrane. The modulus value may be computed using the following equation: $E_m = (F/A_m)/(\Delta L/L)$ L – unstretched length of the membrane (m), ΔL – change in length of the membrane because of applied force (m), and A_m – area of the membrane $= 2t_m W_s$, W_s – width of circumferential strip of membrane.	—

unconsolidated-undrained compressive strength:

$$\sigma_{1c} - \sigma_3$$

$$\sigma_{1c} = (\sigma_1 - \sigma_3)_f - \Delta(\sigma_1 - \sigma_3)_m$$

$(\sigma_1-\sigma_3)_f$ – measured principal stress difference (kPa), and

σ_{1c} – calibrated principal difference (kPa).

Same as in GB (I).

$$c_u = \frac{(\sigma_1 - \sigma_3)_f}{2}$$

Same as in GB (I).

Subtract the membrane correction from the calculated maximum deviatoric stress.

Plotting:
1. Plot the deviatoric stress against the axial strain.
2. Plot the shear stress against the normal stress; take $\dfrac{\sigma_{1f} + \sigma_{3f}}{2}$ as the center of the circle, and $\dfrac{\sigma_{1f} - \sigma_{3f}}{2}$ as the radius, draw the damage stress circle, and draw the envelope of the damage stress circle under different ambient pressures to calculate the shear strength parameters.

Unconsolidated-undrained compressive strength:
1. When there is a peak value, take the peak of the deviatoric stress as the strength.
2. When there is no peak value, the deviatoric stress at 15% axial strain is taken.

Table 9.44 Comparison of consolidated undrained triaxial test

Item	GB	ASTM	BS
Designation	GB 50123-1999	ASTM D4767-11 (CU) ASTM D7181-11 (CD)	BS 1377-8: 1990
Scope	Fine-grained soil and coarse-grained soils with particle diameter ≤20 mm.	All kinds of soil	All kinds of soil
Specimen requirements	Shape: Cylinder Diameter 1. The diameter of the specimen shall be 35–101 mm. For fissured soil, the diameter shall not exceed 60 mm. 2. When the diameter is less than 100 mm, the maximum particle diameter shall be less than one-tenth of the specimen diameter. 3. When the diameter of the specimen exceeds 100 mm, the maximum particle diameter shall be less than 1/5 of the diameter of the specimen. Height-to-diameter ratio: 2–2.5	Specimen shall have a minimum diameter of 33 mm. The largest particle size shall be smaller than one sixth of the specimen diameter.	The diameter of the specimen shall be 38–110 mm. The size of the largest soil particle shall not be greater than one-fifth of the specimen diameter. 2
Back pressure	Increment: 30 kPa Difference between the chamber pressure and the back pressure:—— Saturation: If the pore pressure coefficient B is greater than 0.98, the specimen can be considered to be saturated.	The increment may range from 35 kPa up to 140 kPa, depending on the magnitude of the desired effective consolidation stress. shall not exceed 35 kPa Specimens shall be considered saturated if B is equal to or greater than 0.95, or if B remains unchanged with additional back pressure increments.	The increment shall not exceed 50 kPa. Shall not be greater than the desired effective test pressure, or 20 kPa, whichever is less, and shall not be less than 5 kPa. Same as ASTM
Shearing rate	Silty: 0.1–0.5%/min; Clay: 0.05–0.1%/min.	Determine the time t_{50} (when the specimen has a degree of consolidation of 50%), and calculate the strain rate during shearing: $$\varepsilon = \frac{\varepsilon_f}{10 t_{50}}$$ Where, ε_f is the axial strain at failure, and if ε_f is greater than 4%, then take 4%.	Same as ASTM

Stop the test	15–20% strain.	Normally at 15% strain. The deviatoric stress has dropped 20% after peak point. 5% additional axial strain occurs after peak deviatoric stress.	Same as ASTM.

Calculation

Height of Specimen After Consolidation:

$$h_c = h_0 \left(1 - \frac{\Delta V}{V_0} \right)^{2/3}$$

h_c – height of specimen after consolidation (mm),
h_0 – initial height of specimen (mm),
ΔV – change in volume of specimen during consolidation (cm³), and
V_0 – initial volume of specimen (cm³).

Cross-sectional Area of the Specimen After Consolidation:

$$A_c = A_0 \left(1 - \frac{\Delta V}{V_0} \right)^{2/3}$$

A_c – cross-sectional area of the specimen after consolidation (cm²), and
A_0 – initial cross-sectional area of the specimen (cm²).

The Axial Strain:

$$\varepsilon_1 = \frac{\Delta h_1}{h_0} \times 100$$

ε_1 – axial strain (%),
Δh_1 – change of height during shearing (mm), and
h_0 – initial height of the specimen (mm).

$h_c = h_0 - \Delta h_0$

h_0 – initial height of specimen (mm), and
Δh_0 – change in height of specimen at end of consolidation (mm).

$$\Delta V_{sat} = 3V_0 \left(\frac{\Delta h_s}{h_0} \right)$$

$$A_c = \frac{V_0 - \Delta V_{sat} - \Delta V_c}{h_c}$$

ΔV_{sat} – change in volume of specimen during saturation (mm³),
ΔV_c – change in volume of specimen during consolidation (mm³),
V_0 – initial volume of specimen (mm³), and
Δh_s – change in height of the specimen during saturation (mm).

(Continued)

Table 9.44 (Continued)

Item	GB	ASTM	BS

GB

$$A_a = \frac{A_0}{1-\varepsilon_1}$$

A_a – area of cross section of the specimen normal to its axis (cm²), and
A_0 – initial area of specimen normal to its axis (cm²).

Principal Stress Difference:

$$\sigma_1 - \sigma_3 = \frac{CR}{Aa} \times 10$$

σ_1-σ_3 – principal stress difference (kPa),
σ_1 – maximum principal stress (kPa),
σ_3 – minor principal stress (kPa),
C – dynamometer rate coefficient (N/0.01 mm or N/mV),
R – reading of the force-measuring device (0.01 mm), and
10 – unit conversion factor.

Membrane Correction:—

ASTM

Principal Stress Difference:

$$\sigma_1 - \sigma_3 = \frac{P}{Aa}$$

σ_1-σ_3 – measured deviatoric stress (kPa), and
P – measured applied axial load (kN).

Use the following equation to correct the deviatoric stress for the effect of the rubber membrane if the error in deviatoric stress because of the strength of the membrane exceeds 5%:

$$\Delta(\sigma_1 - \sigma_3)_m = \frac{4E_m t_m \varepsilon_1}{D}$$

$\Delta(\sigma_1$-$\sigma_3)_m$ – membrane correction to be subtracted, from the measured principal stress difference,
D – diameter of the specimen (m),
E_m – Young's modulus for the membrane material (kPa),
t_m – thickness of the membrane (m), and
ε_1 – axial strain.

BS

A membrane correction, which shall be applied to allow for the restraining effect of the membrane. The curve in Figure 9.4 gives the correction apply to a specimen.
Initially 38 mm diameter enclosed in a membrane 0.2 mm thick. For other conditions the correction obtained from Figure 9.4 shall be multiplied by:

$$\frac{38}{D} \times \frac{t}{0.2}$$

The Young's Modulus of the Membrane Material:—

The Young's modulus of the membrane material may be determined by hanging a 15 mm (0.5 in.) circumferential strip of membrane using a thin rod, placing another rod through the bottom of the hanging membrane, and measuring the force per unit strain obtained by stretching the membrane. The modulus value may be computed using the following equation:

$$E_m = (F / A_m) / (\Delta L / L)$$

L – unstretched length of the membrane (m),

ΔL – force applied to stretch the membrane,

F – change in length of the membrane because of the force (m),

A_m – area of the membrane = 2 tm W_s, mm^2 or cm^2; where W_s width of circumferential strip of membrane.

Correction for Filter-Paper Strips—

For vertical filter-paper strips which extend over the total length of the specimen, apply a filter-paper strip correction to the computed values of the principal stress difference (deviatoric stress), if the error in principal stress difference (deviatoric stress) because of the strength of the filter-paper strips exceeds 5%:

$$\Delta(\sigma_1 - \sigma_3)_{fp}:$$

For values of axial strain above 2%:

$$\Delta(\sigma_1 - \sigma_3)_{fp} = \frac{K_{fp} P_{fp}}{A_c}$$

For values of axial strain of 2% or less:

$$\Delta(\sigma_1 - \sigma_3)_{fp} = \frac{50\varepsilon_1 K_{fp} P_{fp}}{A_c}$$

K_{fp} – load carried by filter-paper strips per unit length of perimeter covered by filter-paper (kN/mm or kN/m), and

P_{fp} – perimeter covered by filter-paper (mm).

(Continued)

Table 9.44 (Continued)

Item	GB	ASTM	BS
	Drain Correction:—	—	When vertical side drains are fitted, an additional correction, s_{dr}, shall be applied for strains exceeding 0.02 (2%). The value for σ_{dr} shall be taken from Table 9.45
	Corrected Principal Stress Difference:—	$\sigma_{lc} = (\sigma_1 - \sigma_3)_f - \Delta(\sigma_1 - \sigma_3)_m - \Delta(\sigma_1 - \sigma_3)_{fp}$ $(\sigma_1 - \sigma_3)_f$ – principal stress to failure (kPa), and σ_{lc} – corrected deviatoric stress (kPa).	Subtract the membrane correction and the drain correction from the deviatoric stress.
	$\sigma_1' = \sigma_1 - u$ σ_1' – effective major principal stress (kPa), and u – pore water pressure (kPa). $\sigma_3' = \sigma_3 - u$ σ_3' – effective minor principal stress (kPa). Effective principal stress ratio: $\dfrac{\sigma_1'}{\sigma_3'} = 1 + \dfrac{\sigma_1' - \sigma_3'}{\sigma_3'}$ $B = \dfrac{u_0}{\sigma_3}$ B – initial pore compression coefficient, and u_0 – pore-water pressure at the given axial load (kPa). $A_f = \dfrac{u_f}{B(\sigma_1 - \sigma_3)}$ A_f – pore-pressure coefficient at failure, and u_f – pore-pressure generated by the main stress difference when the sample is at failure (kPa).	—	Same as in GB $A = \dfrac{u - \mu_0}{\sigma_1 - \sigma_3}$ A – pore-pressure coefficient, and u_0 – pore-pressure in the specimen at the beginning of compression (kPa).

$\varphi' = \sin^{-1}\tan\alpha$ —   Same as in GB.

ϕ' – effective internal friction angle (°), and
α – Inclination angle of the stress path (°).

$$c' = \dfrac{d}{\cos\varphi'}$$ —   Same as in GB.

c' – effective cohesion (kPa), and
d – intercept of the failure line of stress path (kPa)

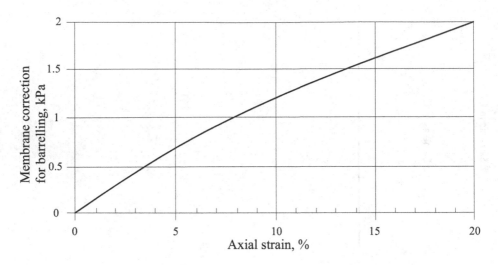

Figure 9.4 Membrane correction in BS

Table 9.45 Drain correction in BS

Specimen diameter	Correction
38	10
50	7
70	5
100	3.5
150	2.5

Note: Corrections for specimens of intermediate diameters may be obtained by interpolation.

Chapter 10

Comparison of rock tests

This chapter analyzes and compares rock tests in Chinese Standard (GB), American Standard (ASTM) and ISRM suggested methods (International Society of Rock Mechanics). This chapter covers both physical and mechanical tests of rocks. The physical tests include moisture test, water absorption, density, particle density, freezing-thawing, expansion and disintegration. Mechanical tests consist of acoustic wave, Brazilian splitting, uniaxial compression, triaxial compression, direct shear of structural plane, point load and rebound tests.

10.1 Physical tests on rocks

10.1.1 Water content

The water content of rock refers to the ratio of mass of water in rock to mass of rock solid, which indirectly reflect the characteristics of voids in rock. The commonly used method for determining water content of rock is to dry the sample in an oven to a constant mass, and then calculate the ratio of the lost mass to the final mass. The drying method is included in GB, ASTM and ISRM. The main difference among these three methods is the criterion to cease the test. GB asks for drying the rock specimen for a certain time period, while ASTM and ISRM require a constant mass to be achieved. A detailed comparison is given in Table 10.1.

10.1.2 Grain density

Grain density of rock refers to the ratio of the mass of grain to its volume, which depends on the density and content of minerals in rock. Commonly used methods for determining rock grain density are pycnometer method and water weighing method. The pycnometer method requires the specimen to be made into small particles. The water weighing method requires a block specimen and the volume of specimen are calculated by the difference between the mass of the dry specimen and the mass of the saturated specimen. The water weighing method has a relatively low accuracy compared to the pycnometer method since the closed/isolated voids in rock are not considered, which leads to relatively small values. Both pycnometer and water weighing methods are included in GB. A detailed comparison of the test methods in GB and ISRM is given in Table 10.2.

10.1.3 Dry density

Dry density of rock refers to the mass of rock of unit volume. Measuring dry density of rocks involves a variety of methods for precisely determining the volume of the specimen. For specimens with regular shape, the volume method is usually measured with a caliper.

Table 10.1 Comparison of water content tests

Item	GB	ASTM	ISRM
Designation	GB 50266-2013	ASTM D2216-10	Suggested methods for determining water content, porosity, density, absorption and related properties and swelling and slake-durability index properties
Scope	All kinds of rocks	Soil, rock and solid conglomerates	Same as GB
Apparatus	Drying oven, desiccator, balances	In addition to the instrument in GB, there is specimen containers	
Specimen	Sample obtained by blasting cannot be used, and wet drilling is allowed in areas with abundant groundwater.	—	—
	Store: The water content of the sample shall be maintained at the natural water content.	The samples shall be stored prior to testing in non-corrodible airtight containers at a temperature between approximately 3 and 30 °C and in an area that prevents direct contact with sunlight.	The change in water content must not exceed 1%.
	Size: The minimum specimen size shall be more than 10 times the diameter of the largest mineral particle.	—	Same as GB
	Quality: 40–200 g	≥500 g	≥50 g
	Quantity: 5	—	10
Test Operation	Drying Temperature: 105–110 °C	110±5 °C	105±3 °C
	Criterion to cease drying: Dry a specimen for 24 h	At least 12–16 h (break the sample into small particles and place in a container with large surface area). The mass loss is less than 1% in 2 hours.	At least 24 h, and the successive mass determinations at intervals of 4 h yield values differing by less than 0.1% of the sample mass.
	Weighing Accuracy: 0.01 g	0.1 g	0.01% of the sample mass
Calculation	$\omega = \dfrac{m_0 - m_s}{m_s} \times 100$	$\omega = \left[\dfrac{M_{cms} - M_{cds}}{M_{cds} - M_c}\right] \times 100$	
	ω – water content (%); m_0 – wet sample mass (g); m_s – dry sample mass (g).	ω – water content (%); M_{cms} – total mass of weighing box and wet sample (g); M_{cds} – total mass of weighing box and dry sample; M_c – mass of weighing box.	
	Accuracy: 0.01	1% or 0.1%	0.1%

Table 10.2 Comparison of grain density tests

Item	GB	ISRM
Designation/ name	GB 50266-2013	Suggested methods for determining water content, porosity, density, absorption and related properties and swelling and slake-durability index properties
Scope	All kinds of rocks	Rock that swells or decomposes when immersed in water
Specimen preparation	The rock was pulverized into rock powder and sieved with a 0.25 mm sieve. If the specimen contains magnetic minerals, the mortar shall be ceramic or agate.	The specimen was crushed into grains with sizes not exceeding 150 μm.
Drying temperature	105–110 °C	105 °C
Drying time	At least 6 h	Dry to constant mass
Cooling time	No explicit requirements	30 min
Vacuum time	When there is no air bubble in the test solution, continue to pump for no less than 1 h.	20 min
Accuracy	The mass is accurate to 0.001 g and the temperature is accurate to 0.5 °C.	Same as GB
Calculation of particle density	$\rho_s = \dfrac{m_s}{m_1 + m_s - m_2} \rho_{WT}$ ρ_s – grain density (g/cm³); ρ_{WT} – test solution density under test temperature (g/cm³); m_s – dry mass of the rock meal (g); m_1 – total mass of bottle and test solution (g); m_2 – total mass of bottle, test solution, and rock powder (g).	$\rho_s = \dfrac{F-D}{V_f \left(1 - \dfrac{G-F}{E-D}\right)}$ ρ_s – grain density (g/cm³); V_f – Volumetric flask volume, usually 50 cm³; D – mass of volumetric flask (g); E – total mass of bottle and test solution (g); F – total mass of bottle and specimen (g); G – Total mass (g) of specimen, bottle and test solution.
Accuracy	0.01	

For irregular specimens, wax-sealing and buoyancy methods are normally used. These two methods take the difference of weight of saturated specimen in air and in the water to back calculate the volume. A detailed comparison of dry density test in different standards is given in Table 10.3.

10.1.4 Absorption

Rock absorption reflects ability of rock absorbing water under certain conditions (pressure and temperature).

Table 10.3 Comparison of dry density tests

Item	GB			ISRM			
Designation	GB 50266-2013			Suggested methods for determining water content, porosity, density, absorption and related properties and swelling and slake-durability index properties			
Method to measure the volume	Caliper	Wax-sealing	Water weighing	Saturated & caliper	Saturated and buoyancy method	Mercury	Mercury & Boyle's Law
Scope	Rock that can be machined	Rock that does not disintegrate, dissolve, or swell in contact with water.	All kinds of rocks	non-friable, coherent rocks that can be machined and do not appreciably swell or disintegrate when oven dried or immersed in water.	The method shall only be used for rocks that do not appreciably swell or disintegrate when oven-dried and immersed in water.	rock material that is liable to swell or disintegrate if being immersed in water.	The method shall only be used for rocks that do not shrink appreciably during oven-drying.
Specimen size	The minimum size shall not be less than 50 mm;	with a side length of 40–60 mm		The minimum size of each specimen shall be such that its mass is at least 50 g (a cube with a side length of 27 mm is sufficient) and its minimum size is at least 10 times the maximum grain size.			
Specimen shape	Regular	Round rock block	Same as wax-sealing method	Cylinder or prism	The sample is washed in water to remove dust.	The size shall be consistent with the chamber.	
Number of specimens	3			At least 3	At least 10	At least 10	At least 3

Calculation

$$\rho_d = \dfrac{m_s}{AH}$$

A – sample cross-sectional area (cm²);
H – sample height (cm);
m_s – dry mass of the specimen (g).

$$\rho_d = \dfrac{m_s}{m_1 - m_2 - \dfrac{m_1 - m_s}{\rho_w}}\,\rho_p$$

$$\rho = \dfrac{m}{m_1 - m_2 - \dfrac{m_1 - m}{\rho_w}}\,\rho_p$$

ρ – the wet density of specimen (g/cm³);
m – mass of the wet specimen (g);
m_1 – mass of the specimen with wax (g);
m_2 – mass of the waxed specimen in water (g);
ρ_w – density of water (g/ cm³);
ρ_p – density of wax (g/ cm³);
ω – water content (%).

Volume method as in GB.

$$\rho_d = \dfrac{m_s}{m_p - m_w}$$

m_s – Drying sample mass (g);
m_p – mass after forced saturation of the test piece (g);
m_w – mass of the forced saturated test piece in water (g).

Water weighing method as in GB.

$$\rho_d = \dfrac{M_s}{V}$$

M_s – dry mass (g);
V – volume of specimen (cm³), which is determined by the mercury displacement.

Free water absorption of rock is generally determined by measuring the mass of absorbed water and that of dry rock under normal atmospheric pressure and at room temperature. The saturated water absorption is usually determined by boiling or vacuum the specimen to get it saturated. The free water absorption test is included in GB and ASTM and ISRM, while the saturated water absorption test is only stipulated in GB.

GB requires the specimen to be dried prior to being immersed in water. ASTM requires the specimen of natural state to be immersed in water. GB asks for a regular cylinder specimen, as in uniaxial compression test. There is no special requirement for shape and size of specimen in ASTM. In GB, the specimen is dried for a certain time period, which may not ensure a 100% drying. See specific comparisons of GB and ASTM for absorption tests in Table 10.4.

10.1.5 Slake durability

The slake durability of rock reflects the ability of rock to resist disintegration under dry-wet cycles. In GB, ASTM and ISRM, the rock is experienced two wet-dry cycles for the measurement. The three standards require very much similar specimen and procedures. Noteworthy, in GB and ISRM, the temperature is required to be maintained at 20 °C, whereas in ASTM, there is no specific requirement for temperature. A detailed comparison of the tests in different standards is given in Table 10.5.

10.2 Mechanical tests on rocks

10.2.1 Swelling properties

Rocks containing hydrophilic and easily swelling minerals (i.e. montmorillonite) can absorb water to lead its volume swelling. The rock swelling indexes are composed of an unconfined swelling strain index, a swelling strain index and a swelling pressure index.

In GB, the free swelling test is to measure the swelling ratio of each side of the specimen when the specimen is immersed freely in water. The confined swelling test is to measure the axial swelling of the specimen. The swelling pressure test is to determine the axial swelling pressure needed to maintain the height of the specimen when the laterally confined specimen is immersed in water. Detailed comparisons of these tests in GB and ISRM are given in Tables 10.6–10.8.

10.2.2 Uniaxial compressive strength

Uniaxial compressive strength of rock refers to the maximum axial stress that the rock with no confining can withstand. In the tests for rock uniaxial compressive strength, cylindrical specimens are normally used. It is found that the size of specimen has influence on the tested results. The larger the height-to-diameter ratio of the specimen, the less the measured strength. When the height-to-diameter ratio of the specimen is greater than 2, the size effect could be ignored. In addition, the diameter of specimen shall be 10 times greater than the maximum size of mineral particles in the specimen. Loading rate also effects the test results, as the higher the loading rate, the higher the strength will be achieved. The uniaxial compression test for rocks is included in GB, ASTM and ISRM. Main differences among these standards is given in Table 10.9.

Table 10.4 Comparison of absorption tests

Item	GB	ASTM	ISRM
Designation	GB 50266-2013	ASTM D6473-15	Suggested methods for determining water content, porosity, density, absorption and related properties and swelling and slake-durability index properties
Scope	Rock that does not disintegrate, is insoluble, and does not shrink and expand in contact with water.	This test is appropriate for breakwater stone, armor stone, riprap, and gabion sized rock materials.	Rock that does not disintegrate in contact with water.
Apparatus	Drying oven, Desiccator, balance, water tank, water weighing device and vacuum pumping or boiling equipment.	Balance, specimen container, water weighing device, water tank, drying oven and Desiccator.	Balance, specimen container, dehydrated silica gel.
Specimen	Shape: Generally, it is regular, and it is possible to use round and irregular specimens when it is difficult to prepare. Size and Mass: The minimum size shall not be less than 50 mm; the error of the height, diameter or side length of the specimen shall not exceed 0.3 mm; the error of non-parallelism on both ends of the specimen shall not exceed 0.05 mm; the end sides of the specimen shall be perpendicular to the axis of the specimen; the maximum deviation shall not be more than 0.25. No requirement for mass. Quantity: 3	No special requirements for size. The mass of specimen shall be at least 1 kg.	—
		At least 5 specimens equal in size. At least 8 specimens unequal in size.	Each piece shall be sized to have a mass of more than 50 g or a minimum size of at least ten times the maximum particle size, take whichever is bigger. At least 10
Water absorption	Dry the specimen first, and then submerge it in water.	Submerge the specimen at its natural state.	Dry first, and then submerge in water.

(Continued)

Table 10.4 (Continued)

Item	GB	ASTM	ISRM
	Drying: Dry the specimen at 105–110 °C for 24 h.	Dry the specimen at least for 24 h at a temperature of 110 ± 5 °C. Constant mass will be considered to have been achieved when weight loss is less than 0.1% in four hours of drying.	The specimen shall be placed in a container filled with dehydrated silica gel at room temperature and left for 24 hours.
Saturated water absorption	Water immersion requirements: Absorption for 48 h Saturating the specimen by boiling or vacuum pumping	24±4h in 20–30 °C water Not included	Absorption for 1 h not included
Accuracy	Weighing accuracy: 0.01 g	5 g or 0.1% of the specimen mass (whichever is less)	0.5 g
Water absorption calculation	$\omega_s = \dfrac{m_0 - m_s}{m_s} \times 100$ ω_s – absorption (%); m_0 – saturated dry mass of specimen surface (g); m_s – dry mass of the specimen (g).		$lv= (B-A)/A \times 100\%$ lv – void index of rock. A – mass of rock after drying (g); B – mass (g) of specimen after water absorption.
Other Calculations	$\omega_{sa} = \dfrac{m_p - m_s}{m_s} \times 100$ $\rho_d = \dfrac{m_s}{m_p - m_w}\rho_w$ $\rho_s = \dfrac{m_s}{m_s - m_w}\rho_w$ ω_{sa} – saturation water absorption (%); ρ_d – rock mass density (g/cm3); ρ_s – grain density (g/cm3); m_p – mass after forced saturation of the specimen (g); m_s – dry mass of the specimen (g); m_w – mass of the saturated specimen in water (g); ρ_w – density of water (g/cm3).	Bulk specific gravity=A/ (B-C) Bulk specific gravity (SSD) = B/(B-C) Apparent specific gravity = A/ (A-C) A – mass of oven-dry test specimen in air, g; B – mass of saturated-surface dry test specimen in air, g; C – buoyant mass of submerged test specimen in water, g.	—

Table 10.5 Comparison of slake durability tests

Comparison	GB	ASTM	ISRM
Designation	GB 50266-2013	ASTM D4644-16	Suggested methods for determining water content, porosity, density, absorption and related properties and swelling and slake-durability index properties
Scope	Rocks tend to disintegrate in water.	Shale or other weak rocks	—
Specimen storage	—	Store in a non-corrosive airtight container at a temperature of 3 to 30 °C, and direct sunlight shall be avoided.	—
Specimen shape	Break off any possible existing sharp corners.		Same as in GB, while the maximum diameter shall not exceed 3 mm.
Specimen number	10		
Mass of each specimen	40–60 g		
Height of filling water	20 mm below the drum axis		
Speed of rotation	20 r/min		
Time of rotation	10 min		
Temperature	20±2 °C	No proper requirements, but measure the water temperature before and after every cycle.	20 °C
Accuracy of balance	0.01 g	0.1 g	0.5 g
Calculation	$I_{d2} = \dfrac{m_r}{m_s} \times 100$ I_{d2} – slake durability index after 2 cycles (%); m_s – mass of oven-dried specimen (g); m_r – residue oven-dried mass after 2 cycles (g). Keep 3 valid digits	$I_{d1} = \left[\dfrac{W_{f1} - C}{W_i - C}\right] \times 100$ $I_{d2} = \left[\dfrac{W_{f2} - C}{W_i - C}\right] \times 100$ I_{d1}, I_{d2} – slake durability index after first and second cycles, respectively (g). W_i – mass of drum plus oven-dried specimen before the first cycle (g). W_{f1}, W_{f2} – mass of drum plus oven-dried specimen retained after the first and the second cycles respectively (g), C – mass of drum (g). 0.1%	

Table 10.6 Comparison of determining the swelling strain developed in an unconfined rock specimen

Item	GB	ISRM
Designation	GB 50266-2013	Suggested methods for determining water content, porosity, density, absorption and related properties and swelling and slake-durability index properties
Scope	Rocks that don't disintegrate in water.	Rocks that don't change geometry in slaking.
Specimen shape	Cylinder or square	Cylinder or rectangle
Specimen size	The diameter of the cylindrical specimen shall be 48–65 mm, the height shall be equal to the diameter, and both ends shall be parallel to each other. The side length of cubic specimen shall be 48–65 mm.	The minimum size of the specimen shall be greater than 15 mm and ten times the maximum particle size.
Number of specimens	3	2
Time of immersion	No less than 48 h	—
Test temperature	Variation of water temperature shall not exceed 2 °C.	—
Accuracy of readings	—	At least 0.1% of the original length
Calculation formula	$V_H = \dfrac{\Delta H}{H} \times 100$ $V_D = \dfrac{\Delta D}{D} \times 100$ V_H – axial free swelling ratio (%); V_D – radial free swelling ration (%); ΔH – figure of specimen axial strain (mm); H – height of specimen (mm); ΔD – average figure of specimen radial strain (mm); D – diameter or length of specimen (mm).	Unconfined swelling strain in direction $X = d/L \times 100\%$. Where: X – a direction relative to the bedding or foliation; d – maximum swelling displacement recorded in direction X during the test; L – initial distance between gauge points in direction X.
Calculation accuracy	Keep 3 valid digits	—

Table 10.7 Comparison of determining the swelling strain index for a radially confined rock specimen with axial surcharge

Item	GB	ISRM
Designation	GB 50266-2013	Suggested methods for determining water content, porosity, density, absorption and related properties and swelling and slake-durability index properties
Scope	All kinds of rocks	
Water content of specimen	—	Test at in-situ water content, and place the specimen in a constant humidity environment.
Shape	Cylinder	
Dimension	The height shall be greater than 20 mm, or 10 times greater than the maximum mineral particle size of the rock. The two ends shall be parallel. The diameter of the specimen shall be 50–65 mm, and be 0.0–0.1 mm less than the diameter of the metal collar.	The diameter shall not be less than 4 times its thickness. The thickness shall be greater than 15 mm and 10 times the maximum particle size. The specimen shall be a close fit in the ring.
Number of specimens	3 specimens for each loading direction.	Duplicate specimens shall be prepared from the same sample, one being used for determination of the water content and the other for swelling testing.
Seating pressure	5 kPa	3 kPa
Criterion of stable swelling	Difference between 3 successive readings is less than 0.001 mm.	—
Time of immersion	No less than 48 hours.	—
Test temperature	Variation of water temperature shall not exceed 2 °C.	—
Accuracy of readings	—	At least 0.1% of the original length
Calculation	$V_{HP} = \dfrac{\Delta H_1}{H} \times 100$ V_{HP} – lateral confined swelling ratio (%); ΔH_1 – axial deformation (mm); H – original height of specimen (mm). Keep 3 valid digits	Swelling strain index = $d/L \times 100\%$ d – maximum swelling deformation recorded during the test; L – initial thickness of the specimen. —

Table 10.8 Comparison of determining the swelling pressure index under conditions of zero volume change

Item	GB	ISRM
Designation	GB 50266-2013	Suggested methods for determining water content, porosity, density, absorption and related properties and swelling and slake-durability index properties
Scope	All kinds of rocks	
Water content of specimen	—	Test at in-situ water content, and place specimen in a constant humidity environment.
Shape	Cylinder	
Dimension	The height of the specimen shall be greater than 20 mm, or greater than 10 times the maximum particle size of the rock. The two ends shall be parallel. The diameter of the specimen shall be 50–65 mm, and shall be 0.0–0.1 mm less than the diameter of the metal collar.	The diameter of sample is not less than 2.5 times its thickness. The thickness shall exceed 15 mm or ten times the maximum grain diameter. The specimen shall be a close fit in the ring.
Number of specimens	3 for each loading direction.	Duplicate specimens shall be prepared from the same sample, one being used for water content determination and the other for swell testing.
Seating pressure	10 kPa	—
Criterion of stable swelling	The difference of three consecutive readings is less than 0.001 mm.	—
Time of immersion	No less than 48 hours	—
Test temperature	Change of water shall not exceed 2 °C.	—
Calculation	$P_e = \dfrac{F}{A}$ P_e – swelling pressure at constant volume (MPa); F – axial load (N); A – cross-sectional area of specimen (mm^2). Keep 3 valid digits	Swelling pressure index = F/A F – maximum axial swelling force A – cross-sectional area of the specimen. —

Table 10.9 Comparison of uniaxial compressive strength tests

Item	GB	ASTM	ISRM
Designation	GB 50266-2013	ASTM D7012-14	Suggested methods for determining the uniaxial compressive strength and deformability of rock materials
Scope	Various types of rocks that can be made into cylinders	Complete core	Same as GB
Apparatus	Material testing machine	Compression apparatus, platens (one plain rigid platen, one spherically seated), timing device	A suitable machine shall be used for applying and measuring axial load to the sample, a disc-shaped steel platen, a spherical seat
Specimen shape	Cylinder		
Diameter	It shall be 48–54 mm, no less than 10 times the maximum mineral particle size in the rock.	The diameter of rock test specimens shall be at least 10 times the diameter of the largest mineral grain. for weak rock types, the specimen diameter shall be at least six times the maximum particle diameter. The minimum diameter is 47 mm.	No less than 54 mm, and the diameter of rock test specimens shall be at least 10 times the diameter of the largest mineral grain
Height-to-diameter ratio	2.0–2.5		2.5–3.0
Parallelism of both ends	<0.05 mm	—	<0.02 mm
Error of diameter along the height of the specimen	<0.3 mm	—	Same as GB
Vertical deviation of the end face to the axis of the sample	<0.25°	—	<3.5'
Number of specimens	3	Determined according to test method ASTM E122	At least 5
Loading rate	0.5–1.0 MPa/s	stress rate is the same as GB, or constant strain rate	Same as GB
Loading time	—	2–15 min	5–10 min
Calculation	$\sigma_u = P/A$	—	Same as GB

σ_u – uniaxial compressive strength (MPa);
P – failure load (kN);
A – cross-sectional area (mm^2).

3 valid digits

10.2.3 Durability under freezing and thawing conditions

This test is intended to measure the frost resistance index of rock, indicated by mass loss ratio after freezing-thawing cycles and the freezing-thawing coefficient. The former refers to the ratio of the mass loss after freezing-thawing to the original dry mass. The latter is the ratio of compressive strength of rock specimen after freezing-thawing cycles to that before freezing-thawing. Detailed comparison of GB and ASTM is given in Table 10.10.

10.2.4 Deformability in uniaxial compression

The deformability of rocks can be determined via measuring the axial and radial strains of rock specimen under conditions of uniaxial compression. The parameter indicating deformability of rocks include the elastic modulus and Poisson's ratio. The former is the ratio of stress to strain in the regime of elastic deformation. The latter is the ratio of transverse strain to axial strain of the specimen.

In the test, the stress-strain relationship curve can be obtained by measuring the load and corresponding strain of the specimen during the compression process. GB, ASTM and ISRM have similar requirements for carrying out uniaxial compression test. The main differences are specimen size as shown in Table 10.11.

10.2.5 The strength in triaxial compression

Triaxial compression test is used to determine the shear strength of rocks. GB, ASTM and ISRM all cover the triaxial compression tests of rocks. GB introduces the method which requires a set of specimens being tested under different confining pressures. ASTM and ISRM also describe the test method, which employs a single specimen and tested at multistage confining pressures. A detailed comparison between the three standards is given in Table 10.12.

10.2.6 Splitting tensile strength

Commonly used methods for determining the tensile strength of rocks are direct tensile test and Brazil test. In the Brazil test, the cylindrical specimen is radially compressed until it is failed. These two test methods are all included in GB, ASTM and ISRM. Differences between these three standards are given in Table 10.13.

10.2.7 Point load strength

The point load strength index is normally used to estimate the uniaxial compressive strength of rocks. It can also be used for rock classification. Compared with the uniaxial compression test, the point load strength test is simple and easy to conduct. GB, ASTM and ISRM all cover the point load strength test. There are no many differences among these three standards. A detailed comparison is given in Table 10.14.

Table 10.10 Comparison of determining the durability under freezing and thawing conditions

Item	GB	ASTM
Designation	GB 50266-2013	ASTM D5312/D5312M - 12
Scope	All kinds of rocks that can be made into cylindrical shape.	All kinds of rocks.
Preparation of test samples	Shape: cylinder	slab
	Materials: Rock of drilling core or block	Rock sources from mine, quarry, outcrop, or field boulders.
		In no case shall the specimen be less than 125 mm (5 in.) on a side.
	Dimensions: Diameter of the specimen shall be 48–54 mm; The diameter shall be greater than 10 times the largest mineral particle; The ratio of height to diameter is 2.0–2.5, and the parallelism error of the end faces shall not exceed 0.05 mm.	
	Number: 6	At least 5
Special solutions	—	0.5% isopropanol alcohol/water solution.
Freezing temperature	-20±2 °C	-18±2.5 °C
Freezing time	4h	At least 12 hours
Thawing temperature	20±2 °C	32±2.5 °C
Thawing time	4 h	8–12 h
Freezing-thawing cycles	Generally, 25 cycles.	—
Quality measured after the freezing and thawing cycle	The quality after drying the specimen.	Quality of largest piece of each slab.
Weighing accuracy	0.01 g	0.1% of the total mass of the specimen.
Data processing	$M = \dfrac{m_p - m_{fm}}{m_s} \times 100$ $K_{fm} = \dfrac{\bar{R}_{fm}}{\bar{R}_w}$ M – mass loss rate (%); K_{fm} – freezing-thawing coefficient; m_p – mass of the saturated sample before freezing and thawing (g); m_{fm} – Mass of specimen after freezing-thawing (g); m_s – Dry mass of the specimen before test (g); $\bar{R}_{fm}$ – Average uniaxial compressive strength of rock after freezing-thawing (MPa); $\bar{R}_w$ – Average uniaxial compressive strength before freezing-thawing (MPa). Accurate to 0.01.	Quantitative examination: For each slab perform the following calculation: % loss = (A-B)/A × 100 A – oven-dried mass of the specimen prior to test (g); B – oven-dried mass of the largest remaining piece of each slab after test (g). 0.1%

Table 10.11 Comparison of determining the deformability in uniaxial compression

Item	GB	ASTM	ISRM
Designation	GB 50266-2013	ASTM D7012-14	Suggested methods for determining the uniaxial compressive strength and deformability of rock materials
Scope	Rocks that can be made into cylinders.	Complete core	Rocks that Same as GB.
Specimen shape	Cylinder		
Diameter	It shall be 48–54 mm. Greater than 10 times the maximum mineral particle size in the rock.	Greater less than 47 mm The diameter shall be at least 10 times larger than the largest mineral grain. For weak rocks, the diameter shall be no less than at least 6 times the biggest particle.	No less than 54 mm. No less than 10 times the biggest mineral particle size in the rock.
Height-to-diameter ratio	2.0–2.5		2.5–3.0
Parallelism of both ends	<0.05 mm	—	<0.02 mm
Error of diameter along the height of the specimen	<0.3 mm	—	Same as GB
Vertical deviation of end faces to the axis of the specimen	<0.25°	—	<3.5'
Loading rate	0.5–1.0MPa/s		
Loading time	—	2–15 min	5–10 min

Accuracy	—	The load is in kN and is retained in two decimal places; the gauge reading is accurate to one decimal place.	The height is accurate to 1.0 mm, the diameter is accurate to 0.1 mm, and the maximum load is accurate to 1%.
Calculation	$$E_{Av} = \frac{\sigma_b - \sigma_a}{\varepsilon_{Ib} - \varepsilon_{Ia}}$$ $$\mu_{Av} = \frac{\sigma_{Db} - \sigma_{Da}}{\varepsilon_{Lb} - \varepsilon_{La}}$$ $$E_{50} = \frac{\sigma_{50}}{\varepsilon_{I50}}$$ $$\mu_{50} = \frac{\varepsilon_{d50}}{\varepsilon_{I50}}$$ E_{Av} and μ_{Av} are the average elastic modulus and average Poisson's ratio, respectively; E_{50} and μ_{50} are the secant elastic modulus and the corresponding Poisson's ratio; ε_d and ε_I are axial and radial strain, respectively; Subscripts a, b, 50 correspond to the stress states.	The tangent modulus E_t is the tangent slope measured at 50% of the peak strength; The average elastic modulus E is the average slope of the linear portion in the curve; The secant modulus E_s is the secant slope from zero to 50% of the intensity value;	The volumetric strain $$\varepsilon_v = \varepsilon_a + 2\varepsilon_d$$ ε_a is the axial strain, and ε_d is the radial strain.
	The modulus takes 3 valid digits and the Poisson's ratio is accurate to 0.01.	—	Both keep 3 valid digits.

Table 10.12 Comparison of determining the strength in triaxial compression

Item	GB	ASTM	ISRM
Designation	GB 50266-2013	ASTM D7012-14	Suggested methods for determining the strength of rock materials in triaxial compression: Revised version
Scope	Rocks that can be made into cylinder.	Complete core	Same as GB
Specimen shape	Cylinder		
Diameter	The diameter of specimen is 48–54 mm, which should be 0.96–1.00 times the bearing platen, and 10 times larger than the largest mineral grain.	The diameter shall be greater than 47 mm and at least 10 times the largest mineral grain.	The diameter shall be ≥54 mm, and no less than 10 times the biggest particle.
Height-to-diameter ratio	2.0–2.5		2.0–3.0
Parallelism of both ends	<0.05 mm	—	<0.02 mm
Error of diameter along the height of the specimen	<0.3 mm	—	Same as in GB
Vertical deviation of the end faces to the axis of the specimen	<0.25°	—	<3.5'
Number of specimens	5	At least 9	At least 5
Temperature	—	Raise to the desired temperature at a rate not exceeding 2 °C/min.	—
Axial loading rate	0.5–1.0 MPa/s		
Axial loading time	—	2–15 min	5–15 min
Measurement accuracy	—	The load is in kN and is retained in two decimal places; the gauge reading is accurate to one decimal place.	Height is accurate to 1.0 mm, and diameter is accurate to 0.1 mm, maximum load is accurate to 1%.

Table 10.13 Comparison of splitting tensile strength tests

Item	GB	ASTM	ISRM
Designation	GB 50266-2013	ASTM D3967-16	Suggested methods for determining tensile strength of rock materials
Specimen shape	Cylinder		
Diameter	Diameter of specimen shall be 48–54 mm, 10 times greater than the largest particle in the specimen.	Diameter of the specimen shall be at least 10 times greater than the largest grain. A diameter of 54 mm generally satisfies.	Diameter of the specimen shall be approximately 54 mm.
Height-to-diameter ratio	0.5–1.0	0.2–0.75	0.5
Accuracy	Parallelism error of the two ends of the specimen shall be less than 0.05 mm; Diameter error is less than 0.3 mm.	Parallelism error is less than 0.5 mm; Both ends shall be perpendicular to the mandrel and the deviation must not exceed 0.5°.	Diameter error is less than 0.025 mm; Parallelism error is less than 0.25 mm; Both ends shall be perpendicular to the mandrel and the deviation must not exceed 0.25°.
Number	3	At least 10	Generally 10
Preloading	—	The loading platen is lowered slowly until the top platen is almost to direct contact with the specimen with few or no loads.	—
Loading rate	0.3–0.5 MPa/s	0.05–0.35 MPa/s	Shall be at a constant rate, recommended as 200 N/s
Failure time	—	1–10 min	15–30 s
Calculation	$\sigma_t = \dfrac{2P}{\pi Dh}$ σ_t – tensile strength (MPa); P – failure load (N); D – specimen diameter (mm); h – specimen height (mm).	Flat bearing platen: $\sigma_t = \dfrac{2P}{\pi Dh}$ Curved bearing platen: $\sigma_t = \dfrac{1.272P}{\pi tD}$ σ_t – tensile strength (MPa); P – maximum applied load (N); t – specimen thickness (mm); D – Specimen diameter (mm).	$\sigma_t = \dfrac{0.636P}{\pi Dt} \approx \dfrac{2P}{\pi Dt}$ σ_t – tensile strength (MPa); P – maximum applied load (N); t – specimen thickness (mm); D – specimen diameter (mm).
	Keep 3 valid digits		

Table 10.14 Comparison of point load strength tests

Item	GB	ASTM	ISRM
Designation	GB 50266-2013	ASTM D5731-16	Suggested method for determining point load strength
Scope	All kinds of rocks	Medium strength rock with compressive strength ≥15MPa	Same as GB
General requirements	—	The specimen diameter shall be 30–85 mm, preferably 50 mm.	—
Core	Radial test: Length-to-diameter ratio > 1. Axial test: Length-to-diameter ratio is 0.3–1.	In addition to the requirements in GB, the diameter is no less than 4 times the largest particle.	Same as GB
Block and irregular specimen	The ratio of the distance between the two loading points to the average width is preferably 0.3 to 1.0. The length of the specimen in the loading direction shall be 30 to 85 mm, preferably 50 mm.		
Number of specimens	Core: 5–10. Blocks and irregular samples: 15–20	Core or block: At least 10. Irregular: At least 20. More is needed if the rock is anisotropic.	At least 10. More is needed if the rock specimen is anisotropic.
Loading time	10–60s		
Measurement requirements	—	Diameter is accurate to ±2%, and width is accurate to ±5%.	
Radial test	$D_e^2 = D^2$ D_e – equivalent core diameter (mm); D – core diameter (mm).		
Other tests	$D_e^2 = \dfrac{4WD}{\pi}$ W – width (or average width) (mm) of the smallest section crossing the two loading points.		
Severe platen penetration	$D_e^2 = DD' = \dfrac{4WD'}{\pi}$ D' – distance between the loading points (mm) at which the specimen breaks.		
Uncorrected rock point load strength index	$I_s = \dfrac{P}{D_e^2}$ I_s – uncorrected rock point load strength (MPa); P – failure load (N).		

Size correction

Find the P_{50} value from the $D_e^2 - P$ curve where D_e^2 is 2500 mm². Calculate the point load strength index:

$$I_{s(50)} = \frac{P_{50}}{2500}$$

$I_{s(50)}$ – Size-corrected rock point-load strength (MPa).

$$I_{s(50)} = FI_s$$

F – size correction factor:

$$F = \left(\frac{D_e}{50}\right)^m$$

m is normally 0.4–0.45.

The $D_e^2 - P$ curve is on double logarithmic scales.

Anisotropy index

$$I_{a(50)} = \frac{I'_{s(50)}}{I''_{s(50)}}$$

$I_{a(50)}$ – anisotropy index of point load strength;
$I'_{s(50)}$ – point load strength index (MPa). Loading direction is parallel to the weak surface (Figure 10.1a);
$I''_{s(50)}$ – point load strength index (MPa). Loading direction is perpendicular to the weak surface (Figure 10.1b).

Estimation of uniaxial compressive strength

UCS = $KI_{s(50)}$
Refer to Table 10.15 for K.

UCS = $KI_{s(50)}$
K is between 20 and 25.

Estimation of tensile strength

Tensile strength = 0.8 $I_{s(50)}$

Accuracy

Keep 3 valid digits

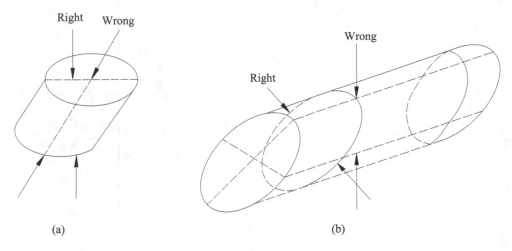

(a)   (b)

Figure 10.1 Correct loading to test the anisotropy of point load strength

Table 10.15 K values in ASTM

Core Size (mm)	21.5	30	42	50	54	60
K	18	19	21	23	24	24.5

10.2.8 The sound velocity by ultrasonic pulse transmission technique

This test is to measure the propagation velocity of longitudinal and transverse waves in rock. Longitudinal and transverse wave velocities can be used to calculate the dynamic elastic parameters of rocks and the integrity index of rock mass. Both GB and ISRM standardize the wave velocity test of rocks. A detailed comparison of these two standards is given in Table 10.16.

10.2.9 Rebound hardness

Rebound hardness can be used to estimate uniaxial compressive strength. The rebound test requires the specimen to be fixed to a backing plate and then to be hammered. GB, ASTM and ISRM are mainly compared in this section with the differences listed in Table 10.18.

Table 10.16 Comparison of determining the sound velocity by ultrasonic pulse transmission technique

Item	GB	ISRM
Designation	GB 50266-2013	Upgraded ISRM suggested method for determining sound velocity by ultrasonic pulse transmission technique
Scope	Rocks that can be made into regular specimens	
Apparatus	Ultrasonic tester, a longitudinal and transverse waves transducer, a test frame.	Ultrasonic testing system components including a signal generator, an oscilloscope, amplifiers and filters, a data acquisition unit interfacing with the apparatus.
Size of specimen	The diameter of cylinder specimen shall be 48–54 mm; The ratio of height to diameter shall be 2.0–2.5.	As specified in Table 10.17.
Coupling agent	Longitudinal waves: petroleum jelly or butter Transverse waves: Solid materials such as aluminum foil, copper foil or phenyl salicylate.	Phenyl salicylate, high-vacuum grease, glycerin, putty, petroleum jelly, oil. A high viscosity medium (e.g., epoxy resin)
Requirements of transducer	The transmit frequency shall be $$f \geq \frac{2v_p}{D}$$ f – transmit frequency of transducer (Hz); v_p – velocity of rock longitudinal (m/s); D – diameter of specimen (m).	Specified in Table 10.17.
Seating load	0.05 MPa	Maintain a small coupling stress of 10 kPa.
Calibration of transducer	The wave delay shall be zero.	The system needs a calibration each time a new pair of transducers are used.
Accuracy of distance	1 mm	0.01 mm
Accuracy of time	0.1 μs	—
Calculation	$$V_p = \frac{L}{t_p - t_0} \quad V_s = \frac{L}{t_s - t_0}$$ V_p – velocity of the longitudinal wave (m/s); V_s – velocity of the transverse wave (m/s); L – travel path length (m); t_p – travel time for P-waves (s); t_s – travel time of S-waves (s); t_0 – zero delay of the system (s). Keep 3 valid digits	$$V_p = L/t_p$$ $$V_s = L/t_s$$ V_p – velocity of the longitudinal wave (m/s); V_s – velocity of transverse wave (m/s); L – travel path length (m); t_p – travel time of P-waves; t_s – travel time of S-waves. —

Table 10.17 ISRM requirements for specimens of different shapes

Slab	Block	Bar

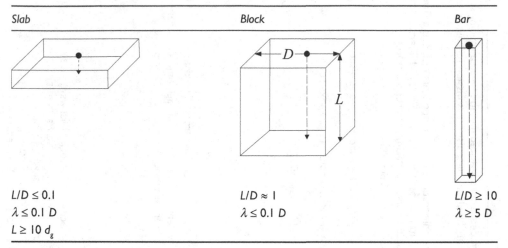

Slab	Block	Bar
$L/D \leq 0.1$	$L/D \approx 1$	$L/D \geq 10$
$\lambda \leq 0.1\ D$	$\lambda \leq 0.1\ D$	$\lambda \geq 5\ D$
$L \geq 10\ d_g$		

Source: d_g: average grain size (equivalent spherical diameter)

Table 10.18 Comparison of rebound hardness tests

Item	China's standard	ASTM	ISRM
Source	In-situ tests for engineering geology	ASTM D5873-14	ISRM suggested method for determination of the Schmidt hammer rebound hardness: Revised version
Scope	Rocks with no significant defects	Rocks with uniaxial compressive strength of 1 to 100 MPa.	N-type Schmidt hammer is more suitable for field testing; L-type is more suitable for porous, weathered weak rocks.
Requirements of core specimens	Length-to-diameter ratio is 2 or 2.5; Length is no less than 10 cm.	Diameter is no less than 54 mm and the length is no less than 15 cm.	For L-type hammers, the diameter should be ≥ 54.7 mm; For N-type hammers, the diameter should be ≥ 84 mm.
Calculation	The first three and last three values of the measurement need be discarded. Take the remaining values to get their average.	Calculate the average of the ten readings obtained for each specimen. Discard the data that differ from the average by more than 7 units.	Use all 20 readings to get the average.

Part 4

Appendices

Reference value of basic parameters of soil

Soil is a collection of crushed minerals formed by intense physical, chemical and biological weathering of rocks. It is usually divided into general and special soils in engineering geology and geotechnical engineering. Widely distributed general soils can be divided into inorganic and organic soils. The original sedimentary inorganic soils can be further divided into sand, gravel, cohesive soil and silty soil. Special soils, such as loess, swelling soil, frozen soil, red clay, soft soil, filled soil, mixed soil, saline soil, contaminated soil, weathered rock and residual soil, are soils with special physical structures and mechanical, physical and engineering properties. The variations of these properties are due to the differences in environment, climate, rainfall, geological origin and history, composition of soil materials, secondary changes and other complex factors.

Many soil types exhibit complex engineering properties. Engineering applications (e.g. foundations, embankments, highway and railway subgrades, slopes and tunnels) utilise different kinds of soil. Soil parameters must be reasonable and reliable to properly evaluate the stability of construction foundation, subgrade and tunnel, the feasibility of site selection (buildings or tunnel roads) and the economic and reasonable design of foundations in terms of geotechnical engineering, design and construction. By referring to soil classification standards, prospect and design specifications, soil mechanics and other geotechnical engineering-related materials, this work summarises the basic soil parameters and divides them into the engineering classification of soils, state division of soils, physical and mechanical parameters and chemical parameters. The typical or reference value range of each parameter is also provided.

The engineering classification of soils is based on the individual characteristics of soil particles (e.g. size and shape). These particles are divided into boulders, cobbles, gravel, sand, silt and clay grains. Table I-1 summarises the soil particle size ranges defined by the Chinese National Standard (GB/T), British Standard (BS), American Society for Testing Materials (ASTM) and International Society of Soil Science (ISSS). Table I-2 summarises the unified soil classification system by fraction and gradation.

The state division of soils includes plasticity index (Table I-3), sensitivity (Table I-4), activity related to swelling (Table I-5), compressibility (Table I-6) and overconsolidation ratio of soil (Table I-7). Soil hardness classification by undrained shear strength and pre-consolidation pressure is show in Table I-8. Table I-9 presents the classification of frost heave capacity of soil. The classification of sand comprises degree of saturation (Table I-10), compactness determined by standard penetration number and relative density (Table I-11) and void ratio (Table I-12). Table I-13 presents the compactness classification of gravelly soil by number of cone compaction. The classification of cohesive soils includes activity number (Table I-14), consistency (Table I-15), sensitivity (saturated cohesive soil) (Table I-16), compressibility (Table I-17) and unconfined compressive strength (Table I-18). For silt, the classification consists of compactness (Table I-19) and humidity condition (Table I-20). Tables I-21 and I-22 present the classification of loess collapsibility and the expansion potential of swelling soil, respectively. The classification of frozen areas according to the freezing index is shown in Table I-23 whilst the classification by thaw compressibility is shown in Table I-24.

The physical index parameter values of soils include density, specific gravity, water content, porosity and void ratio (Table I-25). The mechanical index parameters include void ratio, water content, liquid limit, liquidity index, plasticity index, bearing capacity, modulus of compression, cohesion and internal friction angle (Table I-26). The other physical and mechanical parameters are average physical and mechanical index values (including density, natural water content, void ratio, plastic limit, cohesion, angle of internal friction and modulus of deformation) (Table I-27), consistency of soil-based dry and wet conditions (Table I-28), coefficient of earth pressure at rest (Table I-29) and coefficient of lateral pressure and Poisson's ratio (Table I-30). Soil permeability is the property that demonstrates how soil is permeated by liquid (e.g. water in soil). The parameters related to soil permeability are the coefficient of permeability (Table I-31), viscosity of water (Table I-32) and pore pressure coefficient (pore pressure coefficients A (Table I-33) and B (Table I-34)). For sandy and gravel soils, the physical and mechanical parameters include the reference range of void ratio, porosity and dry weight of common sandy and gravel soils (Table I-35); reference range of elastic modulus (Table I-36); typical values of maximum and minimum porosity of various particle size for some nonclay minerals (Table I-37); typical values of porosity and void ratio for various particle shape at loosest and densest state (Table I-38); and typical values of angle of internal friction of some minerals (Table I-39). The range of the allowable bearing capacity of gravelly soil is shown in Table I-40. For cohesive soil, physical and mechanical parameters include the physical parameters of clay minerals (e.g. liquid, plastic and shrinkage limit; plasticity index; specific surface area and activity number) (Table I-41) and the reference values of modulus of resilience (Table I-42). The range of the basic physical parameters of silt is shown in Table I-43 (e.g. density, specific gravity, void ratio and porosity). The values of loess' shear strength index with different dry density and water content levels are shown in Table I-44. The values of the collapsible loess bearing capacity of Q_3 and $Q_4{}^1$ are shown in Table I-45. The range of physical and mechanical parameters of swelling soil are shown in Table I-46 (e.g. degree of saturation, free swelling, swelling rate, swelling pressure and plastic and shrinkage limits). The basic physical and mechanical parameters of frozen soil are shown in Table I-47 (e.g. liquidity index and plasticity index, water content and modulus of tensile). Coefficient of influence of soil types on freezing depth are presented in Table I-48.

The chemical parameters of soils include the reference range of organic matter (Table I-49), classification of saline soil by salt (Table I-50) content and chemical composition (Table I-51) and solubility of soluble and medium salts in water (Table I-52).

1.1 Engineering classification of soils

Table 1.1 Soil classification by particle size distribution[1–4]

Classification system	Classes			Particle size (mm)
	Over coarse-grains	Boulders		>200
		Cobbles		60–200
	Coarse grains	Gravel	Coarse	20–60
			Medium	5–20
Chinese National Standard (GB/T)			Fine	2–5
		Sand	Coarse	0.5–2
			Medium	0.25–0.5
			Fine	0.075–0.25
	Grain grains	Silt		0.005–0.075
		Clay		≤0.005
	Boulders			>200
	Cobbles			63–200
	Gravel		Coarse	20–63
			Medium	6.3–20
			Fine	2–6.3
British Standard (BS)	Sand		Coarse	0.63–2
			Medium	0.2–0.63
			Fine	0.063–0.2
	Silt		Coarse	0.02–0.063
			Medium	0.0063–0.02
			Fine	0.002–0.0063
	Clay			≤0.002
	Boulders			>300
	Cobbles			75–300
	Gravel		Coarse	19–75
			Fine	4.75–19
American Society for Testing Materials (ASTM)	Sand		Coarse	2–4.75
			Medium	0.425–2
			Fine	0.075–0.425
	Silt			0.005–0.075
	Clay			≤0.005
	Gravel			>2
	Coarse sand			0.2–2
International Society of Soil Science (ISSS)	Fine sand			0.02–0.2
	Silt			0.002–0.02
	Clay			≤0.002

Table 1.2 Unified soil classification system[3][5][6]

Classification system	Classes		Criteria			Group name
Fraction	Over coarse-grained soils	Over coarse-grained soil	Over coarse-grain>75%		Boulder> cobble	Boulder
					Boulders≤ cobble	Cobble

(Continued)

Table 1.2 (Continued)

Classification system	Classes	Criteria			Group name
		Mixed over coarse-grained soil	50%<over coarse-grain≤75%	Boulder>cobble	Boulder-cobble mixture
				Boulder≤cobble	Cobble-boulder mixture
		Over coarse-grained mixed soil	15%<over coarse-grain≤50%	Boulder>cobble	Bouldery soil
				Boulder≤cobble	Cobbley soil
	Gravel (2 mm<d≤60 mm, gravel fraction content>50%)	Clean gravels	Clean gravels less than 5% fines	C_u≥5 and C_c=1–3	Well-graded gravel
				Not meeting the two criteria for well-graded gravel	Poorly graded gravel
		Fine-grained gravels	Gravel with 5–15% fine		Gravel with fine
			Gravel with 15–50% fine	Silt≤50% fine	Clayey gravel
				Silt>50% fine	Silty gravel
	Sand (gravel fraction content≤50%)	Clean sands	Clean sands less than 5% fines	C_u≥5 and C_c=1–3	Well-graded sand
				Not meeting the two criteria for well graded sand	Poorly graded sand
		Fine-grained sands	Sand with 5–15% fine		Sand with fine
			Sand with 15–50% fine	Silt≤50% fine	Clayey sand
				Silt >50% fine	Silty sand
Gradation	Gravel		Particles larger than 200 mm >50%	Round and subround	Boulder
				Angular	Block
			Particles larger than 20 mm>50%	Round and subround	Cobble
				Angular	Aggregate
			Particles larger than 2 mm >50%	Round and subround	Round gravel
				Angular	Breccia
	Sand		Particles larger than 2 mm account for 25–50%	–	Gravelly sand

Classification system	Classes	Criteria		Group name
		Particles larger than 0.5 mm >50%	–	Coarse sand
		Particles larger than 0.25 mm>50%	–	Medium sand
		The weight of particles larger than 0.075 mm>85% of total weight	–	Fine sand
		Particles larger than 0.075 mm>50%	–	Silty sand
	Silt	Particles less than 0.005 mm ≤10%	–	Sandy silt
		Particles less than 0.005 mm >10%	–	Clayey silt

1.2 The state division of soil

Table 1.3 Soil classification by plasticity index[5]

Soil type	Clay	Silty clay	Silt
Plasticity index, I_p	$I_p > 17$	$17 \geq I_p > 10$	$I_p \leq 10$

Table 1.4 Classification of clays by sensitivity[7]

Classification system	Classes	Sensitivity, S_t
Terzaghi method	Slightly sensitive	2–4
	Medium sensitive	4–7
	Very sensitive	7–8
Skempton method	Insensitive	<2
	Medium sensitive	2–4
	Very sensitive	4–8
	Slightly quick	8–16
	Medium quick	16–32
	Very quick	32–64
	Extra quick	>64

Table 1.5 Soil activity related to swelling[8]

Activity	Low	Moderate	High	Very high
Plasticity index, I_p	Less than 18	12–32	22–48	Over 35

Table I.6 Soil classification by compressibility[5]

Compressibility of soil	Low	Moderate	High
Coefficient of compressibility, a (MPa^{-1})	<0.1	$0.1 \leq a < 0.5$	≥ 0.5
Compression index, C_c	<0.2	$0.2 \leq C_c < 0.4$	>0.4
Modulus of compression, E_s (Mpa)	>16	$4 < E_s \leq 16$	≤ 4
Void ratio, e	<0.6	$0.6 < e \leq 1$	>1

Table I.7 Soil classification by overconsolidation ratio[5]

Soil type	Underconsolidated soil	Normally consolidated soil	Overconsolidated soil
Over consolidation ratio, OCR	<1	1	>1

Note: When OCR=1.0–1.2, it can be regarded as normally consolidated soil.

Table I.8 Soil hardness classification by undrained shear strength and pre-consolidation pressure[9,10]

Hardness	Undrained shear strength (kPa)	Pre-consolidation pressure (kPa)
Very soft	<20	0–50
Soft	20–40	50–100
Firm	40–75	100–200
Stiff	75–150	200–400
Very stiff	150–300	400–800
Hard	>300	800–1600

Table I.9 Classification of frost heave capacity of soil[11]

Frost heave capacity	Non	Low	Moderate	High	Very high
Influence coefficient, ψ_{zw}	1.00	0.95	0.90	0.85	0.80

Table I.10 Degree of saturation of sand in various state[29]

Classes	Dry	Humid	Damp	Moist	Wet	Saturated
Degree of saturation, S_r (%)	0	1–25	26–50	51–75	76–99	100

Table I.11 Compactness classification of sand by standard penetration number and relative density[6,12]

Compactness	Very loose	Loose	Medium	Dense
Standard penetration number, N	$N \leq 10$	$10 < N \leq 15$	$15 < N \leq 30$	$N > 30$
Relative density, D_r	$D_r \leq 1/3$	-	$2/3 \geq D_r > 1/3$	$D_r > 2/3$

Table I.12 Compactness classification of sand by void ratio (e)[6,12]

Compactness	Very loose	Loose	Medium	Dense
Gravelly sand, coarse sand, medium sand	e>0.85	$0.75 < e \leq 0.85$	$0.60 \leq e \leq 0.75$	e<0.60
Fine sand, silty sand	e>0.95	$0.85 < e \leq 0.95$	$0.70 \leq e \leq 0.85$	e<0.70

Table I.13 Compactness classification of gravelly soil by number of cone compaction[6]

Compactness	Very loose	Loose	Medium	Dense	Very dense
Drop number of $N_{63.5}$ hammer	$N_{63.5} \leq 5$	$5 < N_{63.5} \leq 10$	$10 < N_{63.5} \leq 20$	$N_{63.5} > 20$	-
Drop number N_{120} hammer	$N_{120} \leq 3$	$3 < N_{120} \leq 6$	$6 < N_{120} \leq 11$	$11 < N_{120} \leq 14$	$N_{120} > 14$

Table I.14 Classification of cohesive soil by activity number[12]

Classes	Activity number, A
Inactive clay	$A \leq 0.75$
Normal clay	$0.75 < A \leq 1.25$
Active clay	$A > 1.25$

Table I.15 Consistency of cohesive soil[12]

Description	Nonplastic	Slightly plastic	Medium plasticity	High plasticity	Very high plasticity
Liquidity index, I_L	$I_L \leq 0$	$0 < I_L \leq 0.25$	$0.25 < I_L \leq 0.75$	$0.75 < I_L \leq 1.0$	$I_L > 1.0$

Table I.16 Sensitivity of saturated cohesive soil[5]

Description	Slightly sensitive	Medium sensitive	Very sensitive
Sensitivity, S_t	$1 < S_t \leq 2$	$2 < S_t \leq 4$	$S_t > 4$

Table I.17 Compressibility of cohesive soil[12]

Compressibility	Low	Moderate	High
Coefficient of compressibility, a_{1-2}	$a_{1-2} < 0.1$	$0.5 > a_{1-2} \geq 0.1$	$a_{1-2} \geq 0.5$
Coefficient of compressibility, a_{1-3}	$a_{1-3} < 0.1$	$1 > a_{1-3} \geq 0.1$	$a_{1-3} \geq 1$
Modulus of compressibility, E_s	$E_s > 15$	$5 < E_s \leq 15$	$E_s \leq 5$

Note:

1) a_{1-2} – coefficient of compressibility at pressure of 100–200kPa, (MPa^{-1});

2) a_{1-3} – coefficient of compressibility at pressure of 100–300kPa, (MPa^{-1});

3) E_s – Modulus of compressibility, MPa.

Table I.18 Consistency of clay in terms of unconfined compressive strength[12]

Consistency	Unconfined compressive strength, q_u (kPa)
Very soft	$q_u < 30$
Soft	$60 > q_u \geq 30$
Firm	$120 > q_u \geq 60$
Stiff	$240 > q_u \geq 120$
Very stiff	$q_u \geq 240$

Table I.19 Compactness of silt by void ratio (e)[6]

Compactness	Loose	Medium	Dense
Void ratio	e >0.90	0.90≥ e ≥0.75	e <0.75

Table I.20 Humidity condition of silt[6]

Classes	Humid	Damp	Moist
Water content, ω (%)	<20	20–30	>30

Table I.21 Classification of collapsibility of loess[5,12]

Classes	Non	Slight	Medium	High
Coefficient of collapsibility, δ_s	δ_s <0.015	0.015≤ δ_s ≤0.03	0.03< δ_s ≤0.07	δ_s >0.07

Table I.22 Expansion potential of swelling soil[12]

Free swelling, δ_{ef} (%)	Potential for volume change
40≤ δ_{ef} <65	Low
65≤ δ_{ef} <90	Moderate
δ_{ef} ≥90	High

Table I.23 Classification of frozen area by frozen index[13]

Classes	Non-frozen	Slightly frozen	Medium frozen	High frozen
Frozen index	≤50	50–800	800–2000	≥2000

Table I.24 Classification of frozen soil by thawing compressibility[12]

Classes	Non	Slight	Medium	High	Very high
Thaw compressibility, a_0	<1%	1–3%	3–10%	10–25%	>25%

I.3 Physical and mechanical parameters of soil

Table I.25 Typical values of basic physical parameters of soil[5,14–18]

Soil type		Nature density, ρ (g/cm^3)	Specific gravity, G_s	Water content, ω (%)	Saturated water content, ω_{sat} (%)	Dry density, ρ_d (g/cm^3)	Saturated density, ρ_{sat} (g/cm^3)	Void ratio, e	Porosity, n (%)
Cohesive soil	Organic soft clay	1.8–2.0	2.72–2.76	Hard cohesive soil: ω≤30; Saturated soft clay: ω≥60	70	0.93	1.58	1.94	66
	Organic clay				110	0.69	1.43	3	75
	Boulder clay				9	2.11	2.32	0.25	20
	Ice soft clay				45	1.21	1.17	1.2	55
	Ice hard clay				22	1.7	2.07	0.59	37

Soil type		Nature density, ρ (g/cm³)	Specific gravity, G_s	Water content, ω (%)	Saturated water content, ω_{sat} (%)	Dry density, ρ_d (g/cm³)	Saturated density, ρ_{sat} (g/cm³)	Void ratio, e	Porosity, n (%)
Sand	Homogeneous loose sand	1.6–2.0	2.65–2.70	Dry coarse sand: ≈ 0; Saturated sand: 40	32	1.44	1.89	0.85	46
	Homogeneous dense sand				19	1.74	2.09	0.52	34
	Heterogeneous loose sand				25	1.59	1.99	0.67	40
	Heterogeneous dense sand				16	1.86	2.16	0.43	30
Loess		1.4–1.65	-	10–20	12–61	1.27	-	0.85–1.24	50

Table I.26 Physical and mechanical parameters range of soil[12,14,16,19]

Soil type	Physical and mechanical parameter								
	Void ratio, e	Liquidity index, I_L	Water content, ω (%)	Liquid limit, ω_L (%)	Plasticity index, I_p	Bearing capacity (kPa)	Modulus of compressibility, E_s (MPa)	Cohesion, c (kPa)	Angle of internal friction, φ (°)
Xiashu system clay	0.6–0.9	<0.8	15–25	25–40	10–18	300–800	>15	40–100	22–30
Ordinary clay	0.55–1.0	0–1.0	15–30	25–45	5–20	100–450	4–15	10–50	15–22
Recently deposited clay	0.7–1.2	0.25–1.2	24–36	30–45	6–18	80–140	2–7.5	10–20	7–15
Silty soil Coastal Inland Montane	1.0–2.0	>1.0	36–70	30–65	10–25	40–100 50–110 30–80	1–5 2–5 1–6	5–15	4–10
Typical red clay	1.1–1.7		20–75	50–110	30–50	180–380	>3	40–90	8–20
Loess	0.85–1.24		10–20	19–34	7–13	110–230	1.7–5	0–42	5–31
Swelling soil	0.5–0.8		20–30	38–55	18–35			Big difference before and after immersion	

Table I.27 The average physical and mechanical parameters value of soil[12]

Soil type		Density, ρ (g/cm³)	Nature water content, ω (%)	Void ratio, e	Plastic limit, ω_p	Cohesion, c (kPa)		Angle of internal friction, φ (°)	Modulus of deformation, E (MPa)
						Standard value	Calculated value		
Sand	Coarse sand	2.05	15–18	0.4–0.5		2	0	42	46
		1.95	19–22	0.5–0.6		1	0	40	40
		1.90	23–25	0.6–0.7		0	0	38	33

(Continued)

Table 1.27 (Continued)

Soil type		Density, ρ (g/cm³)	Nature water content, ω (%)	Void ratio, e	Plastic limit, ωp	Cohesion, c (kPa) Standard value	Cohesion, c (kPa) Calculated value	Angle of internal friction, φ (°)	Modulus of deformation, E (MPa)
	Medium sand	2.05	15–18	0.4–0.5		3	0	40	46
		1.95	19–22	0.5–0.6		2	0	38	40
		1.90	23–25	0.6–0.7		1	0	35	33
	Fine sand	2.05	15–18	0.4–0.5		6	0	38	37
		1.95	19–22	0.5–0.6		4	0	36	28
		1.90	23–25	0.6–0.7		2	0	32	24
	Silty sand	2.05	15–18	0.4–0.5		8	5	36	14
		1.95	19–22	0.5–0.6		6	3	34	12
		1.90	23–25	0.6–0.7		4	2	28	10
Silt		2.10	15–18	0.4–0.5	<9.4	10	6	30	18
		2.00	19–22	0.5–0.6		7	5	28	14
		1.95	23–25	0.6–0.7		5	2	27	11
		2.10	15–18	0.4–0.5	9.5–12.4	12	7	25	23
		2.00	19–22	0.5–0.6		8	5	24	16
		1.95	23–25	0.6–0.7		6	3	23	13
Cohesive soil	Silty clay	2.10	15–18	0.4–0.5	12.5–15.4	42	25	24	45
		2.00	19–22	0.5–0.6		21	15	23	21
		1.95	23–25	0.6–0.7		14	10	22	15
		1.90	26–29	0.7–0.8		7	5	21	12
		2.00	19–22	0.5–0.6	15.5–18.4	50	35	22	39
		1.95	23–25	0.6–0.7		25	15	21	18
		1.90	26–29	0.7–0.8		19	10	20	15
		1.85	30–34	0.8–0.9		11	8	19	13
		1.80	35–40	0.9–1.0		8	5	18	8
		1.95	23–25	0.6–0.7	18.5–22.4	68	40	20	33
		1.90	26–29	0.7–0.8		34	25	19	19
		1.85	30–34	0.8–0.9		28	20	18	13
		1.80	35–40	0.9–1.0		19	10	17	9
	Clay	1.90	26–29	0.7–0.8	22.5–26.4	82	60	18	28
		1.85	30–34	0.8–0.9		41	30	17	16
		1.75	35–40	0.9–1.1		36	25	16	11
		1.85	30–34	0.8–0.9	26.5–30.4	94	65	16	23
		1.75	35–40	0.9–1.1		47	35	15	14

Note:

1) Average specific gravity: sand, 2.65; silt, 2.70; silty clay, 2.71; clay, 2.74;

2) Modulus of deformation of coarse and medium sands are applicable to uniformity coefficient c_u=3; when c_u>5, the values listed in the table are reduced by 2/3, and the median value of c_u is determined by the interpolation method;

3) Used for the foundation to determine the calculation, the calculated φ value is less than the standard value.

Table 1.28 Reference range of consistency of soil foundation in dry and wet state[13]

Classes Soil type	Humid	Damp	Moist	Wet
Sand	$\omega_c \geq 1.20$	$1.20 > \omega_c \geq 1.00$	$1.00 > \omega_c \geq 0.85$	$\omega_c < 0.85$
Clay	$\omega_c \geq 1.10$	$1.10 > \omega_c \geq 0.95$	$0.95 > \omega_c \geq 0.80$	$\omega_c < 0.80$
Silt	$\omega_c \geq 1.05$	$1.05 > \omega_c \geq 0.90$	$0.90 > \omega_c \geq 0.75$	$\omega_c < 0.75$

Note: ω_c is the average consistency within a depth of 80 cm below the surface of the roadbed.

Table I.29 Reference value of coefficient of earth pressure at rest[11,17,20]

Soil type		Coefficient of earth pressure at rest, K_0
Sand	Loose sand	0.4–0.45
	Dense sand	0.45–0.5
	Sand loam	0.35
Clay	Slightly medium plasticity clay	0.4–0.5
	Silty clay	
	Medium-high plasticity clay	0.5–0.6
	High plasticity clay	0.6–0.75
	Very high plasticity clay	0.75–0.8
	Clay loam	0.45
	Normally consolidated clay	0.5–0.6
	Overconsolidated soil	1–4
Compacted fill		0.8–1.5
Silt		0.4–0.5
Hard soil		0.2–0.4
Gravel, cobble		0.20

Table I.30 Reference value of coefficient of lateral pressure and Poisson's ratio[5,15]

Soil type		Coefficient of lateral pressure	Poisson's ratio
Gravelly soil		0.18–0.33	0.15–0.25
Egg gravelly soil		0.18–0.25	0.15–0.20
Gravel		0.25–0.33	0.20–0.25
Sand		0.33–0.43	0.25–0.30
Silt		0.43	0.30
Silty clay	Nonplastic	0.33	0.25
	Medium plasticity	0.43	0.30
	High and very high plasticity	0.53	0.35
Clay	Nonplastic	0.33	0.25
	Medium plasticity	0.53	0.35
	High and very high plasticity	0.72	0.42

Table I.31 Reference range of coefficient of permeability of various soil[4,7,12,15,16,21,22]

Soil type	Coefficient of permeability, k (cm/s)	Soil type	Coefficient of permeability, k (cm/s)
Clay	$<10^{-6}$	Fine sand	$10^{-5}–10^{-3}$
Silty clay	$5\times10^{-9}–10^{-6}$	Medium sand	$5\times10^{-5}–2.4\times10^{-2}$
Clay silt	$6.0\times10^{-5}–6.0\times10^{-4}$	Coarse sand	$2\times10^{-4}–6\times10^{-2}$
Loess	$3.0\times10^{-4}–6.0\times10^{-4}$	Gravelly sand	$6\times10^{-2}–1.8\times10^{-1}$
Silty sand	$10^{-6}–1.2\times10^{-3}$	Soft soil	$<1\times10^{-7}$
Red clay	$1.0\times10^{-5}–1.0\times10^{-3}$	Pebble	$1\times10^{-2}–6\times10^{-2}$
Cobble	$10^{-3}–5\times10^{-3}$	Silt	$5\times10^{-8}–10^{-3}$
Gravel	>1	Sand	$10^{-3}–1$
Compacted silt	$10^{-6}–10^{-5}$	Homogeneous coarse sand	$7\times10^{-2}–8\times10^{-2}$
Homogeneous medium sand	$4\times10^{-2}–6.0\times10^{-2}$	–	

Table I.32 Viscosity of water, η (mPa·s)[5]

Temperature, °C	η	Temperature, °C	η	Temperature, °C	η
0.0	1.794	13.0	1.206	21.5	0.974
5.0	1.561	13.5	1.190	22.0	0.963
5.5	1.493	14.0	1.175	22.5	0.952
6.0	1.470	14.5	1.160	23.0	0.941
6.5	1.449	15.0	1.144	24.0	0.919
7.0	1.428	15.5	1.130	25.0	0.899
7.5	1.407	16.0	1.115	26.0	0.879
8.0	1.387	16.5	1.101	27.0	0.859
8.5	1.367	17.0	1.088	28.0	0.841
9.0	1.347	17.5	1.074	29.0	0.823
9.5	1.328	18.0	1.061	30.0	0.806
10.0	1.310	18.5	1.048	31.0	0.789
10.5	1.292	19.0	1.035	32.0	0.773
11.0	1.274	19.5	1.022	33.0	0.757
11.5	1.256	20.0	1.010	34.0	0.742
12.0	1.239	20.5	0.998	35.0	0.727
12.5	1.223	21.0	0.986	–	–

Table I.33 Pore pressure coefficient, A[4][5]

Soil type (saturated)	A	Note
Very loose fine sand	2–3	
Sensitive clay	1.5–2.5	
Normally consolidated clay	0.7–1.3	Used for checking the value of soil failure
Slightly overconsolidated clay	0.3–0.7	
Severely overconsolidated clay	−0.5–0	
Highly sensitive soft clay	>1	
Normally consolidated clay	0.5–1	Used to calculate the value of foundation deformation
Overconsolidated clay	0.25–0.5	
Severely overconsolidated clay	0–0.25	
Highly Sensitive clay	0.75–1.5	
Compacted sandy clay	0.5–0.75	Value at failure

Table I.34 Pore pressure coefficient, B[23]

Soil type	Description	Void ratio	B value of soil (saturated)
Soft soil	Normally consolidated clay	2	0.9998
Firm soil	Compacted silt and clay	0.6	0.9988
Stiff soil	Overconsolidation (stiff) clay, medium dense sand	0.6	0.9877
Hard soil	Dense sand and stiff clay under high confining pressure	0.4	0.913

Table I.35 Reference ranges of void ratio, porosity and dry weight of common sandy and gravel soils[4,10]

Soil type	Void ratio, e	Porosity, n (%)	Dry weight, γ_d (kN/m³)
Gravel	0.3–0.6	–	16–20
Coarse sand	0.35–0.75	17–49	15–19
Fine sand	0.4–0.85		14–19
Standard Ottawa sand	0.5–0.8	33–44	14–17
Gravelly sand	0.2–0.7	–	15–22
Silty sand	0.4–1	23–47	13.7–20
Silty sand and sandy gravel	0.15–0.85	12–46	14–22.9
Micaceous sand	0.4–1.2	29–55	11.9–18.9

Table I.36 Reference ranges of elastic modulus of sand[4]

Soil type	Elastic modulus, E (MN/m²)
Loose sand	10.35–24.15
Silty sand	10.35–17.25
Medium dense sand	17.25–27.60
Dense sand	34.5–55.2
Gravelly sand	69.0–172.5

Table I.37 Typical values of maximum and minimum porosities of soils consisting of nonclay minerals[12]

Particle size (mm)	Quartz	Orthoclase	White mica	Quartz	Orthoclase	White mica
	Maximum porosity, n_{max} (%)			Minimum porosity, n_{min} (%)		
2–1	47.63	47.50	87.00	37.90	45.46	80.46
1–0.5	47.10	51.98	95.18	40.61	47.88	75.20
0.5–0.25	46.98	54.76	83.71	41.09	49.18	72.16
0.25–0.1	52.47	58.46	82.74	44.82	51.62	66.30
0.1–0.06	54.60	61.22	82.98	45.31	52.72	68.98
0.06–0.01	55.99	62.53	–	45.68	–	65.33

Table I.38 Typical values of porosity and void ratio of soils consisting of various shapes of particles[12]

Particle shape	Loosest state		Densest state	
	Maximum porosity, n_{max} (%)	Maximum void ratio, e_{max}	Minimum porosity, n_{min} (%)	Minimum void ratio, e_{min}
Angular quartz sand (d=0.25–0.70 mm)	50.1	1	44	0.79
Alluvium sand (d=0.1–2.7 mm)	41.6	0.71	33.9	0.51
Round hillock sand	45.8	0.85	38.9	0.64
Theoretical spheroids of uniform size	47.6	0.91	25.9	0.35

Table I.39 Typical values of angle of internal friction of some minerals[12]

Mineral	Angle of internal friction of the following particle size (mm), φ (°)				
	2–1	1–0.5	0.5–0.25	0.25–0.1	0.06–0.01
Mica	28	26	17.5	19	27
Feldspar	–	–	39	–	17
Angular quartz	66	56	46	27	15
Round quartz	61	–	27	28	18.5

Table I.40 Allowable bearing capacity of gravelly soil, $[\sigma_0]$ (kPa)[12]

Soil	Compactness			
	Very loose	Loose	Medium dense	Dense
Cobble	500–300	650–500	1000–650	1200–1000
Gravelly soil	400–200	550–400	800–550	1000–800
Round gravel	300–200	400–300	600–400	850–600
Angular gravel	300–200	400–300	500–400	700–500

Table I.41 Physical index of common clay minerals[4,24,25]

Mineral	Mineral component	Liquid limit (%)	Plastic limit (%)	Shrinkage limit (%)	Plasticity index	Specific surface area (m²/g)	Activity number
Kaolinite	Na	53	32	26.8	21	10–20	0.3–0.5
	K	49	29	–	20		
	Ca	38	27	24.5	11		
	Mg	54	31	28.7	23		
	Fe	59	37	29.2	22		
Illite	Na	120	53	15.4	67	65–100	0.5–1.3
	K	120	60	17.5	60		
	Ca	100	45	16.8	55		
	Mg	95	46	14.7	49		
	Fe	110	49	15.3	61		
Montmorillonite	Na	710	54	9.9	656	Up to 840	1–7
	K	660	98	9.3	562		
	Ca	510	81	10.5	429		
	Mg	410	60	14.7	350		
	Fe	290	75	10.3	215		
Attapulgite	H	270	150	7.6	150	–	0.5–1.2
Halloysite (4H$_2$O)	–	–	–	–	–	–	0.5
Halloysite (2H$_2$O)	–	–	–	–	–	–	0.1
Allophane	–	–	–	–	–	–	0.5–1.2

Table I.42 Modulus of resilience of clay[26]

Load level	Unit pressure, P (kPa)	Modulus of resilience, E_e (kPa)
1	25	22670
2	50	22670
3	100	22375
4	150	22571
5	200	22522

Table 1.43 Physical parameters of silt[5,14,15]

Soil type	Density, ρ (g/cm³)	Specific gravity, G_s	Void ratio, e	Porosity, n (%)
Silt	1.35	2.65–2.71	0.4–1.2	30–60

Table 1.44 Shear strength parameters of loess[12]

Dry density, ρ_d (g/cm³)	Water content, ω (%)	Angle of internal friction, φ (°)	Cohesion, c (kPa)	Dry density, ρ_d (g/cm³)	Water content, ω (%)	Angle of internal friction, φ (°)	Cohesion, c (kPa)
1.25–1.27	3.9	39°20′	70	1.42–1.44			
	8.6	33°50′	52		18.3	29°20′	40
	14.5	31°20′	32		21	27°	26
	19.2	30°10′	21		23.3	26°30′	20
	23.8	26°20′	6		25.6	25°50′	10
	27.9	26°	2				
1.36–1.38	6.1	36°50′	80	1.48–1.50			
	9.5	35°	65		7.8	37°10′	157
	12.8	31°20′	46		10	33°	120
	15.1	29°	35		14.4	28°20′	80
	20.6	28°20′	20		18.5	26°30′	52
	25.4	26°30′	10		24.4	26°	20
	26.5	25°20′	5				
1.42–1.44				1.53–1.55	14.3	36°10′	132
					17.7	34°30′	100
	7	34°10′	96		21.6	31°20′	70
	12.1	28°50′	58		23.9	26°10′	42
	15.8	28°30′	46		25.6	25°40′	31
					26.8	25°10′	26

Table 1.45 Basic value of Q_3, $Q_4{}^1$ collapsible loess bearing capacity, f_0[12]

ω_L/e \ ω (%) \ f_0 (kPa)	≤13	16	19	22	25
22	180	170	150	130	110
25	190	180	160	140	120
28	210	190	170	150	130
31	230	210	190	170	150
34	250	230	210	190	170
37	–	250	230	210	190

Note:

1) When natural water content of soil less than plastic limit, the bearing capacity of can be determined by plastic limit;

2) e, void ratio; ω, water content; ω_L, liquid limit.

Table 1.46 Parameter ranges of swelling soil[14]

Soil type	Degree of saturation	Free swelling (%)	Swelling ratio (%)	Swelling pressure (kPa)	Plastic limit (%)	Shrinkage limit (%)
Swelling soil	>0.85	40–58	1–4	10–110	20–35	11–18

Table I.47 Parameter ranges of frozen soil[27,28]

Soil type	Liquidity index, I_L	Water content, ω (%)	Plasticity index, I_p	Modulus of tensile, E_t (MPa)
Frozen soil	0.12–0.65	10–49	8.3–36.9	1.5–15

Table I.48 Coefficient of soil type on freezing depth[11]

Soil type	Coefficient, ψ_{zs}
Cohesive soil	1.0
Fine sand, silty sand, silt	1.2
Medium, coarse, gravelly sand	1.3
Block gravelly soil	1.4

1.4 Chemical parameters of soil

Table I.49 Reference range of organic matter[5]

Soil type	Non-organic soil	Organic soil	Cumulosoil			Peat
			Low	Moderate	High	
Organic matter (%)	$O_m <5$	$5 \leq O_m \leq 10$	$10< O_m \leq 25$	$25<O_m \leq 40$	$40<O_m \leq 60$	$O_m >60$

Table I.50 Classification of saline soil by salt content[12]

Soil type	Average saline soil (%)		
	Chlorine saline soil and sub chlorine saline soil	Sulphate saline soil and sulfite saline soil	Alkaline saline soil
Low saline soil	0.5–1	–	–
Moderate saline soil	1–5	0.5–2	0.5–1
High saline soil	5–8	2–5	1–2

Table I.51 Classification of saline soil by chemical composition[12]

Soil type	$c\,(Cl^-)\,/\,2c\,(SO4^{2-})$	$2\,c\,(CO_3^{2-}) + c\,(HCO_3^-)/c\,(Cl^-) + 2\,c\,(SO_4^{2-})$
Chlorine saline soil	>2	–
Sub chlorine saline soil	2–1	–
Sulfite saline soil	1–0.3	–
Sulphate saline soil	<0.3	–
Alkaline saline soil	–	>0.3

Table I.52 Solubility of soluble and medium salts in water[12]

Mineral	Molecular formula	Relative density	Solubility (g/L)
Gypsum	$CaSO_4 \cdot 2H_2O$	2.3–2.4	2.0
Anhydrite	$CaSO_4$	2.9–3.0	2.1
Mirabilite	$Na_2SO_4 \cdot 10H_2O$	1.48	448.0
Thenardite	Na_2SO_4	2.68	398 (40 °C)
Glauberite	$Na_2SO_4 \cdot CaSO_4$	2.70–2.85	-
Epsomite	$MgSO_4 \cdot 7H_2O$	1.75	262
Hexahydrite	$MgSO_4 \cdot 6H_2O$	1.76	308
Halite	$NaCl$	2.1–2.2	264
Sylvinite	KCl	1.98	340

Reference value of basic parameters of rock

Rocks are solid aggregates composed of minerals or debris with stable shapes. A rock mass is a geological body characterised by discontinuity, heterogeneity and anisotropy and is formed by a network of rock blocks and structural planes. For general rock mass engineering, a laboratory rock test is used to estimate physical and mechanical parameters. However, this method is subjective and ignores the influence of structural plane properties, distribution, groundwater effects and others. Improper rock parameters can also mislead engineering practice. This part refers to relevant geotechnical engineering manuals and materials of rock mechanics. The basic parameters of rocks are summarised and classified into three categories: engineering classification, physical mechanics and dynamic parameters. The typical values of the parameters are given.

The engineering classification of rocks or rock masses have an important reference value for the evaluation of rock integrity, weathering degree and quality rating. It covers rock material strength (Table II-1), rock quality designation (Table II-2), integrality (Table II-3), rocks mass weathering zones (Table II-4) and slaking durability (Table II-5).

The physical and mechanical parameters of rocks are important basis for evaluating the stability of rock masses. General physical parameters include density, particle density, porosity, absorption, water saturation and softening and water saturation coefficient. The parameter ranges are shown in Table II-6. The coefficient of permeability, which is determined by the number, size and connectivity of pores, is an important mechanical index for rock permeability. The specific parameters are shown in Table II-7. This coefficient is important in solving practical problems related to seepage (e.g. removal of water seepage from rock caverns, impermeability evaluation of reservoirs and extraction of oil and gas stored in rock fissures). In rock mechanics, rock strength refers to the ability of rocks to resist damage. When the load reaches or exceeds this strength, the rock is destroyed. The range of rock strength index parameters mainly includes basic strength indices (e.g. compressive and tensile strength, angle of internal friction, cohesion, modulus of deformation and Poisson's ratio) (Table II-8), unconfined compressive strength (Table II-9), point load strength index (Table II-10), structural plane shear strength parameters (Table II-11) and shear strength parameters (Table II-12). The allowable bearing pressure of rocks is important not only in evaluating the stability of rock foundations but also in determining the bearing pressure of foundations. The parameter range is shown in Table II-13.

The properties of a rock mass under dynamic loads are the dynamic characteristics. The study of a rock mass' dynamic properties can provide a theoretical basis for the dynamic measurements of various physical and mechanical parameters. These parameters (Table II-14), which include elastic wave velocity, dynamic modulus of elasticity and dynamic Poisson's ratio, play an important role in the evaluation of dynamic stability in rock mass engineering.

2.1 Engineering classification of rock

Table II.1 Classification of rock material strength[6]

Classes	Extremely weak	Weak	Moderately weak	Moderately strong	Strong
Standard values for saturated uniaxial compressive strength, f_{rk} (MPa)	$f_{rk} \leq 5$	$15 \geq f_{rk} > 5$	$30 \geq f_{rk} > 15$	$60 \geq f_{rk} > 30$	$f_{rk} > 60$

Table II.2 Relation between in situ rock quality and RQD[29]

Rock quality designation, RQD	0–0.25	0.25–0.5	0.5–0.75	0.75–0.9	0.9–1
Rock quality	Very poor	Poor	Fair	Good	Excellent

Table II.3 Classification of integrality of rock mass[6]

Classes	Very highly fractured	Highly Fractured	Moderately fractured	Slightly fractured	Integrated
Index of rock mass integrity	< 0.15	0.35–0.15	0.55–0.35	0.75–0.55	> 0.75

Note: Index of rock mass integrity is square of the ratio of velocity of compressive wave of intact rock to that of rock mass.

Table II.4 Classification of rocks mass weathering zones[12]

Classes	Characteristics	Degree of weathering parameters	
		Wave velocity ratio, K_v	Coefficient of weathering, K_f
Unweathered	Fresh rock; May show slight discoloration along discontinuities.	0.9–1.0	0.9–1.0
Slightly weathered	Original rock texture is generally unchanged but a few weathered fissures; Fresh rock colors generally retained but stained near joint surfaces.	0.8–0.9	0.8–0.9
Moderately weathered	Original rock texture is partly destroyed, secondary minerals and weathered fissures retained along the joint surface, and rock mass is cut into rock blocks; Difficult to crumble with picks, core drilling can be crumbled.	0.6–0.8	0.4–0.8

(Continued)

Table II.4 (Continued)

Classes	Characteristics	Degree of weathering parameters	
		Wave velocity ratio, K_v	Coefficient of weathering, K_f
Highly weathered	Most of original rock texture is destroyed, the mineral composition changes significantly; Weathered fissure developed, rock mass broken; Can be crumbled by picks, but the dry drill cannot crumble.	0.4–0.6	< 0.4
Completely weathered	Original rock texture preserved, but still recognizable; residual structural strength retained, can be crumbled by picks and dry drill.	0.2–0.4	-
Residual soil	Residual soil derived from *in situ* weathering; Original rock texture completely destroyed: 100% soil; Easy to be crumbled by picks and dry drill, and can be shaped.	< 0.2	-

Note:

1) Wave velocity ratio, K_v, is the ratio of compression wave velocity of weathered rock to fresh rock;

2) Coefficient of weathering, K_f, is the ratio of saturated uniaxial compressive strength of weathered rock to fresh rock;

3) Degree of weathering in rock, in addition to the field characteristics and quantitative index are listed in the table, it can also be classified according to local experience.

4) Granitic rock can be divided according to the standard penetration test, N≥50 is highly weathered; 50> N≥30 is completely weathered; N < 30 is residual soil;

5) Mudstone and semi-diagenesis may not be classified by degree of weathering.

Table II.5 Gamble's slake durability classification[30]

Group name	Retained after one 10 min cycle (dry weight basis, %)	Retained after two 10 min cycle (dry weight basis, %)
Very low durability	< 60	< 30
Low durability	60–85	30–60
Medium durability	85–95	60–85
Medium high durability	95–98	85–95
High durability	98–99	95–98
Very high durability	> 99	> 98

2.2 Physical and mechanical parameters of rock

Table II.6 Reference range of general physical parameters in rock[12,30–42]

Rock type	Particle density, ρ_s (g/cm³)	Bulk density, ρ (g/cm³)	Porosity, n (%)	Absorption, ω_a (%)	Water saturation, ω_{sa} (%)	Softening coefficient, η_c	Water saturation coefficient
Granite	2.50–2.84	2.3–2.8	0–1	0.1–4	0.84	0.72–0.97	0.55
Diorite	2.60–3.10	2.52–2.96	0.2–0.5	0.3–5.0	0.07–0.15	0.60–0.80	0.59
Quartz diorite				0.32	0.54		0.59
Dioritic porphyrite	2.6–2.84	2.4–2.8	2.1–5.0	0.4–1.7	0.42	0.78–0.81	0.83
Diabase	2.60–3.10	2.53–2.97	0.3–5.0	0.8–5.0	0.1–9.85	0.33–0.90	
Gabbro	2.70–3.20	2.55–2.98	0.3–4.0	0.5–4.0	0.22	0.44–0.90	0.59
Andesite	2.40–2.80	2.30–2.70	1.1–4.5	0.3–4.5		0.81–0.91	
Porphyrite	2.6–2.9	2.4–2.86	2.1–5.0	0.07–0.65		0.78–0.81	
Basalt	2.48–3.30	1.8–3.10	0.3–21.8	0.3	0.39	0.30–0.95	0.69
Basaltic porphyrite		2.4–2.75		0.31–3.91	0.38–4.21		
Marlstone	2.70–2.80	2.10–2.70	16.0–52.0	0.5–8.16		0.44–0.86	0.35
Dolomite limestone				0.74	0.92		0.8
Dolomite	2.60–2.90	2.10–2.70	0.3–25.0	0.1–3.0	0.05–1.54	0.83	0.8
Gneiss	2.6–3.1	2.30–3.00	0.3–2.4	0.1–0.7		0.75–0.97	
Quartz schist	2.60–2.80	2.10–2.70	0.7–3.0	0.1–0.3		0.44–0.84	
Quartz sandstone	2.64–2.65	2.58–2.59		0.41–0.46		0.65–0.97	
Chlorite schist	2.80–2.90	2.10–2.85	0.8–2.1	0.1–0.6		0.53–0.69	
Phyllite	2.81–2.96	2.71–2.86	0.4–3.6	0.5–1.8		0.67–0.96	
Argillite argillyte	2.70–2.85	2.30–2.80	0.1–0.5	0.1–0.3		0.39–0.52	
Griotte	2.7–2.87	2.60–2.70	0.1–6.0	0.1–0.8		0.8–0.98	
Quartzite	2.53–2.84	2.40–3.3	0.1–8.7	0.1–1.5		0.94–0.96	
Syenite	2.50–2.90	2.5–3.0	1.38	0.47–1.94			
Porphyry	2.3–2.8	2.70–2.74	0.29–2.75			0.65–0.88	0.82
Tuff	≈2.6	0.75–1.4	25	0.5–7.5		0.52–0.86	
Schist	2.6–2.9	2.3–3.01	0.02–1.85	0.1–0.2	0.14	0.49–0.80	0.92
Mica schist			8–2.1	0.13	1.31	0.53–0.69	0.13
Serpentinite	2.4–2.8	≈2.6	0.1–2.5	0.2–2.5			
Killas	2.7–2.84	2.6–2.7	0.1–0.45	0.1–0.3	0.33–2.39	0.52–0.82	
Shale	2.57–2.77	2.3–2.62	0.4–33.5	0.5–3.2	0.49–0.79	0.24–0.55	
Sandstone	1.8–2.75	2.2–2.71	0.7–34	0.2–9.0	11.99	0.44–0.97	0.6
Siltite	2.66–2.8	2.26–2.71		0.13–3.04	0.16–3.97		
Limestone	2.48–2.76	1.8–2.6	0.53–27.0	0.1–4.45	0.25	0.58–0.94	0.36
Conglomerate	2.67–2.71	2.4–2.7	0.8–10.0	1.0–5.0		0.5–0.96	
Trachyte	2.4–2.7	2.3–2.7	8.5				
Liparite	2.65	2.53–2.62	4–6	0.31–2.04	0.39–2.07	0.75–0.95	
Amphibolite	2.98–3.08	2.93–2.95		0.1–0.11	0.12–0.13		

Table II.7 Reference range of coefficient of permeability in rock[33,34,43,44]

Rock type	Pore condition	Coefficient of permeability, k (cm/s)
Granite	Medium dense, microfissures	1.1×10^{-12}–9.5×10^{-11}
	Contain microfissures	1.1×10^{-11}–2.5×10^{-11}
	Microfissures and partly coarse fissures	2.8×10^{-9}–7×10^{-8}
Limestone	Dense	3×10^{-12}–6×10^{-10}
	Microfissures, pores	2×10^{-9}–3×10^{-6}
	Relatively developed pores	9×10^{-5}–3×10^{-4}
Gneiss	Dense	$<10^{-13}$
	Microfissures	9×10^{-8}–4×10^{-7}
	Developed microfissures	2×10^{-6}–3×10^{-5}
Diabase	Dense	$<10^{-13}$
Basalt	Dense	$<10^{-13}$
	Weak developed fissures	1×10^{-3}–1.9×10^{-3}
Sandstone	Medium dense	10^{-13}–2.5×10^{-10}
	Developed pores	5.5×10^{-6}
Siltite		10^{-9}–10^{-8}
Fine sandstone		2×10^{-7}
Slate		7×10^{-11}–1.6×10^{-10}
Shale	Developed microfissures	2×10^{-10}–8×10^{-9}
Pelitic shale	Microfissures	3×10^{-4}
	Weathered, medium fissures	4×10^{-4}–5×10^{-4}
Schist	Developed microfissures	10^{-9}–5×10^{-5}
Black schist	Have fissures	10^{-4}–3×10^{-4}
Quartzite	Microfissures	1.2×10^{-10}–1.8×10^{-10}
Karst limestone	General condition	10^{-2}–10^{-6}
Rhyolite porphyry	Dense	$<10^{-13}$
Andesitic porphyrite	Microfissures	8×10^{-11}
Stiff mudstone		6×10^{-7}–2×10^{-6}
Mudstone		10^{-9}–10^{-13}
Dolomite		1.6×10^{-7}–1.2×10^{-5}
Permeable volcanic rock		10^{-2}–10^{-7}
Tuff breccia		1.5×10^{-4}–2.3×10^{-4}
Tuff		6.4×10^{-4}–4.4×10^{-3}
Conglomerate		2.7×10^{-8}

Table II.8 Reference range of strength structural parameters in rock[12,34,36,39,41,45–47]

Rock type	Compressive strength, σ_c (MPa)	Tensile strength, σ_t (MPa)	Angle of internal friction, φ (°)	Cohesion, c (MPa)	Elastic modulus, E ($\times10^4$ MPa)	Poisson's ratio, μ
Granite	100–250	7–25	45–60	14–50	5–10	0.2–0.3
Liparite	180–300	15–30	45–60	10–50	5–10	0.1–0.25
Diorite	100–250	10–25	53–55	10–50	7–15	0.1–0.3
Andesite	100–250	10–20	45–50	10–40	5–12	0.2–0.3
Gabbro	180–300	15–36	50–55	10–50	7–15	0.12–0.2
Diabase	200–350	15–35	55–60	25–60	8–15	0.1–0.3
Basalt	150–300	10–30	48–55	20–60	6–12	0.1–0.35

Rock type	Compressive strength, σ_c (MPa)	Tensile strength, σ_t (MPa)	Angle of internal friction, φ (°)	Cohesion, c (MPa)	Elastic modulus, E ($\times 10^4$ MPa)	Poisson's ratio, μ
Quartzite	150–350	10–30	50–60	20–60	6–20	0.1–0.25
Shale	10–100	2–10	10–30	3–20	2–8	0.2–0.4
Gneiss	50–200	5–20	30–50	3–5	1–10	0.22–0.35
Phyllite	10–100	2–10	15–30	3–20	1–8	0.2–0.4
Schist						
Slate	60–200	7–15	45–60	2–60	2–8	0.2–0.3
Sandstone	20–200	4–25	27–50	8–40	1–10	0.2–0.3
Quartz sandstone	68–102.5	1.9–3.0	75		0.39–1.25	0.05–0.25
Conglomerate	10–150	2–15	35–50	8–50	2–8	0.2–0.3
Limestone	50–200	5–20	30–50	10–50	5–10	0.2–0.35
Dolomite	80–250	15–25	35–50	20–50	4–8	0.2–0.35
Griotte	100–250	7–20	35–50	15–30	1–9	0.2–0.35
			25–30			
Syenite	80–100	2.3–2.8	62–66	1.0–3.0	4.8–5.3	0.18–0.26
	120–250	3.4–5.7				
Marlstone	50–100	12–98	20–21	0.07–0.44	0.4–0.7	0.3–0.4
Gypsum			29.4		0.1–0.8	0.3
Dolerite	200–350	15–35				
Mudstone			23	0.01	0.8–3.5	
Porphyry	160	5.4	85		6.6–7.0	0.16
Tuff	120–150	3.4–7.1	75–87		2.2–11.4	0.02–0.29
Volcanic breccia / agglomerate	120–250	3.4–7.1	80–87		1.0–11.4	0.05–0.16

Table II.9 Reference range of unconfined compressive strength in rock[41,47]

Rock type	Unconfined compressive strength, q_u (MN/m^2)
Sandstone	70–140
Shale	35–70
Limestone	105–210
Griotte	60–70
Tuff	11.3
Granite	140–210
Dolomite	87–90

Table II.10 Reference range of point load strength index in rock[48]

Rock type	Point load strength index, $I_{s\,(50)}$ (MPa)
Tertiary sandstone and clay rock	0.05–1
Coal	0.2–2
Limestone	0.25–8
Mudstone, Shale	0.2–8
Volcanic flow rock	3–15
Dolomite	6–11

Table II.11 Reference range of structural plane shear strength parameters[34][36]

Structural plane type	Angle of internal friction, φ (°)	Cohesion, c (MPa)
Structure surfaces of mud	10–20	0–0.05
Loess rock layers	20–30	0.05–0.10
Marlstone layers	20–30	0.05–0.10
Tuff layers	20–30	0.05–0.10
Shale layers	20–30	0.05–0.10
Sandstone layers	30–40	0.05–0.10
Conglomerate layers	30–40	0.05–0.10
Limestone layers	30–40	0.05–0.10
Phyllite planes	28	0.12
Talc schist schistosity planes	10–20	0–0.05
Mica schistosity planes	10–20	0–0.05
(Griotte, diorite) fault structure planes	8–37	0.–0.2
Shale joint surfaces (straightness)	18–29	0.10–0.19
Sandstone joint surfaces (straightness)	32–38	0.05–1.0
Limestone joint surfaces (straightness)	35	0.2
Quartz syenite diorite joint surfaces (straightness)	32–35	0.02–0.08
Rough structural surfaces	40–48	0.08–0.30
Gabbro, granite joint surfaces	30–38	0.20–0.40
Granite joint surfaces (roughness)	42	0.4
Limestone unload joint surfaces (roughness)	37	0.04
(Sandstone, granite) rock/ concrete contact surfaces	55–60	0–0.48

Table II.12 Reference range of shear strength of various rock masses[34]

Rock type	Cohesion, c (MPa)		Angle of internal friction, φ (°)
Lignite	0.014–0.03		15–18
Clay rock	Range	0.002–0.18	10–45
	General	0.04–0.09	15–30
Mudstone	0.01		23
Marlstone	0.07–0.44		20–41
Quartzite	0.01–0.53		22–40
Diorite	0.2–0.75		30–59
Gneiss	0.35–1.4		29–68
Gabbro	0.76–1.38		38–41
Shale	Range	0.03–1.36	33–70
	General	0.1–1.4	38–50
Limestone	Range	0.02–3.9	13–65
	General	0.1–1	38–52
Siltite	0.07–1.7		29–59
Arenaceous shale	0.07–0.18		42–63
Sandstone	Range	0.04–2.88	28–70
	General	1–2	48–60
Basalt	0.06–1.4		36–61

Rock type	Cohesion, c (MPa)		Angle of internal friction, φ (°)
Granite	Range	0.1–4.16	30–70
	General	0.2–0.5	45–52
Griotte	Range	1.54–4.9	24–60
	General	3–4	49–55
Quartz diorite	1.0–2.2		51–61
Andesite	0.89–2.45		53–74
Syenite	1–3		62–66

Table II.13 Allowable bearing pressure in rock [σ_0] (kPa)[12]

Rock type	Degree of joint development and spacing (cm)	Very developed joint, 2–20	Developed joint, 20–40	Undeveloped or moderately developed joint, >40
	Rupture degree	Fragment	Broken	Block
Very weak rock		400–800 (200–300)	600–1000 (300–400)	800–1200 (400–500)
Weak rock		800–1200 (500–800)	1000–1500 (700–1000)	1500–3000 (900–1200)
Moderately weak rock		800–1000	1000–1500	1500–3000
Strong rock		1500–2000	2000–3000	>4000 (>3000)

Note:

1) The allowable bearing pressure requires that the compressive stress acting on the base does not exceed the ultimate bearing pressure of foundation, and has sufficient safety, and induced deformation does not exceed the allowable deformation of the building. The load that per unit area of foundation is called the allowable bearing capacity of the foundation meeting the requirements described earlier.

2) Rock has weathered into gravel, sand, soil (weathered deposits), and its allowable bearing pressure can be determined by the corresponding soil types. If there is a certain cementation force between the particles, can be compared with the corresponding soil to increase appropriately.

3) For cave, fault, weak interlayer and soluble rock, etc., It should be determined by individual studies.

4) Low values should be taken when fissure is open or filled with mud.

2.3 Dynamic parameter

Table II.14 Reference range of elastic wave velocity and dynamic elastic parameters in rock[12,31,34,45,47,49]

Rock type	velocity of longitudinal wave, V_l (m/s)	velocity of transverse wave, V_t (m/s)	Dynamic modulus of elasticity, E_d (GPa)	Dynamic Poisson's ratio, μ_d
Basalt	4570–7500	3050–4500	53.1–162.8	0.1–0.22
Andesite	4200–5600	2500–3300	41.4–83.3	0.22–0.23
Diorite	5700–6450	2793–3800	52.8–96.2	0.23–0.34
Dioritic porphyrite	4391–5873	2766–3402	50–84	0.13–0.20
Quartz diorite	5211–6455	3057–3426	65–78	0.17–0.20

(Continued)

Table 11.14 (Continued)

Rock type	velocity of longitudinal wave, V_l (m/s)	velocity of transverse wave, V_t (m/s)	Dynamic modulus of elasticity, E_d (GPa)	Dynamic Poisson's ratio, μ_d
Granite	4000–6500	2370–3800	37.0–113.0	0.13–0.40
Gabbro	5500–7500	3200–4000	63.4–114.8	0.20–0.21
Dunite	6500–7980	4080–4800	128.3–183.8	0.17–0.22
Quartz trachyte	3000–5300	1800–3100	18.2–66.0	0.22–0.24
Diabase	5200–5800	3100–3500	59.5–88.3	0.21–0.22
Rhyolite	4800–6990	2900–4100	40.2–107.7	0.21–0.29
Perlite	5049–5226	3275–3292	58	0.13–0.17
Quartzite	3030–5610	1800–3200	20.4–76.3	0.23–0.26
Quartz porphyry	4949–5051	3431–3711	66–69	0.09–013
Schist	5800–6420	3500–3800	78.8–106.6	0.21–0.23
Gneiss	6000–6700	3500–4000	76.0–129.1	0.22–0.24
Slate	3650–4450	2160–2860	29.3–48.8	0.15–0.23
Griotte	5800–7300	3500–4700	79.7–137.7	0.15–0.21
Phyllite	2800–5200	1800–3200	20.2–70.0	0.15–0.20
Sandstone	1500–5000	915–2400	5.3–37.9	0.20–0.22
Shale	1330–3970	780–2300	3.4–35.0	0.23–0.25
Limestone	2500–6000	1450–3500	12.1–88.3	0.24–0.25
Siliceous limestone	4400–4800	2600–3000	46.8–61.7	0.18–0.23
Argillaceous limestone	2000–3500	1200–2200	7.9–26.6	0.17–0.22
Dolomite	2500–6000	1500–3600	15.4–94.8	0.22
Conglomerate	1500–2500	900–1500	3.4–16.0	0.19–0.22
Concrete	2000–4500	1250–2700	8.85–49.8	0.18–0.21
Anhydrite mudstone	5000–6000	3000–3300	69.6–83.12	0.23–0.315
Clay shale				0.286
Tuff	3000–6800		7.1–11.4	
Volcanic breccia/ agglomerate	3000–6800		7.1–11.4	
Amphibolite			1.8–5.2	

Basic geotechnical engineering terms

This part refers to the standard for soil test method (GB/T 50123-1999), standard for test methods for rock mass engineering (GB/T 50266-2013), British Standards (BS 1377: 1990), American Society for Testing and Materials (ASTM D653–14), and geotechnical engineering-related materials. Common terminologies are collected in geotechnical tests, and the corresponding definitions, symbols, dimensions and units in English are given.

The basic terminologies can be divided into physical (basic properties and structural characteristics of rock or soil mass in solid, liquid and gas phases), mechanical (deformation and strength characteristics of rock or soil mass under external force), thermal (exchange of thermal energy between rock or soil mass and the outside world) and chemical indices (distribution of chemical elements in rock or soil mass). The indices are arranged alphabetically.

3.1 Physical index

Absorption (ω_a, %): Increment of mass because of water entering the pores of rock material. Does not include water adhering to outer surfaces of particles, and is expressed as a percentage of dry mass.

Activity number (A, Dimensionless): The ratio of the plasticity index of a cohesive soil to the percent by mass of particles having an equivalent diameter smaller than 2 μm.

Buoyant density (ρ', ML^{-3}, g/cm³): Difference between the saturated density of soil and the density of water (at 20 °C or project-specified temperature).

Buoyant unit weight (γ', $ML^{-2}T^{-2}$, kN/cm³): Buoyant density multiplied by standard acceleration of gravity (at 20 °C or project-specific temperature).

Coefficient of curvature (C_c, Dimensionless): The ratio $(d_{30})^2 / (d_{10} \cdot d_{60})$, where d_{60}, d_{30}, and d_{10} are the particle sizes corresponding to 60, 30, and 10% finer on the cumulative particle-size distribution curve, respectively.

Compaction: Densification of a soil by means of mechanical manipulation.

Consistency (ω_c, Dimensionless): The ratio of the liquid limit minus the water content to the plasticity index of cohesive soils.

Constrained size (d_{60}, Dimensionless): Diameter at which 60% by weight (dry) particles of a soil are finer.

Degree of compaction (λ_c, Dimensionless): A measure of compaction given by:

$$\lambda_c = \frac{\rho_d}{\rho_{dmax}}$$

λ_c – degree of compaction,
ρ_d – dry unit density to be achieved in the field, and
ρ_{dmax} – in standard compaction test, the maximum value defined by the compaction curve.

Degree of saturation (Sr, %): The ratio of water volume to void volume determined at a given temperature (usually 20 °C).

Density (ρ, ML^{-3}, g/cm^3): Mass of soil or rock per unit volume.

Density of frozen ground (ρ_f, ML^{-3}, g/cm^3): Mass of frozen earth materials of a unit volume.

Discharge velocity (V, LT^{-1}, m/s): Rate of discharge of water through a porous medium per unit area perpendicular to the direction of flow.

Dry density (ρ_d, ML^{-3}, g/cm^3): The ratio of dry mass of soil or rock per unit total volume.

Dry unit weight (γ_d, ML^{-2}T^{-2}, kN/cm^3): The product of dry density and the gravitational acceleration.

Effective size (d_{10}, Dimensionless): Diameter at which 10% by weight (dry) the soil particles are finer.

Electrical resistivity (K, T^{3}I^{2}M^{-1}L^{-3}, S/m): A property of a material that determines the electrical current flowing through a unit volume under a unit electrical potential.

Free swelling (δ_{ef}, %): The ratio of wetting-induced change in volume to the original volume of a loose and dry soil.

Freezing temperature (T, θ, °C): Temperature at which water in soil starts to freeze.

Frequency (f, T^{-1}, Hz): Number of cycles per unit time.

Gradation (G, %): The proportion by mass of various particle sizes of soils.

Linear shrinkage (δ_{si}, %): Percentage change in length of a bar-shaped soil sample when dried from about its shrinkage limit.

Liquid limit (ω_L, %): Water content of a cohesive soil representing transition from the semi-liquid to plastic states.

Liquidity index (I_L, Dimensionless): The ratio of the water content of a cohesive soil at a given condition/state minus its plastic limit to its plasticity index.

Longitudinal wave: Wave in which direction of particle displacement is normal to wave front. The propagation velocity, V_l, calculated as follows:

$$V_1 = \sqrt{(\frac{E}{\rho})[\frac{(1-v)}{(1+v)(1-2v)}]}$$

E – Young's modulus,
ρ – mass density, and
v – Poisson's ratio.

Maximum dry density (ρ_{dmax}, ML^{-3}, g/cm^3): Densest state (in dry condition) of a soil determined using a standard test method.

Maximum-index void ratio (e_{max}, Dimensionless): A reference void ratio at minimum index density/unit weight.

Median size (d_{30}, Dimensionless): Diameter at which 30% by weight (dry) the soil particles are finer.

Minimum-index void ratio (e_{min}, Dimensionless): A reference void ratio at maximum index density/unit weight.

Optimum water content (ω_{opt}, %): The water content at which a soil can be compacted to a maximum dry unit weight by a given compaction effort.

Plastic limit (ω_p, %): The water content of a cohesive soil at the boundary representing the transition from the plastic to semi-solid states.

Plasticity index (I_p, Dimensionless): The range of water content over which a cohesive soil behaves plastically. Numerically, it is the difference between the liquid limit and the plastic limit.

Porosity (n, %): The ratio of volume of voids to the total volume of soil or rock mass.

Relative density (D_r, Dimensionless): A parameter describing the void ratio/density of a soil sample relative to the loosest and densest states for that soil, usually expressed as a percentage. It is defined by either of the following two equations:

(a) $$D_r = \frac{e_{max} - e}{e_{max} - e_{min}} \times 100$$

D_r – relative density in %,
e_{max} – void ratio in loosest state, from minimum dry density,
e – any given void ratio (typically an in-situ test value or that of a test specimen), and
e_{min} – void ratio in densest state, from maximum dry density.

(b) $$D_r = \frac{\rho_{dmax}}{\rho_d} \frac{\rho_{dmax} - \rho_d}{\rho_{dmax} - \rho_{dmin}} \times 100$$

ρ_{dmax} – maximum dry density in kg/m³,
ρ_d – any given dry density (typically an in-situ test value or that of a test specimen) in kg/m³, and
ρ_{dmin} – minimum dry density in kg/m³.

Rock structure: Large-scale features of a rock mass, like bedding, foliation, jointing, cleavage, or brecciation, also, the sum total of such features as contrasted with texture.

Saturated density (ρ_{sat}, ML⁻³, g/cm³): The ratio of fully saturated mass to its total volume.

Saturated unit weight (γ_{sat}, ML⁻²T⁻², kN/cm³): The product of density and the gravitational acceleration.

Specific gravity of soil particle (G_s, Dimensionless): The ratio of (1) density of soil or rock to that of water at a given temperature (usually 20 °C) or (2) mass in air of a given volume of soil or rock to that of an equal volume of distilled/demineralized water at a given temperature.

Shrinkage limit (ω_n, %): The maximum water content at which a reduction in water content will not cause a decrease in soil volume.

Swelling ratio (δ, %): The ratio of vertical swell to test specimen height before swell.

Swell pressure (P_e, ML⁻¹T⁻², kPa): The pressure required to maintain constant volume, i.e. to prevent swelling, when a soil has access to water.

Structural plane: Any surface across which some property of a rock mass is discontinuous. This includes fracture surfaces and bedding planes.

Slake durability index (I_{d2}, %): The ratio of dry mass retained of a collection of shale pieces on a 2.00 mm sieve after two cycles of oven drying and 10 min of soaking in water with a standard tumbling and abrasion action.

Solids or particle density (ρ_s, ML⁻³, g/cm³): Mass of dry solids (particles) of soil or rock per unit volume of solids without any voids.

Transverse wave: Wave in which direction of particle displacement is parallel to wave front. The propagation velocity, V_t, is calculated as follows:

$$V_t = \sqrt{\frac{G}{\rho}} = \sqrt{(\frac{E}{\rho})[\frac{1}{2(1+v)}]}$$

G – shear modulus,
ρ – mass density,
v – Poisson's ratio, and
E – Young's modulus.

Unfrozen water content (ω_u, %): The ratio of either weights or volumes of unfrozen water and dry soil.

Uniformity coefficient (C_u, dimensionless): The ratio of d_{60}/d_{10}, where d_{60} and d_{10} are soil particle diameters corresponding to 60% and 10% finer on the cumulative particle size curve, respectively.

Unit weight (γ, ML⁻²T⁻², kN/cm³): Density multiplied by gravitational acceleration.

Void ratio (e, Dimensionless): The ratio of void volume to solid volume of soil or rock.

Volumetric shrinkage (δ_v, %): The ratio of the change in volume to the original volume when water content is reduced to the shrinkage limit.

Water content (ω, %): The ratio of mass of pore water of soil or rock material, to mass of its solid particles, expressed as a percentage.

Water saturation (ω_{sa}, %): The ratio of water mass that penetrates rock under high pressure (usually 15 MPa) or under vacuum to total volume.

3.2 Mechanical index

Angle of internal friction (φ, °): Angle of the Mohr-Coulomb envelope.

Angle of shear resistance, in terms of effective stress (φ', °): The slope of the Mohr-Coulomb effective stress envelope.

Axial strain (ε_1, %): The ratio of length changes as read from deformation indicator to initial length of a specimen.

Axial stress (σ_1, ML⁻¹T⁻², kPa): The axial force divided by the area on which it acts.

California bearing ratio (CBR, %): The ratio in percent and at a penetration of either 0.1 or 0.2 in. of: (1) stress required to penetrate a soil mass to (2) stress required to penetrate a standard material (crushed aggregate) using standard equipment and procedures.

Coefficient of collapsibility (δ_s, Dimensionless): The coefficient of collapsibility, δ_s, is expressed as follows:

$$\delta_s = \frac{h_1 - h_2}{h_0}$$

h_1 – stable deformation of the sample height under a certain pressure, mm,

h_2 – final height after wetting, mm, and

h_0 – initial height of specimen, mm.

Coefficient of compressibility (a_v, $M^{-1}LT^2$, MPa^{-1}): The secant slope, for a given pressure increment, of the compression curve.

Coefficient of consolidation (C_v, L^2T^{-1}, cm^2/s): A coefficient utilized in the theory of consolidation, containing the physical constants of a soil affecting its rate of volume change.

Coefficient of deformation because of leaching (δ_{wt}, Dimensionless): The ratio of the difference in height to the original height because of the leaching under standard conditions.

Coefficient of permeability (k, LT^{-1}, cm/s): The rate of discharge of water under laminar flow conditions through a unit cross-sectional area of a porous medium under a unit hydraulic gradient and standard temperature conditions (usually 20 °C).

Coefficient of self-weight collapsibility (δ_{zs}, %): The ratio of wetting-induced change in height to the height immediately prior to wetting when the soil pressure to saturated self-weight stress.

Coefficient of shrinkage (λ_n, Dimensionless): In the shrinkage curve, the ratio of the difference between the two water contents in the linear contraction phase to the difference between the corresponding line shrinkages.

Coefficient of viscosity (η, $ML^{-1}T^{-1}$, Pa·s): The shearing stress required to maintain a unit velocity gradient between two parallel layers of a fluid a unit distance apart.

Coefficient of volume compressibility (m_v, $M^{-1}LT^2$, MPa^{-1}): The compression of a soil layer per unit original thickness because of a unit increase in pressure. It is numerically equal to the coefficient of compressibility divided by one plus the original void ratio.

Cohesion (c, $ML^{-1}T^{-2}$, kPa): Shear resistance at zero normal stress.

Cohesion intercept, in terms of effective stress (c', $ML^{-1}T^{-2}$, kPa): The intercept of the Mohr-Coulomb effective stress envelope.

Collapse: Wetting-induced decrease in height of a soil specimen.

Compression curve: A curve representing the relationship between stress and void ratio of a soil as obtained from a consolidation test.

Compression index (C_c, Dimensionless): The slope of the linear portion of the compression curve on a semi-log plot.

Compressive strength (σ_c, $ML^{-1}T^{-2}$, Mpa): Load per unit area at which an unconfined cylindrical specimen of soil or rock will fail in a simple compression test.

Consolidation: The gradual reduction in volume of a soil mass resulting from an increase in compressive stress.

Consolidation ratio (U_z, %): The ratio of the amount of primary consolidation at a given time and distance (location) from a drainage surface to the total amount of primary consolidation obtainable at that point under a given stress increment.

Contraction: Linear strain associated with a decrease in length.

Critical damping: The minimum viscous damping that will allow a displaced system to return to its initial position without oscillation.

Damping: Reduction in the amplitude of vibration of a body or system because of dissipation of energy.

Damping ratio (ζ, Dimensionless): For a system with viscous damping, the ratio of actual to critical damping coefficients.

Deviator stress ($\sigma_1 - \sigma_3$, ML^{-1}T^{-2}, kPa): The difference between the major and minor principal stresses.

Dilatancy: The expansion of cohesionless rocks or soils when subjected to shearing deformation.

Dynamic modulus of elasticity (E_d, ML^{-1}T^{-2}, Mpa): The ratio of stress to strain for a material under dynamic uniaxial loading conditions.

Dynamic Poisson's ratio (μ_d, Dimensionless): The absolute value of the ratio between the diametral and axial deformations under dynamic uniaxial loading conditions.

Elastic modulus (E, ML^{-1}T^{-2}, Mpa): The ratio of the increase in stress on a test specimen to the resulting increase in strain over the range of loading within the elastic domain.

Effective stress (σ', ML^{-1}T^{-2}, kPa): The difference between the total stress, σ, and the pore water pressure, u_w.

Failure criteria: Specification of the mechanical condition under which solid materials fail by fracturing or by deforming beyond some specified limit.

Frost heave: The expansion of volume because of the formation of ice in the underlying soil or rock.

Frost heaving ratio (η_f, %): The ratio of the difference between before and after freezing volumes to the volume before freezing.

Head (h, L, m): Fluids pressure at a point, expressed in terms of the vertical distance the fluid rises.

Hydraulic gradient (i, Dimensionless): The change in total head (head loss, Δh) per unit distance (L) in the direction of flow.

Initial pressure of collapsibility (P_{sh}, ML^{-1}T^{-2}, kPa): The minimum collapsing pressure when the coefficient of collapsibility is up to 0.015.

Modulus of deformation (E, ML^{-1}T^{-2}, Mpa): The ratio of stress to strain for a material under a given loading condition, numerically equal to the slope of the tangent or the secant of a stress-strain curve.

Modulus of resilience (E_e, ML^{-1}T^{-2}, Mpa): The ratio of the vertical pressure to the resilience strain during the process of unloading under the confined conditions.

Over consolidation ratio (OCR, Dimensionless): The ratio of pre-consolidation vertical stress to the current effective overburden stress.

Peak shear strength (τ_p, ML^{-1}T^{-2}, Mpa): The maximum stress determined in complete shear stress versus shear displacement curve under a given constant normal stress.

Point load strength anisotropy index ($I_{a(50)}$, Dimensionless): The strength anisotropy index is defined as the ratio of mean $I_{s(50)}$ values measured perpendicular and parallel to planes of weakness. That is, the ratio of greatest to least point load strength indices that result in the greatest and least point load strength values.

Point load strength index (I_s, ML^{-1}T^{-2}, Mpa): An indicator of strength obtained by subjecting a rock specimen to increasing load applied by a pair of truncated, conical platens, until failure occurs.

Poisson's ratio (μ, Dimensionless): The ratio between linear strain changes perpendicular to and in the direction of a given uniaxial stress change.

Pore pressure coefficients (A and B): Coefficients that relate principal stresses to the pore pressure changes in accordance with the equation:

$$\Delta u = B[\Delta\sigma_3 + A(\Delta\sigma_1 - \Delta\sigma_3)]$$

Δu – change in pore pressure,

$\Delta\sigma_1$ – change in total major principal stress,

$\Delta\sigma_3$ – change in total minor principal stress,

$(\Delta\sigma_1 - \Delta\sigma_3)$ – change in deviator stress, and

A and B – pore pressure coefficients.

Pore water pressure (u, $ML^{-1}T^{-2}$, kPa): The pressure of water in the voids between solid particles.

Preconsolidation pressure (P_c, $ML^{-1}T^{-2}$, kPa): The yield stress of a soil specimen as determined from a standard one-dimensional consolidation test.

Primary consolidation: The reduction in volume of a soil mass caused by the application of a sustained load to the mass and due principally to a squeezing out of water from the void spaces of the mass and accompanied by a transfer of the load from the soil water to the soil solids.

Principal plane: Each of three mutually perpendicular planes through a point, on which the shearing stresses are zero.

Residual strain (ε_r, Dimensionless): Strain associated with a state of residual stress.

Residual strength (S_r, $ML^{-1}T^{-2}$, Mpa): Shear strength, corresponding to a specific normal stress, for which the shear stress remains essentially constant with increasing shear displacement.

Residual stress (σ_r, $ML^{-1}T^{-2}$, Mpa): Stress remaining in a solid under zero external stress after some process that causes the dimensions of the various parts of the solid to be incompatible under zero stress, for example, (1) deformation under the action of external stress when some parts of the body suffer permanent strain, or (2) heating or cooling of a body in which the thermal expansion coefficient is not uniform throughout the body.

Secant modulus (E_s, $ML^{-1}T^{-2}$, Mpa): Slope of the line connecting the origin and a given point on the stress-strain curve (generally taken at a stress equal to half the compressive strength).

Secondary consolidation: Reduction in volume of a soil mass caused by the application of a sustained load and due principally to the adjustment of the internal structure of the soil mass after most of the load has been transferred from the soil water to the soil solids.

Seepage: The infiltration or percolation of water through rock or soil to or from the surface. The term seepage is usually restricted to the very slow movement of groundwater.

Seepage force (J, $ML^{-2}T^{-2}$, kN/m³): The frictional drag of water flowing through voids or interstices, causing an increase in the intergranular pressure.

Sensitivity (S_t, Dimensionless): The ratio of the strength of an undisturbed soil specimen to the strength of the same specimen after remolding.

Shear strength (S, $ML^{-1}T^{-2}$, kPa): The maximum resistance of a material to shearing stresses.

Shear stress (τ, $ML^{-1}T^{-2}$, kPa): A stress component acting parallel to the surface of the plane being considered.

Size-corrected point load strength index ($I_{s(50)}$, $ML^{-1}T^{-2}$, Mpa): The original point load strength index value multiplied by a factor to normalize the value that would have been obtained with diametral test of a sample with a standard diameter (D).

Softening coefficient (η_c, Dimensionless): The ratio of the saturated to the dry compressive strengths of rocks.

Stiffness (K, MT^{-2}, N/m): The ratio of change of force (or torque) to the corresponding change in translational (or rotational) deformation of an elastic element.

Strain (ε, Dimensionless): The change in length per unit of original length in a given direction.

Stress (p, $ML^{-1}T^{-2}$, kPa): The force per unit area.

Swell index (C_s, Dimensionless): Slope of the rebound pressure-void ratio curve on a semi-log plot.

Tangent modulus (E_t, $ML^{-1}T^{-2}$, Mpa): Slope of the tangent to the stress-strain curve at a given stress value (generally taken at a stress equal to half the compressive strength).

Tensile strength (σ_t, $ML^{-1}T^{-2}$, Mpa): The maximum resistance to deformation of a material when subjected to tension by an external force.

Thaw compressibility (a_0, Dimensionless): In the permafrost melting process, the ratio of the change in height to the initial height under the self-weight pressure.

Thaw-settlement coefficient (a_{tc}, $M^{-1}LT^2$, MPa^{-1}): The thaw-settlement coefficient, a_{tc}, expressed as the relative compression deformation per unit change in pressure.

$$a_{tc} = \frac{S_{i+1} - S_i}{P_{i+1} - P_i}$$

S_i – the ratio of change in height to original height under a certain pressure, and
P_i – a reference pressure (MPa).

Time factor (T_v, Dimensionless): Dimensionless factor, utilized in the theory of consolidation, containing the physical constants of a soil stratum influencing its time-rate of consolidation, expressed as follows:

$$T_v = \frac{C_v \cdot t}{H^2}$$

T_v – time factor,
t – elapsed time that the stratum consolidated,
C_v – coefficient of consolidation, and
H – thickness of stratum drained on one side only. If stratum is drained on both sides, its thickness equals $2H$.

Total stress (σ_t, $ML^{-1}T^{-2}$, kPa): The total force per unit area. It is the sum of the pore pressure and effective stresses.

Unconfined compressive strength (q_u, $ML^{-1}T^{-2}$, kPa): The axial load per unit area at which a prismatic or cylindrical specimen will fail in a simple compression test without lateral support.

Undrained shear strength (S_u, $ML^{-1}T^{-2}$, kPa): Shear strength of a soil under undrained conditions, before drainage because of application of stress can take place.

Yield stress (σ_s, $ML^{-1}T^{-2}$, Mpa): The stress beyond which the induced deformation is not fully annulled after complete distressing.

3.3 Thermal index

Mass heat capacity (c, $L^2T^{-2}\theta^{-1}$, J/(kg·K)): The quantity of heat required to change the temperature of a unit mass of a substance one degree, measured as the average quantity over the temperature range specified. It is distinguished from true specific heat by being an average rather than a point value.

Thermal conductivity (λ, $LMT^{-3}\theta^{-1}$, W/m·K): The quantity of heat that will flow through a unit area of a substance in unit time under a unit temperature gradient.

3.4 Chemical index

Organic matter (O_m, %): The amount of organic matter determined by the following:

$$O_m = 100 - \frac{C \times 100}{B}$$

O_m – organic matter, %,
C – ash, g, and
B – oven-dried test specimen, g.

PH value (pH, Dimensionless): An index of the acidity or alkalinity in terms of the logarithm of the reciprocal of the hydrogen ion concentration.

Common physical quantities and unit conversion for geotechnical testing

This part summarises the unit names, symbols and mutual conversion relationships of the 18 common physical quantities related to geotechnical testing based on SI units and recommendations for the use of their multiples and of certain other units (GB 3100–93). Table IV-1 summarises the SI units and symbols of the physical quantities, excluding the physical quantities with combined units, such as velocity (unit: metre per second), density (unit: kilogram per cubic meter), etc.

The conversions of the physical quantity units are listed and organised as follows: length (Table IV-2), area (Table IV-3), volume (Table IV-4), mass (Table IV-5), force (Table IV-6), time (Table IV-7), density (Table IV-8), dynamic viscosity (Table IV-9), kinematic viscosity (Table IV-10), velocity (Table IV-11), temperature (Table IV-12), pressure (Table IV-13), volume flow rate (Table IV-14), work (energy, heat) (Table IV-15), power (heat flow rate) (Table IV-16), mass heat capacity (Table IV-17), thermal conductivity (Table IV-18) and coefficient of heat transfer (Table IV-19). The values in each table represent the conversion ratios between row and column units.

Table IV.1 Common physical quantity units

Physical quantity	Unit	Symbol	Unit	Symbol
Length	kilometer	km	meter	**m**
	decimeter	dm	centimeter	cm
	millimeter	mm	nanometer	nm
	mile	mi	foot	ft
	inch	in	nautical mile	nmi
	yard	yd		
Area	square kilometer	km²	square meter	m²
	hectare	ha	acre	ac
	square mile	mi²	square foot	ft²
	square inch	in²	square yard	yd²
Volume	cubic meter	m³	cubic decimeter	dm³
	cubic centimeter	cm³	liter	L
	milliliter	mL	cubic foot	ft³
	cubic inch	in³		
Mass	ton	t	kilogram	**kg**
	gram	g	pound	lb
	ounce	oz	carat	ct
Force	newton	N	kilogram-force	kgf
	pound-force	lbf	dyne	dyn
Time	hour	h	minute	min
	second	s		
Dynamic viscosity	poise	P	centipoise	cP
Kinematic viscosity	stoke	St	centistoke	cSt
Temperature	degree Celsius	°C	kelvin	**K**
	degree Fahrenheit	°F	degree Rankine	°R
	Réaumur scale	°Ré		
Pressure	bar	bar	Pascal	Pa
	standard atmosphere	atm	technical atmosphere	at
	millimeter of mercury	mmHg		
Work (energy, heat)	joule	J	calorie	cal
	kilocalorie	kcal		
power (heat flow rate)	watt	W	kilowatt	kW
	British thermal unit	Btu		

Note: The units in bold are the basic units of the international system of units (SI).

Table IV.2 Length conversion between units

Unit	Kilometer (km)	Meter (m)	Decimeter (dm)	Centimeter (cm)	Millimeter (mm)	Nanometer (nm)	Mile (mi)	Foot (ft)	Inch (in)	Nautical mile (nmi)	Yard (yd)
Kilometer (km)	1	10^3	10^4	10^5	10^6	10^{12}	6.214×10^{-1}	3.281×10^3	3.937×10^4	5.4×10^{-1}	1.094×10^3
Meter (m)	10^{-3}	1	10	10^2	10^3	10^9	6.214×10^{-4}	3.281	3.937×10^1	5.4×10^{-4}	1.094
Decimeter (dm)	10^{-4}	10^{-1}	1	10	10^2	10^8	6.214×10^{-5}	3.281×10^{-1}	3.937	5.4×10^{-5}	1.094×10^{-1}
Centimeter (cm)	10^{-5}	10^{-2}	10^{-1}	1	10	10^7	6.214×10^{-6}	3.281×10^{-2}	3.937×10^{-1}	5.4×10^{-6}	1.094×10^{-2}
Millimeter (mm)	10^{-6}	10^{-3}	10^{-2}	10^{-1}	1	10^6	6.214×10^{-7}	3.281×10^{-3}	3.937×10^{-2}	5.4×10^{-7}	1.094×10^{-3}
Nanometer (nm)	10^{-12}	10^{-9}	10^{-8}	10^{-7}	10^{-6}	1	6.214×10^{-13}	3.281×10^{-9}	3.937×10^{-8}	5.4×10^{-13}	1.094×10^{-9}
Mile (mi)	1.609	1.609×10^3	1.609×10^4	1.609×10^5	1.609×10^6	1.609×10^{12}	1	5.28×10^3	6.336×10^4	8.69×10^{-1}	1.76×10^3
Foot (ft)	3.048×10^{-4}	3.048×10^{-1}	3.048	3.048×10^1	3.048×10^2	3.048×10^8	1.894×10^{-4}	1	1.2×10^1	1.646×10^{-4}	3.33×10^{-1}
Inch (in)	2.54×10^{-5}	2.54×10^{-2}	2.54×10^{-1}	2.54	2.54×10^1	2.54×10^7	1.578×10^{-5}	8.333×10^{-2}	1	1.371×10^{-5}	2.778×10^{-2}
Nautical mile (nmi)	1.852	1.852×10^3	1.852×10^4	1.852×10^5	1.852×10^6	1.852×10^{12}	1.151	6.076×10^3	7.291×10^4	1	2.025×10^3
Yard (yd)	9.144×10^{-4}	9.144×10^{-1}	9.144	9.144×10^1	9.144×10^2	9.144×10^8	5.682×10^{-4}	3	3.6×10^1	4.937×10^{-4}	1

Table IV.3 Area conversion between units

Unit	Square kilometer (km²)	Square meter (m²)	Hectare (ha)	Acre (ac)	Square mile (mi²)	Square foot (ft²)	Square inch (in²)	Square yard (yd²)
Square kilometer (km²)	1	10^6	10^2	2.471×10^2	3.861×10^{-1}	1.076×10^7	1.55×10^9	1.196×10^6
Square meter (m²)	10^{-6}	1	10^{-4}	2.471×10^{-4}	3.861×10^{-7}	1.076×10^1	1.55×10^3	1.196
Hectare (ha)	10^{-2}	10^4	1	2.471	3.861×10^{-3}	1.076×10^5	1.55×10^7	1.196×10^4
Acre (ac)	4.047×10^{-3}	4.047×10^3	4.047×10^{-1}	1	1.563×10^{-3}	4.356×10^4	6.273×10^6	4.84×10^3
Square mile (mi²)	2.59	2.59×10^6	2.59×10^2	6.4×10^2	1	2.788×10^7	4.014×10^9	3.098×10^6
Square foot (ft²)	9.29×10^{-8}	9.29×10^{-2}	9.29×10^{-6}	2.296×10^{-5}	3.587×10^{-8}	1	1.44×10^2	1.111×10^{-4}
Square inch (in²)	6.452×10^{-10}	6.452×10^{-4}	6.452×10^{-8}	1.594×10^{-7}	2.491×10^{-10}	6.944×10^{-3}	1	7.716×10^{-4}
Square yard (yd²)	8.361×10^{-7}	8.361×10^{-1}	8.361×10^{-5}	2.066×10^{-4}	3.228×10^{-7}	9	1.296×10^3	1

Table IV.4 Volume conversion between units

Unit	Cubic meter (m^3)	Liter (L)	Cubic centimeter (cm^3)	Cubic foot (ft^3)	Cubic inch (in^3)
Cubic meter (m^3)	1	10^3	10^6	3.531×10^1	6.102×10^4
Liter (L)	10^{-3}	1	10^3	3.531×10^{-2}	6.102×10^1
Cubic centimeter (cm^3)	10^{-6}	10^{-3}	1	3.531×10^{-5}	6.102×10^{-2}
Cubic foot (ft^3)	2.832×10^{-2}	2.832×10^1	2.832×10^4	1	1.728×10^3
Cubic inch (in^3)	1.639×10^{-5}	1.639×10^{-2}	1.639×10^1	5.787×10^{-4}	1

Table IV.5 Mass conversion between units

Unit	Ton (t)	Kilogram (kg)	Gram (g)	Pound (lb)	Ounce (oz)	Carat (ct)
Ton (t)	1	10^3	10^6	2.205×10^3	3.527×10^4	5×10^6
Kilogram (kg)	10^{-3}	1	10^3	2.205	3.527×10^1	5×10^3
Gram (g)	10^{-6}	10^{-3}	1	2.205×10^{-3}	3.527×10^{-2}	5
Pound (lb)	4.536×10^{-4}	4.536×10^{-1}	4.536×10^2	1	1.6×10^1	2.268×10^3
Ounce (oz)	2.835×10^{-5}	2.835×10^{-2}	2.835×10^1	6.25×10^{-2}	1	1.417×10^2
Carat (ct)	2×10^{-7}	2×10^{-4}	2×10^{-1}	4.409×10^{-4}	7.055×10^{-3}	1

Table IV.6 Force conversion between units

Unit	Newton (N)	Kilogram-force (kgf)	Pound-force (lbf)	Dyne (dyn)
Newton (N)	1	1.02×10^{-1}	2.248×10^{-1}	10^5
Kilogram-force (kgf)	9.807	1	2.205	9.807×10^5
Pound-force (lbf)	4.448	4.536×10^{-1}	1	4.448×10^5
Dyne (dyn)	10^{-5}	1.02×10^{-6}	2.248×10^{-6}	1

Table IV.7 Time conversion between units

Unit	Hour (h)	Minute (min)	Second (s)
Hour (h)	1	60	3600
Minute (min)	1.667×10^{-2}	1	60
Second (s)	2.778×10^{-4}	1.667×10^{-2}	1

Table IV.8 Density conversion between units

Unit	Kilogram/cubic meter (kg/m^3)	Gram/cubic centimeter (g/cm^3)	Gram/cubic meter (g/m^3)	Pound/cubic foot (lb/ft^3)	Pound/cubic inch (lb/in^3)
Kilogram/cubic meter (kg/m^3)	1	10^{-3}	10^3	6.244×10^{-2}	3.613×10^{-5}
Gram/cubic centimeter (g/cm^3)	10^3	1	10^6	6.244×10^1	3.613×10^{-2}

Unit	Kilogram/cubic meter (kg/m^3)	Gram/cubic centimeter (g/cm^3)	Gram/cubic meter (g/m^3)	Pound/cubic foot (lb/ft^3)	Pound/cubic inch (lb/in^3)
Gram/cubic meter (g/m^3)	10^{-3}	10^{-6}	1	$6.244×10^{-5}$	$3.613×10^{-8}$
Pound/cubic foot (lb/ft^3)	$1.602×10^1$	$1.602×10^{-2}$	$1.602×10^4$	1	$5.787×10^{-4}$
Pound/cubic inch (lb/in^3)	$2.768×10^4$	$2.768×10^1$	$2.768×10^7$	$1.728×10^3$	1

Table IV.9 Dynamic viscosity conversion between units

Unit	Pascal·second ($Pa·s$)	Poise (P)	Centipoise (cP)	Pound-force·second / square foot ($lbf·s/ft^2$)
Pascal·second ($Pa·s$)	1	10	10^3	$2.09×10^{-2}$
Poise (P)	10^{-1}	1	10^2	$2.09×10^{-3}$
Centipoise (cP)	10^{-3}	10^{-2}	1	$2.09×10^{-5}$
Pound-force·Second/square foot ($lbf·s/ft^2$)	$4.788×10^1$	$4.788×10^2$	$4.788×10^4$	1

Table IV.10 Kinematic viscosity conversion between units

Unit	Stoke (St)	Centistoke (cSt)	Square meter/second (m^2/s)	Square centimeter/second (cm^2/s)	Square foot/second (ft^2/s)	Square inch/second (in^2/s)
Stoke (St)	1	10^2	10^{-4}	1	$1.076×10^{-3}$	$1.55×10^{-1}$
Centistoke (cSt)	10^{-2}	1	10^{-6}	10^{-2}	$1.076×10^{-5}$	$1.55×10^{-3}$
Square meter/second (m^2/s)	10^4	10^6	1	10^4	$1.076×10^1$	$1.55×10^3$
Square centimeter/second (cm^2/s)	1	10^2	10^{-4}	1	$1.076×10^{-3}$	$1.55×10^{-1}$
Square foot/second (ft^2/s)	$9.29×10^2$	$9.29×10^4$	$9.29×10^{-2}$	$9.29×10^2$	1	$1.44×10^2$
Square inch/second (in^2/s)	6.452	$6.452×10^2$	$6.452×10^{-4}$	6.452	$6.944×10^{-3}$	1

Table IV.11 Velocity conversion between units

Unit	Meter/second (m/s)	Foot/second (ft/s)	Mile/second (mi/s)	Kilometer/hour (km/h)	Nautical mile/hour (nmi/h)	Mile/hour (mi/h)
Meter/second (m/s)	1	3.281	6.214×10^{-4}	3.6	1.942	2.237
Foot/second (ft/s)	3.048×10^{-1}	1	1.894×10^{-4}	1.097	5.926×10^{-1}	6.818×10^{-1}
Mile/second (mi/s)	1.609×10^{3}	5.28×10^{3}	1	5.794×10^{3}	3.126×10^{3}	3.6×10^{3}
Kilometer/hour (km/h)	2.778×10^{-1}	9.111×10^{-1}	1.726×10^{-4}	1	5.4×10^{-1}	6.214×10^{-1}
Nautical mile/hour (nmi/h)	5.144×10^{-1}	1.688	3.2×10^{-4}	1.852	1	1.151
Mile/hour (mi/h)	4.47×10^{-1}	1.467	2.778×10^{-4}	1.609	8.69×10^{-1}	1

Table IV.12 Temperature conversion between units

Unit	Degree Celsius (°C)	Kelvin (K)	Degree Fahrenheit (°F)	Degree Rankine (°R)	Réaumur scale (°Ré)
Degree Celsius (°C)	1	[°c] = [k] − 273.15	[°c] = ([°f] − 32) / 1.8	[°c] = ([°r] − 491.67) / 1.8	[°c] = [°ré] × 1.25
Kelvin (K)	[k] = [°c] + 273.15	1	[k] = ([°f] + 459.67) / 1.8	[k] = [°r] / 1.8	[k] = [°ré] × 1.25 + 273.15
Degree Fahrenheit (°F)	[°f] = [°c] × 1.8 + 32	[°f] = [k] × 1.8 − 459.67	1	[°f] = [°r] − 459.67	[°f] = [°ré] × 2.25 + 32

Table IV.13 Pressure conversion between units

Unit	Bar (bar)	Pascal (Pa)	Standard atmosphere (atm)	Technical atmosphere (at)	Millimeter of mercury (mmHg)	Pound-force /square inch (lbf/in²)	Pound-force /square foot (lbf/ft²)
Bar (bar)	1	10^5	9.869×10^{-1}	1.0197	7.501×10^2	1.45×10^1	2.089×10^3
Pascal (Pa)	10^{-5}	1	9.869×10^{-6}	1.0197×10^{-5}	7.501×10^{-3}	1.45×10^{-4}	2.089×10^{-2}
Standard atmosphere (atm)	1.01325	1.01325×10^5	1	1.033	7.6×10^2	1.47×10^1	2.116×10^3
Technical atmosphere (at)	9.807×10^{-1}	9.807×10^4	9.678×10^{-1}	1	7.356×10^2	1.422×10^1	2.048×10^3
Millimeter of mercury (mmHg)	1.333×10^{-3}	1.333×10^2	1.316×10^{-3}	1.36×10^{-3}	1	1.934×10^{-2}	2.784
Pound-force/ Square inch (lbf/in²)	6.895×10^{-2}	6.895×10^3	6.805×10^{-2}	7.031×10^{-2}	5.171×10^1	1	1.44×10^2
Pound-force/square foot (lbf/ft²)	4.788×10^{-4}	4.788×10^1	4.725×10^{-4}	4.882×10^{-4}	3.591×10^{-1}	6.944×10^{-3}	1

Table IV.14 Volume flow rate conversion between units

Unit	Cubic meter/ hour (m³/h)	Cubic meter/ minute (m³/min)	Cubic meter/ second (m³/s)	Liter/hour (L/h)	Liter/minute (L/min)	Liter/second (L/s)	Cubic foot/ hour (ft³/h)	Cubic foot/ minute (ft³/min)
Cubic meter/ hour (m³/h)	1	1.667×10^{-2}	2.778×10^{-4}	10^3	1.667×10^1	2.778×10^{-1}	3.531×10^1	5.886×10^{-1}
Cubic meter/ minute (m³/ min)	6×10^1	1	1.667×10^{-2}	6×10^4	10^3	1.667×10^1	2.119×10^3	3.531×10^1
Cubic meter/ second (m³/s)	3.6×10^3	6×10^1	1	3.6×10^6	6×10^4	10^3	1.271×10^5	2.119×10^3
Liter/hour (L/h)	10^{-3}	1.667×10^{-5}	2.778×10^{-7}	1	1.667×10^{-2}	2.778×10^{-4}	3.531×10^{-2}	5.886×10^{-4}
Liter/minute (L/min)	6×10^{-2}	10^{-3}	1.667×10^{-5}	6×10^1	1	1.667×10^{-2}	2.119	3.531×10^{-2}
Liter/second (L/s)	3.6	6×10^{-2}	10^{-3}	3.6×10^3	6×10^1	1	1.271×10^2	2.119
Cubic foot/ hour (ft³/h)	2.832×10^{-2}	4.719×10^{-4}	7.866×10^{-6}	2.832×10^1	4.719×10^{-1}	7.866×10^{-3}	1	1.667×10^{-2}
Cubic foot/ minute (ft³/ min)	1.699	2.832×10^{-2}	4.719×10^{-4}	1.699×10^3	2.832×10^1	4.719×10^{-1}	6×10^1	1

Table IV.15 Work conversion between units

Unit	Joule (J)	Calorie (cal)	Kilogram-force·meter (kgf·m)	Kilowatt·hour (kW·h)	Foot-pound-force (ft·lbf)	British thermal unit (Btu)
Joule (J)	1	2.39×10^{-1}	1.02×10^{-1}	2.778×10^{-7}	7.376×10^{-1}	9.478×10^{-4}
Calorie (cal)	4.186	1	4.27×10^{-1}	1.163×10^{-6}	3.087	3.967×10^{-3}
Kilogram-force·meter (kgf·m)	9.804	2.342	1	2.72×10^{-6}	7.231	9.292×10^{-3}
Kilowatt·hour (kW·h)	3.56×10^{6}	8.6×10^{5}	3.672×10^{5}	1	2.655×10^{6}	3.412×10^{3}
Foot-pound-force (ft·lbf)	1.356	3.239×10^{-1}	1.383×10^{-1}	3.766×10^{-7}	1	1.285×10^{-3}
British thermal unit (Btu)	1.055×10^{3}	2.521×10^{2}	1.076×10^{2}	2.931×10^{-4}	7.782×10^{2}	1

Table IV.16 Power conversion between units

Unit	Watt (W)	Calorie/second (cal/s)	British thermal unit/second (Btu/s)	Foot-pound-force/second (ft·lbf/s)
Watt (W)	1	2.39×10^{-1}	9.478×10^{-4}	7.376×10^{-1}
Calorie/second (cal/s)	4.186	1	3.966×10^{-3}	3.086
British thermal unit/second (Btu/s)	1.055×10^{3}	2.522×10^{2}	1	7.782×10^{2}
Foot-pound-force/second (ft·lbf/s)	1.356	3.24×10^{-1}	1.285×10^{-3}	1

Table IV.17 Mass heat capacity conversion between units

Unit	Joule/(kilogram·kelvin), J/(kg·K)	Calorie/(gram·degree Celsius), cal/(g·°C)	British thermal unit/(pound·degree Fahrenheit), Btu/(lb·°F)	Joule/(pound·degree Celsius), J/(lb·°C)
Joule/(kilogram·kelvin), J/(kg·K)	1	2.388×10^{-4}	2.388×10^{-4}	4.536×10^{-1}
Calorie/(gram·degree Celsius), cal/(g·°C)	4.187×10^{3}	1	1	1.899×10^{3}
British thermal unit/(pound·degree Fahrenheit), Btu/(lb·°F)	4.187×10^{3}	1	1	1.899×10^{3}
Joule/(pound·degree Celsius), J/(lb·°C)	2.204	5.266×10^{-4}	5.266×10^{-4}	1

Table IV.18 Thermal conductivity conversion between units

Unit	Kilocalorie/ (meter·hour·degree Celsius), kcal/(m·h·°C)	Calorie/ (centimeter·second·degree Celsius), cal/(cm·s·°C)	Watt/ (meter·kelvin), W/(m·K)	Joule/ (centimeter·second·degree Celsius), J/(cm·s·°C)	British thermal unit/ (foot·hour·degree Fahrenheit), Btu/(ft·h·°F)
Kilocalorie/ (meter·hour·degree Celsius), kcal/(m·h·°C)	1	2.778×10^{-3}	1.16	1.16×10^{-2}	6.72×10^{-1}
Calorie/ (centimeter·second·degree Celsius), cal/(cm·s·°C)	3.6×10^{2}	1	4.186×10^{2}	4.186	2.42×10^{2}
Watt/(meter·kelvin), W/(m·K)	8.598×10^{-1}	2.39×10^{-3}	1	10^{-2}	5.78×10^{-1}
Joule/ (centimeter·second·degree Celsius), J/(cm·s·°C)	8.598×10^{1}	2.39×10^{-1}	10^{2}	1	5.78×10^{1}
British thermal unit/ (foot·hour·degree Fahrenheit), Btu/(ft·h·°F)	1.49	4.13×10^{-3}	1.73	1.73×10^{-2}	1

Table IV.19 Coefficient of Heat Transfer conversion between units

Unit	Watt/(square meter·kelvin), W/(m²·K)	Kilocalorie/(square meter·hour·degree Celsius), kcal/(m²·h·°C)	Calorie/(square centimeter·second·degree Celsius), cal/(cm²·s·°C)	British thermal unit/(square foot·hour·degree Fahrenheit), Btu/(ft²·h·°F)
Watt/(square meter·kelvin), W/(m²·K)	1	8.598×10^{-1}	2.388×10^{-5}	1.761×10^{-1}
Kilocalorie/(square meter·hour·degree Celsius), kcal/(m²·h·°C)	1.162	1	2.778×10^{-5}	2.048×10^{-1}
Calorie/(square centimeter·second·degree Celsius), cal/(cm²·s·°C)	4.187×10^{4}	3.6×10^{4}	1	7.373×10^{3}
British thermal unit/(square foot·hour·degree Fahrenheit), Btu/(ft²·h·°F)	5.678	4.882	1.356×10^{-4}	1

References

[1] British Standards Institution. (1990) *Methods of Test for Soils for Civil Engineering Purposes: General Requirements and Sample Preparation* (BS 1377-1:2016). BSI, London.

[2] British Standards Institution. (1990) *Methods of Test for Soils for Civil Engineering Purposes: Classification Tests* (BS 1377-2:1990). BSI, London.

[3] The Ministry of Water Resources of the People's Republic of China. (2007) *Standard for Engineering Classification of Soil* (GB/T 50145-2007). China Planning Press, Beijing, China.

[4] Das, B.M. (2008) *Advanced Soil Mechanics*. 3rd ed. Taylor & Francis, London and New York.

[5] Zhang, K.G. & Liu, S.Y. (2010) *Soil Mechanics*. China Architecture & Building Press, Beijing, China.

[6] Ministry of Housing and Urban-Rural Development of the People's Republic of China. (2001) *Code for Investigation of Geotechnical Engineering* (GB/T 50021-2001). China Architecture & Building Press, Beijing, China.

[7] Lin, Z.Y. (1994) *Geotechnical Engineering Test Monitoring Manual*. China Architecture & Building Press, Beijing, China.

[8] Ng, C.W.W. & Menzies, B.K. (2007) *Advanced Unsaturated Soil Mechanics & Engineering*. Taylor & Francis, London and New York.

[9] Craig, R.F. (2004) *Craig's Soil Mechanics Seventh Edition Solutions Manual*. Taylor & Francis, London and New York.

[10] Mitchell, J.K. & Soga, K. (2005) *Fundamentals of Soil Behavior*. 3rd ed. John Wiley & Sons, Hoboken, NJ.

[11] Ministry of Housing and Urban-Rural Development of the People's Republic of China (2002) *Code for Design of Building Foundation* (GB 50007-2002). China Architecture & Building Press, Beijing, China.

[12] Chang, S.B. & Zhang, S.M. (2006) *Engineering Geology Manual*. China Architecture & Building Press, Beijing, China.

[13] CCCC Highway Consultants Co., Ltd. (2006) *Specifications for Design of Highway Asphalt Pavement* (JTJ D50-2006). China Communications Press, Beijing, China.

[14] Chen, X.Z. & Ye, J. (2013) *Soil Mechanics and Foundation*. Tsinghua University Press, Beijing, China.

[15] Fang, Y. (2002) *Soil Mechanics*. China University of Geosciences Press, China.

[16] Hou, Z.X. (2007) *Special Soil Foundation*. China Building Materials Industry Press, China.

[17] Lu, T.H. (2002) *Soil Mechanics*. Hohai University Press, China.

[18] Zhao, C., Shao, M., Jia, X., Nasir, M. & Zhang, C. (2016) Using pedotransfer functions to estimate soil hydraulic conductivity in the loess plateau of China. *Catena*, 143, 1–6.

[19] Han, J.G. (2014) *Soil Mechanics and Foundation Engineering*. Chongqing University Press, China.

[20] Gong, W.H. (2007) *Soil Mechanics*. Huazhong University of Science and Technology Press, China.

[21] He, Y.N., Han, L.J. & Wang, Y.S. (2010) *Rock Mechanics*. China University of Mining and Technology Press, China.

[22] Schofield, A.N. & Wroth, C.P. (1968) *Critical State Soil Mechanics*. Cambridge University, Cambridge, UK.

[23] Black, D.K. & Lee, K.L. (1973) Saturating laboratory samples by back pressure. *Journal of the Soil Mechanics & Foundations Division*, 99(SM1), 75–93.

[24] Knappett, J. (2012) *Craig's Soil Mechanics*. 8th ed. Spon Press.

[25] Lambe, W.T. & Whitman, R.V. (1969) *Soil Mechanics*. John Wiley & Sons, Hoboken, New Jersey, USA.

[26] Research Institute of Highway Ministry of Transport. (2007) *Test Methods of Soils for Highway Engineering* (JTJ E40-2007). China Communications Press, Beijing, China.

[27] Kurz, D., Alfaro, M. & Graham, J. (2017) Thermal conductivities of frozen and unfrozen soils at three project sites in northern Manitoba. *Cold Regions Science & Technology*, 140, 30–38.

[28] Ming, F., Li, D., Zhang, M. & Zhang, Y. (2017) A novel method for estimating the elastic modulus of frozen soil. *Cold Regions Science & Technology*, 141, 1–7.

[29] Das, B.M., Emeritus, D. & Sobhan, K. (2012) *Principles of Geotechnical Engineering*. 8th ed. PWS, Stanford, CT.

[30] Wang, W.M., Yang, G.S., Zhang, X.D. & Ye, H.D. (2010) *Rock Mechanics*. China University of Mining and Technology Press, China.

[31] Shen, M.R. & Chen, J.F. (2006) *Rock Mechanics*. Tongji University Press, China.

[32] Cai, M.F. (2002) *Rock Mechanics and Engineering*. Science Press, China.

[33] Ling, X.C. & Cai, D.S. (2002) *Rock Mass Mechanics*. Harbin Institute of Technology Press, China.

[34] Liu, Y.R. & Tang, H.M. (2009) *Rock Mass Mechanics*. Chemical Industry Press, China.

[35] Lin, Z.Y. (2005) *Geotechnical Test Monitoring Manual*. China Architecture & Building Press, Beijing, China.

[36] Lin, Z.Y. (1996) *Geotechnical Engineering Survey and Design Manual*. Liaoning Science and Technology Press, China.

[37] Miao, C. (2015) *Study on the Influence of Rock Mass Alteration Characteristics and Engineering Geological Characteristics - a Case Study of Rock Mass in Dagangshan Dam Area*. Chengdu University of Technology, Chengdu.

[38] Wu, D.L., Huang, Z.H. & Zhao, M.J. (2002) *Rock Mechanics*. Xinjiang University Press, China.

[39] Zhang, Z.T., Jing, F. and Yang, H.L. (2009) *Engineering Practical Rock Mechanics*. China Water & Power Press, China.

[40] Zhu, Y.M. (2012) Research on application of gabbro artificial aggregate in hydraulic concrete. *Concrete*, (6), 96–101 (In Chinese).

[41] Goodman, R.E. (1989) *Introduction to Rock Mechanics*. 2nd ed. John Wiley & Sons, Hoboken, NJ.

[42] Mogi, K. (2013) *Experimental Rock Mechanics*. Taylor & Francis, London and New York.

[43] Zhao, W. (2010) *Rock Mechanics*. Central South University Press, China.

[44] Wang, Z.T., Zhou, H.Q. & Xie, Y.S. (2007) *Mining Rock Mass Mechanics*. China University of Mining and Technology Press, China.

[45] China Institute of Water Resources and Hydropower. (1991) *Manual of Rock Mechanics Parameters*. China Water & Power Press, China.

[46] Zhang, S.L. & Zhang, Y.F. (1997) Experimental study of the physical-mechanical property of marl-the rock material used for constructing tailing dam. *Gold*, (2), 26–30 (In Chinese).

[47] Das, B.M. (1984) *Principles of Foundation Engineering*. Brooks/Cole Engineering Division, Thomson, Canada.

[48] Broch, E. & Franklin, J.A. (1972) The point load strength test. *International Journal of Rock Mechanics and Mining Sciences*, 9(6), 669–676.

[49] Zheng, L.H., Shan, Y.M., Yi, S., Wang, Q.Q. & Liu, Y. (2013) Analysis of cream mudstone's mechanical properties in Tarim X area. *Science Technology and Engineering*, 13(28), 8409–8414 (In Chinese).

Index